MATHEMATICS RESEARCH DEVELOPMENTS

Selected Topics of Invariant Measures in Polish Groups

MATHEMATICS RESEARCH DEVELOPMENTS

MATHEMATICS RESEARCH DEVELOPMENTS

Selected Topics of Invariant Measures in Polish Groups

Gogi Pantsulaia

nova publishers
New York

Library of Congress Cataloging-in-Publication Data

Pantsulaia, Gogi, author.
Selected topics of invariant measures in Polish groups / Gogi Pantsulaia (Department of Mathematics, Georgian Technical University, Tbilisi, Georgia).
pages cm
Includes bibliographical references and index.
ISBN 978-1-62948-831-8 (hardcover)
1. Polish spaces (Mathematics) 2. Invariant measures. I. Title.
QA611.28.P35 2014
514'.325--dc23

2013047177

Published by Nova Science Publishers, Inc. † New York

Contents

Preface

This book explores a number of new applications of invariant quasi-finite diffused Borel measures in Polish groups for a solution of various problems stated by famous mathematicians (for example, Carmichael, Erdos, Fremlin, Darji and so on). By using natural Borel embeddings of an infinite-dimensional function space into the standard topological vector space of all real-valued sequences (endowed with the Tychonoff topology), a new approach for the construction of different translation-invariant quasi-finite diffused Borel measures with suitable properties and for their applications in a solution of various partial differential equations in an entire vector space is proposed.

Introduction

The present monograph is devoted to selected topics of invariant measures in Polish groups and consists of seventeen chapters:

New concepts of Lebesgue measure on $\mathbb{R}^\infty$ are proposed and some of their realizations in the *ZFC* theory are given in Chapter 1. Also, it is shown that Bakers both measures [2], [3], Mankiewicz and Preiss-Tiśer generators [92] and the measure of [91] are not α-standard Lebesgue measures on $\mathbb{R}^\infty$ for $\alpha = (1,1,\cdots)$.

In Chapter 2, in terms of the "Lebesgue measure" λ [2], we consider concepts of the uniformly distribution and the Riemann integrability in infinite-dimensional rectangles in R^∞ and prove infinite-dimensional versions of Lebesgue [86] and Weyl [144] famous results, respectively.

Let $(\alpha_n^{(k)})_{n\in N}$ be an infinite sequence of different integer numbers for every $k \in N$. Then a set of all sequences $(x_k)_{k\in N}$ in R^∞ for which a sequence of increasing sets $(Y_n((x_k)_{k\in N}))_{n\in N}$ is λ-uniformly distributed on the $\prod_{k\in N}[a_k,b_k])$, where

$$Y_n((x_k)_{k\in N}) = \prod_{k=1}^{n}((\cup_{j=1}^{n}\{<\alpha_j^{(k)}x_k>(b_k-a_k)\})+a_k)\times \prod_{k\in N\setminus\{1,\dots,n\}}\{a_k\}$$

and λ is "Lebesgue measure" constructed by R. Baker [2], is of λ-measure zero and, hence shy in R^∞.

In Chapter 3, we extend change of variable formula for ordinary Lebesgue measures on $\mathbb{R}^{\mathbb{N}}$ and give a generalization of the certain result obtained in [2].

By using a technique of invariant measures developed by Oxtoby [88] for a Polish group, which is dense in itself, in Chapter 4 we prove an existence of a quasi-finite left (or right) invariant generator of left (or right)-shy sets in an entire group and establish that no any element of the class of all quasi-finite generators of left (or right)-shy sets possesses a uniqueness property. We get the validity of an analogous result for quasi-finite generators of two-sided-shy sets in Polish groups, which are dense in itself and are equipped with two-sided invariant metrics. These results allows us to answer to Questions 2.1-2.2 posed in [99].

In Chapter 5 we give an extension of the notion of generators of shy sets in Polish topological vector spaces [96] to all Polish groups (see, Definition 5.2.1) and study their various properties (existence, uniqueness, non-σ-finiteness, etc). By a new approach supported on the technique of generators of shy sets, we extend Randal Dougherty criterion of shyness for subsets (cf. [31], Proposition 12, p. 87) in the product of unimodular Polish groups that are not compact (see, Theorem 5.4.1 and Example 5.4.1)

In Chapter 6, a new class of translation-invariant quasi-finite Borel measures (the so called, ordinary and standard "Lebesgue Measures") in an infinite-dimensional separable Banach space X is constructed and some its properties are studied.

Let α be an infinite parameter set and let $(\alpha_i)_{i\in I}$ be its any partition such that α_i is a non-empty finite subset for every $i \in I$. For $j \in \alpha$, let μ_j be a σ-finite Borel measure defined on a Polish metric space (E_j, ρ_j). In Chapter 7, we introduce a new concept of a standard $(\alpha_i)_{i\in I}$-product of measures $(\mu_j)_{j\in\alpha}$ and prove its existence(see, Theorem 7.1). As consequence, we construct "a standard $(\alpha_i)_{i\in I}$-Lebesgue measure" on the Borel σ-algebra of subsets of $\mathbb{R}^{\alpha}$ for every infinite parameter set α such that it is invariant under a group generated by shifts and canonical permutations (see, Remark 7.3).

Our method of construction of measures presented in Chapter 8, is a concept of strict standard and strict ordinary products of an infinite family of (no only σ-finite) measures, which allows us to construct Mankiewicz and Preiss-Tišer generators in $\mathbf{R}^{\infty}$. We show that if $f : \mathbf{R}^{\infty} \to R$ is a Lipschitz function and $R^{(N)}$ is a group of all eventually zero sequences, then f is λ-almost everywhere $R^{(N)}$-Gâteaux differentiable on $\mathbf{R}^{\infty}$ in Solovay model, where λ denotes the strict standard Lebesgue measure in $\mathbf{R}^{\infty}$.

In Chapter 9, by using a structure of the Fourier differential operator in $\mathbb{R}^{\infty}$, under some general restrictions we describe a new and essentially different approach for a construction of a particular solution of the non-homogeneous differential equation of the higher order with real constant coefficients and give its representation in an explicit form.

By using a structure of the Fourier differential operator in $\mathbb{R}^{\infty}$, a new and essentially different approach for a solution of the old functional problem posed by R. D. Carmichael in 1936 is described in Chapter 10. More precisely, under some natural restrictions, we express in an explicit form the general solution of the linear inhomogeneous differential equation of infinite order with real constant coefficients. In addition, we construct an invariant measure for the corresponding differential equation. Also, we describe a certain approach for a solution of an initial value problem for a special class of linear inhomogeneous partial differential equations of infinite order with real constant coefficients and describe behaviors of corresponding dynamical systems in terms of partial analogs of the Lebesgue measure.

The solution of a certain generalization of the infinite order of the von Foerster-Lasota differential equation [A. L. Dawidowicz, A. Poskrobko, On chaotic and stable behaviour of the von Foerster-Lasota equation in some Orlicz spaces, Proc. Est. Acad. Sci., 57(2) (2008), 6169] is obtained in an explicit form and the behaviour of corresponding motions in $\mathbb{R}^{\infty}$ is studied in terms of ordinary and standard Lebesgue measures in Chapter 11 .

The structure of uniformly distributed sequences on $[-1/2, 1/2]$ is studied in terms of Yamasaki measure μ [145] in Chapter 12. In particular, it is shown that μ-almost every element of $\mathbb{R}^{\infty}$ is uniformly distributed on $[-1/2, 1/2]$.

The expansion of a real-valued square integrable (with respect to the infinite-dimensional Lebesgue measure λ [2]) function on an infinite-dimensional rectangle $\mathcal{R}$ (with $\lambda(\mathcal{R}) < +\infty$) in $\mathbb{R}^{\infty}$ into an infinite-dimensional multiple trigonometric series is given in Chapter 13. An analogous question for real-valued square integrable (with respect to the same measure) functions on $\mathbb{R}^{\infty}$ is considered. More precisely, it is shown that every element f of the Hilbert space $L_2(\mathbb{R}^{\infty}, \mathcal{M}, \lambda)$ [3] admits such an expansion into an infinite dimensional multiple trigonometric series converging to f.

It is proved in Chapter 14 that the duality principle between the measure and category is valid with respect to the sentence P defined by: *For every two Polish groups G_1 and G_2, and for every Haar null set $Y \subset G_1$ we have*

$$(\forall X)(X \subseteq G_2 \rightarrow Y \times X \textit{ is Haar null in } G_1 \times G_2).$$

By using that approach which differs from the approach of M. P. Cohen (2012), it is shown that every infinite product of unimodular Polish groups that are not compact decomposes into the disjoint union of a Haar null set and a meagre set, which gives a partial positive answer to a question of Darji (2011). We show that Baker's (2004) product of Haar measures defined on the product of unimodular Polish groups that are not compact is concentrated on a meagre set which is not covered by a countable family of compact sets. It is shown also that a question from David Fremlins problem list (2012) asking *whether one can simplify the definition of a Haar null set in $\mathbb{R}$ by leaving out the Borel set* is independent of $ZF + DC$ theory. Similar result is obtained for a question of M. Elekes and J. Steprâns (2012) *asking whether there exists an atomless singular Borel probability measure in $\mathbb{R}$ which reflects the positive measure of every subset in $\mathbb{R}$.* By using Erdös-Sierpiński duality principle, it proved that dual of the Sierpiński set answers negatively on a certain topological analogue of M. Elekes and J. Steprâns (2012). As consequence, we get that Bartoszynski's(2002) and Burke-Miller's (2005) result is independent of $ZF + DC$ theory.

In Chapter 15 it is shown that for every translation-invariant quasi-finite diffused Borel measure μ in an uncountable non-locally compact Polish group G that is dense in himself there does not exist a positive constant c such that each Borel set E with the μ-measure bigger than c contains no three points such that an area defined by that points is equal to one. This answers negatively to a certain modification of P. Erdős problem [P. Erdős, Set-theoretic, measure-theoretic, combinatorial, and number- theoretic problems concerning point sets in Euclidean space, *Real Anal. Exchange,* 4(2), (1978/79), 113–138].

In Chapter 16, a certain version of the Erdős problem is studied. More precisely, it is proved that there does not exist a finite constant c such that each plane set with an outer Lebesgue measure greater than c contains the vertices of a triangle of area 1. It is shown that a sentence *"each plane set E with Lebesgue outer measure $+\infty$ contains the vertices of a triangle of area 1"* is independent from the theory $(ZF)\&(DC)$. The Erdős problem is studied for the shy-measure in an infinite-dimensional separable Banach space and it is established that any number from the interval $[0,1[$ is Erdős constant for such a measure. There is constructed such an example of a thick (in the sense of shyness) subset of ℓ_2 which does not contain vertices of a triangle of area 1.

In Chapter 17 (Appendix) we give proofs of Mycielski-Swierczkowski , Martin-Solovay and Liouville theorems.

Acknowledgment. The preparation of the present monograph partially was supported by the Sh. Rustaveli National Science Foundation(Grants: no. $31/24$, no. $31/25$).

Chapter 1

On Ordinary and Standard Lebesgue Measures in $\mathbb{R}^\infty$

We discuss the problem of the existence of an analog of the Lebesgue measure for the vector space of all real-valued sequences $\mathbb{R}^\infty = \prod_{i=1}^{\infty} \mathbb{R}$ equipped with Tychonoff topology.

R. Baker [2] introduced the notion of "Lebesgue measure" in $\mathbb{R}^\infty$ as follows: a measure λ being the completion of a translation invariant Borel measure in $\mathbb{R}^\infty$ is called a "Lebesgue measure" in $\mathbb{R}^\infty$ if for any measurable rectangle $\prod_{i=1}^{\infty}(a_i,b_i)$, $-\infty < a_i < b_i < +\infty$ with $0 \leq \prod_{i=1}^{\infty}(b_i - a_i) < +\infty$, the following equality

$$\lambda\Big(\prod_{i=1}^{\infty}(a_i,b_i)\Big) = \prod_{i=1}^{\infty}(b_i - a_i)$$

holds, where

$$\prod_{i=1}^{\infty}(b_i - a_i) := \lim_{n\to\infty}\prod_{i=1}^{n}(b_i - a_i).$$

Subsequently, R. Baker [3] extended his notion of "Lebesgue measure" in $\mathbb{R}^\infty$ as follows : a measure λ being the completion of a translation invariant Borel measure on $\mathbb{R}^\infty$ is called a "Lebesgue measure" if for any measurable rectangle $\prod_{i=1}^{\infty} R_i$, $R_i \in \mathcal{B}(\mathbb{R})$ with $0 \leq \prod_{i=1}^{\infty} m(R_i) < \infty$, the following equality

$$\lambda\Big(\prod_{i=1}^{\infty} R_i\Big) = \prod_{i=1}^{\infty} m(R_i)$$

holds, where m denotes a linear Lebesgue measure in $\mathbb{R}$.

To propose a new concept of "Lebesgue measure" on $\mathbb{R}^\infty$ we attract our attention to the following two simple facts.

Fact 1.1. *Let μ be a probability measure defined on a measure space (E,S). Then the product measure $\mu^{\mathbb{N}}$ defined on $(E^{\mathbb{N}},S^{\mathbb{N}})$ has the following essential property: if f is any permutation of $\mathbb{N}$ and $A_f((x_k)_{k\in\mathbb{N}}) = (x_{f(k)})_{k\in\mathbb{N}}$ for $(x_k)_{k\in\mathbb{N}} \in E^{\mathbb{N}}$, then $\mu^{\mathbb{N}}(A_f(X)) = \mu^{\mathbb{N}}(X)$ for every $X \in S^{\mathbb{N}}$.*

Fact 1.2. *The n-dimensional Lebesgue measure ℓ_n in $\mathbb{R}^n$ has the following property: if f is any permutation of $\{1,\dots,n\}$ and*

$$A_f((x_k)_{1\le k\le n}) = (x_{f(k)})_{1\le k\le n} \; ((x_k)_{1\le k\le n} \in \mathbb{R}^n),$$

then $\ell_n(A_f(X)) = \ell_n(X)$ for every $X \in \mathcal{B}(\mathbb{R}^n)$.

In view of these facts we can say that Baker's measures [2], [3] have no essential property of a product - measure to be an invariant under the group of all canonical permutations[1] of $\mathbb{R}^\infty$.

Indeed, if we consider the following infinite-dimensional rectangular set

$$X = \prod_{k=1}^{\infty}[0, e^{\frac{(-1)^k}{k}}],$$

then for every non-zero real number a there exists a permutation f_a of $\mathbb{N}$ such that $\lambda(A_{f_a}(X)) = a$, where λ is any Baker's measure [2], [3].

To introduce new concepts of the Lebesgue measure in $\mathbb{R}^\infty$, we need some definitions.

Let $(\beta_j)_{j\in\mathbb{N}} \in [0,+\infty]^N$.

Definition 1.1. We say that a number $\beta \in [0,+\infty]$ is an ordinary product of numbers $(\beta_j)_{j\in\mathbb{N}}$ if

$$\beta = \lim_{n\to\infty}\prod_{i=1}^{n}\beta_i.$$

An ordinary product of numbers $(\beta_j)_{j\in\mathbb{N}}$ is denoted by $(\mathbf{O})\prod_{i\in\mathbb{N}}\beta_i$.

Definition 1.2. A standard product of the family of numbers $(\beta_i)_{i\in\mathbb{N}}$ is denoted by $(\mathbf{S})\prod_{i\in\mathbb{N}}\beta_i$ and defined as follows:

$(\mathbf{S})\prod_{i\in\mathbb{N}}\beta_i = 0$ if $\sum_{i\in\mathbb{N}^-}\ln(\beta_i) = -\infty$, where $N^- = \{i : ln(\beta_i) < 0\}$ [2], and $(\mathbf{S})\prod_{i\in\mathbb{N}}\beta_i = e^{\sum_{i\in\mathbb{N}}\ln(\beta_i)}$ if $\sum_{i\in\mathbb{N}^-}\ln(\beta_i) \neq -\infty$.

Let $\alpha = (n_k)_{k\in\mathbb{N}} \in (\mathbb{N}\setminus\{0\})^{\mathbb{N}}$. We set

$$F_0 = [0,n_0]\cap\mathbb{N},\ F_1 = [n_0{+}1, n_0{+}n_1]\cap\mathbb{N},\ \dots, F_k = [n_0{+}\cdots{+}n_{k-1}{+}1, n_0{+}\cdots{+}n_k]\cap\mathbb{N},\dots.$$

Definition 1.3. We say that a number $\beta \in [0,+\infty]$ is an ordinary α-product of numbers $(\beta_i)_{i\in\mathbb{N}}$ if β is an ordinary product of numbers $(\prod_{i\in F_k}\beta_i)_{k\in\mathbb{N}}$. An ordinary α-product of numbers $(\beta_i)_{i\in\mathbb{N}}$ is denoted by $(\mathbf{O},\alpha)\prod_{i\in\mathbb{N}}\beta_i$.

Definition 1.4. We say that a number $\beta \in [0,+\infty]$ is a standard α-product of numbers $(\beta_i)_{i\in\mathbb{N}}$ if β is a standard product of numbers $(\prod_{i\in F_k}\beta_i)_{k\in\mathbb{N}}$. A standard α-product of numbers $(\beta_i)_{i\in\mathbb{N}}$ is denoted by $(\mathbf{S},\alpha)\prod_{i\in\mathbb{N}}\beta_i$.

---

[1] Let f be any permutation of $\mathbb{N}$. A mapping $A_f : \mathbb{R}^\infty \to \mathbb{R}^\infty$ defined by $A_f((x_k)_{k\in\mathbb{N}}) = (x_{f(k)})_{k\in\mathbb{N}}$ for $(x_k)_{k\in\mathbb{N}} \in \mathbb{R}^\infty$ is called a canonical permutation of $\mathbb{R}^\infty$.

[2] We set $ln(0) = -\infty$

Definition 1.5. Let $\alpha = (n_k)_{k\in\mathbb{N}} \in (\mathbb{N}\setminus\{0\})^{\mathbb{N}}$. Let $(\alpha)\mathcal{OR}$ be the class of all infinite-dimensional measurable α-rectangles $R = \prod_{i\in\mathbb{N}} R_i (R_i \in \mathcal{B}(\mathbb{R}^{n_i}))$ for which an ordinary product of numbers $(m^{n_i}(R_i))_{i\in\mathbb{N}}$ exists and is finite.

We say that a measure λ being the completion of a translation-invariant Borel measure is an ordinary α-Lebesgue measure in $\mathbb{R}^\infty$(or, shortly, O(α)LM) if for every $R \in (\alpha)\mathcal{OR}$ we have

$$\lambda(R) = (\mathbf{O})\prod_{k\in\mathbb{N}} m^{n_k}(R_k).$$

Definition 1.6. Let $\alpha = (n_k)_{k\in\mathbb{N}} \in (\mathbb{N}\setminus\{0\})^{\mathbb{N}}$. Let $(\alpha)\mathcal{SR}$ be the class of all infinite-dimensional measurable α-rectangles $R = \prod_{i\in\mathbb{N}} R_i (R_i \in \mathcal{B}(\mathbb{R}^{n_i}))$ for which a standard product of numbers $(m^{n_i}(R_i))_{i\in\mathbb{N}}$ exists and is finite.

We say that a measure λ being the completion of a translation-invariant Borel measure is a standard α-Lebesgue measure in $\mathbb{R}^\infty$(or, shortly, S(α)LM) if for every $R \in (\alpha)\mathcal{SR}$ we have

$$\lambda(R) = (\mathbf{S})\prod_{k\in\mathbb{N}} m^{n_k}(R_k).$$

Proposition 1.1. *Note that for every $\alpha = (n_k)_{k\in\mathbb{N}} \in (\mathbb{N}\setminus\{0\})^{\mathbb{N}}$ the following strict inclusion*

$$(\alpha)\mathcal{OR} \subset (\alpha)\mathcal{SR}$$

holds.

Proof. Suppose that $R = \prod_{i\in\mathbb{N}} R_i \in (\alpha)\mathcal{OR}$. This means that

$$0 \le \lim_{n\to\infty}\prod_{k=1}^{n}(m^{n_k}(R_k)) < +\infty.$$

There are possible three cases:

1) $\sum_{i=1}^{\infty} \ln(m^{n_k}(R_k))$ is convergent to $-\infty$;
2) $\sum_{i=1}^{\infty} \ln(m^{n_k}(R_k))$ is conditionally convergent to any finite real number;
3) $\sum_{i=1}^{\infty} \ln(m^{n_k}(R_k))$ is absolutely convergent to any finite real number;

Conditions 1) and 2) imply that

$$(\mathbf{S})\prod_{k\in\mathbb{N}} m^{n_k}(R_k) = 0.$$

The condition 3) implies that

$$0 < (\mathbf{S})\prod_{k\in\mathbb{N}} m^{n_k}(R_k) < +\infty. \qquad \square$$

The main purpose of the present chapter is to give a new construction of translation-invariant Borel measures in $\mathbb{R}^\infty$ which will be different from the construction [3] in the sense that it does not apply the metric properties of $\mathbb{R}^\infty$. It will be an adaptation of a construction from general measure theory which will allows us to construct interesting examples of analogs of a Lebesgue measure on the entire space.

Let (E,S) be a measurable space and let $\mathcal{R}$ be any subclass of the σ-algebra S. Let $(\mu_B)_{B\in\mathcal{R}}$ be such a family of σ-finite measures that for $B \in \mathcal{R}$ we have $\mathrm{dom}(\mu_B) = S \cap \mathcal{P}(B)$, where $\mathcal{P}(B)$ denotes the power set of the set B.

Definition 1.7. A family $(\mu_B)_{B\in\mathcal{R}}$ is called to be consistent if

$$(\forall X)(\forall B_1,B_2)(X\in S\ \&\ B_1,B_2\in\mathcal{R}\rightarrow\mu_{B_1}(X\cap B_1\cap B_2)=\mu_{B_2}(X\cap B_1\cap B_2)).$$

Convention 1.1 In the proof of Lemma 1.1, we use the convention that $\mathbb{N}=\{1,2,\cdots\}$ and $\cup_{k=1}^{0}A_k=\emptyset$ for an arbitrary family of subsets $(A_k)_{k\in\mathbb{N}}$.

The following assertion plays a key role in our future investigation.

Lemma 1.1. *Let $(\mu_B)_{B\in\mathcal{R}}$ be a consistent family of σ-finite measures. Then there exists a measure $\mu_{\mathcal{R}}$ on (E,S) such that*

(i) *$\mu_{\mathcal{R}}(B)=\mu_B(B)$ for every $B\in\mathcal{R}$;*

(ii) *if there exists a non-countable family of pairwise disjoint sets $\{B_i\ :\ i\in I\}\subseteq\mathcal{R}$ such that $0<\mu_{B_i}(B_i)<\infty$, then the measure $\mu_{\mathcal{R}}$ is non-σ-finite;*

(iii) *if G is a group of measurable transformations of E such that $G(\mathcal{R})=\mathcal{R}$ and*

$$(\forall B)(\forall X)(\forall g)\big((B\in\mathcal{R}\ \&X\in S\cap\mathcal{P}(B)\ \&\ g\in G)\rightarrow\mu_{g(B)}(g(X))=\mu_B(X)\big),$$

where $\mathcal{P}(B)$ denotes a power set of the set B, then the measure $\mu_{\mathcal{R}}$ is G-invariant.

Proof. If $X\in S$ is covered by any countable family $(A_n)_{n\in\mathbb{N}}$ of elements of $\mathcal{R}$, then we put

$$\mu_{\mathcal{R}}(X)=\sum_{n\in\mathbb{N}}\mu_{A_n}((A_n\setminus\cup_{k=1}^{n-1}A_k)\cap X).$$

We set that $\mu_{\mathcal{R}}(X)=+\infty$ if X is not covered by any countable family of elements of $\mathcal{R}$.

Let us show the correctness of the definition of the functional $\mu_{\mathcal{R}}$.

If X is not covered by any countable family of elements of $\mathcal{R}$, then the correctness is obvious.

Now let us use Convention 1.1 and let X be covered by two countable families $(A_n)_{n\in\mathbb{N}}$ and $(B_n)_{n\in\mathbb{N}}$ of elements of $\mathcal{R}$. We have to show that

$$\sum_{n\in\mathbb{N}}\mu_{A_n}\big((A_n\setminus\cup_{k=1}^{n-1}A_k)\cap X\big)=\sum_{n\in\mathbb{N}}\mu_{B_n}\big((B_n\setminus\cup_{k=1}^{n-1}B_k)\cap X\big).$$

Indeed, we have

$$\sum_{n\in\mathbb{N}}\mu_{A_n}\big((A_n\setminus\cup_{k=1}^{n-1}A_k)\cap X\big)=\sum_{n\in\mathbb{N}}\mu_{A_n}\big((A_n\setminus\cup_{k=1}^{n-1}A_k)\cap\big(\cup_{m\in\mathbb{N}}(B_m\setminus\cup_{l=1}^{m-1}B_l)\big)\cap X\big)=$$

$$=\sum_{n\in\mathbb{N}}\mu_{A_n}\big(\cup_{m\in\mathbb{N}}\big((A_n\setminus\cup_{k=1}^{n-1}A_k)\cap(B_m\setminus\cup_{l=1}^{m-1}B_l)\big)\cap X\big)=$$

$$=\sum_{n\in\mathbb{N}}\sum_{m\in\mathbb{N}}\mu_{A_n}\big((A_n\setminus\cup_{k=1}^{n-1}A_k)\cap(B_m\setminus\cup_{l=1}^{m-1}B_l)\cap X\big)=$$

$$=\sum_{m\in\mathbb{N}}\sum_{n\in\mathbb{N}}\mu_{A_n}\big((A_n\setminus\cup_{k=1}^{n-1}A_k)\cap(B_m\setminus\cup_{l=1}^{m-1}B_l)\cap X\big)=$$

$$=\sum_{m\in\mathbb{N}}\sum_{n\in\mathbb{N}}\mu_{B_m}\big((A_n\setminus\cup_{k=1}^{n-1}A_k)\cap(B_m\setminus\cup_{l=1}^{m-1}B_l)\cap X\big)=$$

$$=\sum_{m\in\mathbb{N}}\mu_{B_m}\big(\cup_{n\in\mathbb{N}}(A_n\setminus\cup_{k=1}^{n-1}A_k)\cap(B_m\setminus\cup_{l=1}^{m-1}B_l)\cap X\big)=$$

$$= \sum_{m\in\mathbb{N}} \mu_{B_m}\big((B_m \setminus \cup_{l=1}^{m-1} B_l) \cap X\big).$$

Thus the correctness is proved.

Let us prove that the functional $\mu_{\mathcal{R}}$ is σ-additive.

Let $(X_k)_{k\in\mathbb{N}}$ be a countable family of pairwise disjoint elements of S.

Case I. Each X_k is covered by a countable family of elements of $\mathcal{R}$. Then so will be their union. Let $(A_m)_{m\in\mathbb{N}}$ be such a family of elements of $\mathcal{R}$ that covers $\cup_{k\in\mathbb{N}} X_k$. We have

$$\mu_{\mathcal{R}}(\cup_{k\in\mathbb{N}} X_k) = \sum_{n\in\mathbb{N}} \mu_{A_n}((A_n \setminus \cup_{k=1}^{n-1} A_k) \cap (\cup_{k\in\mathbb{N}} X_k)) =$$
$$= \sum_{n\in\mathbb{N}} \sum_{k\in\mathbb{N}} \mu_{A_n}((A_n \setminus \cup_{k=1}^{n-1} A_k) \cap X_k) =$$
$$= \sum_{k\in\mathbb{N}} \sum_{n\in\mathbb{N}} \mu_{A_n}((A_n \setminus \cup_{k=1}^{n-1} A_k) \cap X_k) =$$
$$= \sum_{k\in\mathbb{N}} \mu_{\mathcal{R}}(X_k).$$

Case II. Let us assume that not every element of the family $(X_k)_{k\in\mathbb{N}}$ is covered by a countable family of elements of $\mathcal{R}$. Then so will be their union and we get

$$\mu_{\mathcal{R}}(\cup_{k\in\mathbb{N}} X_k) = +\infty = \sum_{k\in\mathbb{N}} \mu_{\mathcal{R}}(X_k).$$

Proof of the item (i)**.** We set $A_k = B$ for $k \in \mathbb{N}$. Then the family $(A_k)_{k\in\mathbb{N}}$ covers B and by the definition of $\mu_{\mathcal{R}}$ we have

$$\mu_{\mathcal{R}}(B) = \mu_B(B) + \mu_B((B \setminus B) \cap B) + \cdots = \mu_B(B).$$

The proof of the item (ii) is obvious and we omit it.

Proof of the item (iii)**.** Let G be a group of measurable transformations of E such that $G(\mathcal{R}) = \mathcal{R}$ and

$$(\forall B)(\forall X)(\forall g)\big((B \in \mathcal{R}\ \& X \in B \cap S\ \&\ g \in G\big) \to \mu_{g(B)}(g(X)) = \mu_B(X)).$$

We are to show that the measure $\mu_{\mathcal{R}}$ is G-invariant.

Let $X \in S$ be covered by a countable family $(A_n)_{n\in\mathbb{N}}$ of elements of $\mathcal{R}$. Then $g(X)$ will be covered by $(g(A_n))_{n\in\mathbb{N}}$ which is a countable family of elements of $\mathcal{R}$.

We have

$$\mu_{\mathcal{R}}(g(X)) = \sum_{n\in\mathbb{N}} \mu_{g(A_n)}\big((g(A_n) \setminus \cup_{k=1}^{n-1} g(A_k)) \cap g(X)\big) =$$
$$= \sum_{n\in\mathbb{N}} \mu_{g(A_n)}(g\big((A_n \setminus \cup_{k=1}^{n-1} A_k) \cap X\big)) =$$
$$= \sum_{n\in\mathbb{N}} \mu_{A_n}\big((A_n \setminus \cup_{k=1}^{n-1} A_k) \cap X\big) = \mu_{\mathcal{R}}(X).$$

If X is not covered by any countable family of elements of $\mathcal{R}$, then the same is true for $g(X)$ and we get

$$\mu_{\mathcal{R}}(g(X)) = \mu_{\mathcal{R}}(X) = +\infty. \qquad \square$$

Lemma 1.2. *Let* $\alpha = (n_i)_{i\in\mathbb{N}} \in (\mathbb{N}\setminus\{0\})^{\mathbb{N}}$. *We set* $\mathcal{R} = (\alpha)O\mathcal{R}$. *Suppose that* $R = \prod_{i\in\mathbb{N}} R_i \in \mathcal{R}$ *for which* $R_i \in \mathcal{B}(\mathbb{R}^{n_i})$ *for* $i \in \mathbb{N}$.

For $X \in \mathcal{B}(R)$, *we set* $\mu_R(X) = 0$ *if*

$$(\mathbf{O})\prod_{i\in\mathbb{N}} m^{n_i}(R_i) = 0,$$

and

$$\mu_R(X) = (\mathbf{O})\prod_{i\in\mathbb{N}} m^{n_i}(R_i) \times \Big(\prod_{i\in\mathbb{N}} \frac{m^{n_i}{}_{R_i}}{m^{n_i}(R_i)}\Big)(X)$$

otherwise, where $\frac{m^{n_i}{}_{R_i}}{m^{n_i}(R_i)}$ *is a Borel probability measure defined on* R_i *as follows*

$$(\forall X)\Big(X \in \mathcal{B}(R_i) \to \frac{m^{n_i}{}_{R_i}}{m^{n_i}(R_i)}(X) = \frac{m^{n_i}(Y \cap R_i)}{m^{n_i}(R_i)}\Big).$$

Then the family of measures $(\mu_R)_{R\in\mathcal{R}}$ *is consistent.*

Proof. Let $R_1 = \prod_{i=1}^{\infty} R_i^{(1)}$ and $R_2 = \prod_{i=1}^{\infty} R_i^{(2)}$ be two elements of the class $\mathcal{R}$.

Without loss of generality it can be assumed that $0 < (\mathbf{O})\prod_{i\in\mathbb{N}} m^{n_i}(R_i^{(1)}) < \infty$ and $0 < (\mathbf{O})\prod_{i\in\mathbb{N}} m^{n_i}(R_i^{(2)}) < \infty$.

We will show that $\mu_{R_1}(X) = \mu_{R_2}(X)$ for every $X \in \mathcal{B}(R_1 \cap R_2)$. In this case it is sufficient to show that $\mu_{R_1}(Y) = \mu_{R_2}(Y)$ for every elementary measurable rectangle $Y = \prod_{i=1}^{\infty} Y_i$ in $R_1 \cap R_2$. Note here that under an elementary measurable rectangle $Y = \prod_{i=1}^{\infty} Y_i$ in $R_1 \cap R_2$ we assume a subset of $R_1 \cap R_2$ such that $Y_i \in \mathcal{B}(R_i^{(1)} \cap R_i^{(2)})$ for every $i \in \mathbb{N}$ and, in addition, there exists a natural number n such that $Y_i = R_i^{(1)} \cap R_i^{(2)}$ for $i \geq n$.

For every $i \in \mathbb{N}$ and for every $Y_i \in \mathcal{B}(R_i^{(1)} \cap R_i^{(2)})$ we have

$$m^{n_i}(Y_i \cap R_i^{(1)} \cap R_i^{(2)}) = m^{n_i}(Y_i \cap R_i^{(1)}) = m^{n_i}(Y_i \cap R_i^{(2)}).$$

The latter relation implies that

$$(\mathbf{O})\prod_{i=1}^{\infty} m^{n_i}(Y_i \cap R_i^{(1)} \cap R_i^{(1)}) = \lim_{n\to\infty}\prod_{i=1}^{n} m^{n_i}(Y_i \cap R_i^{(1)} \cap R_i^{(1)}) =$$

$$= \lim_{n\to\infty}\prod_{i=1}^{n} m^{n_i}(Y_i \cap R_i^{(1)}) = (\mathbf{O})\prod_{i\in\mathbb{N}} m^{n_i}(Y_i \cap R_i^{(1)}).$$

Analogously, we have

$$(\mathbf{O})\prod_{i\in\mathbb{N}} m^{n_i}(Y_i \cap R_i^{(1)} \cap R_i^{(1)}) = \lim_{n\to\infty}\prod_{i=1}^{n} m^{n_i}(Y_i \cap R_i^{(1)} \cap R_i^{(1)}) =$$

$$= \lim_{n\to\infty}\prod_{i=1}^{n} m^{n_i}(Y_i \cap R_i^{(2)}) = (\mathbf{O})\prod_{i\in\mathbb{N}} m^{n_i}(Y_i \cap R_i^{(2)}).$$

Hence we get

$$\mu_{R_1}(\prod_{i=1}^{\infty} Y_i) = (\mathbf{O})\prod_{i\in\mathbb{N}} m^{n_i}(Y_i \cap R_i^{(1)}) = (\mathbf{O})\prod_{i\in\mathbb{N}} m^{n_i}(Y_i \cap R_i^{(1)} \cap R_i^{(1)}) =$$

$$= (\mathbf{O})\prod_{i\in\mathbb{N}} m^{n_i}(Y_i \cap R_i^{(2)}) = \mu_{R_2}\Big(\prod_{i=1}^{\infty} Y_i\Big).$$

Since a class $\mathcal{A}(R_1 \cap R_2)$ of all finite disjoint unions of elementary measurable rectangles in $R_1 \cap R_2$ is a ring, and since, by definition, the class $\mathcal{B}(R_1 \cap R_2)$ of Borel measurable sets of $R_1 \cap R_2$ is a minimal σ-ring generated by the ring $\mathcal{A}(R_1 \cap R_2)$, we claim (cf. [46], Theorem B, p. 27) that the class of all those sets of $R_1 \cap R_2$ for which this equality holds coincides with the class $\mathcal{B}(R_1 \cap R_2)$.

The consistency of the family of measures $(\mu_R)_{R\in\mathcal{R}}$ is proved. □

Lemma 1.3. *Let $\alpha = (n_i)_{i\in\mathbb{N}} \in (\mathbb{N}\setminus\{0\})^{\mathbb{N}}$. We set $\mathcal{R} = (\alpha)\mathcal{SR}$. Suppose that $R = \prod_{i\in\mathbb{N}} R_i \in \mathcal{R}$) for which $R_i \in \mathcal{B}(\mathbb{R}^{n_i})$ for $i \in \mathbb{N}$ and $R \in (\alpha)\mathcal{SR}$.*

For $X \in \mathcal{B}(R)$, we set $\mu_R(X) = 0$ if

$$(\mathbf{S})\prod_{i\in\mathbb{N}} m^{n_i}(R_i) = 0,$$

and

$$\mu_R(X) = (\mathbf{S})\prod_{i\in\mathbb{N}} m^{n_i}(R_i) \times \Big(\prod_{i\in\mathbb{N}} \frac{m^{n_i}{}_{R_i}}{m^{n_i}(R_i)}\Big)(X)$$

otherwise, where $\frac{m^{n_i}{}_{R_i}}{m^{n_i}(R_i)}$ is a Borel probability measure defined in R_i as follows

$$(\forall X)\Big(X \in \mathcal{B}(R_i) \to \frac{m^{n_i}{}_{R_i}}{m^{n_i}(R_i)}(X) = \frac{m^{n_i}(Y \cap R_i)}{m^{n_i}(R_i)}\Big).$$

Then the family of measures $(\mu_R)_{R\in\mathcal{R}}$ is consistent.

The proof of Lemma 1.3 can be obtained by the scheme applied in the proof of Lemma 1.2.

Let us consider some corollaries of Lemmas 1.1–1.3.

Theorem 1.1. *For every $\alpha = (n_i)_{i\in\mathbb{N}} \in (\mathbb{N}\setminus\{0\})^{\mathbb{N}}$, there exists a Borel measure μ_α in $\mathbb{R}^\infty$ which is $O(\alpha)LM$.*

Proof. By Lemma 1.2, the class of measures $(\mu_R)_{R\in(\alpha)\mathcal{OR}}$ is consistent. Since the class $(\alpha)\mathcal{OR}$ is translation-invariant and the condition (iii) in Lemma 1.1 has been satisfied with respect to the group of all translations of $\mathbb{R}^\infty$, by Lemma 1.1 we claim that $\mu_\alpha := \lambda_{(\alpha)\mathcal{OR}}$ is O(α)LM. □

Theorem 1.2. *For every $\alpha = (n_i)_{i\in\mathbb{N}} \in (\mathbb{N}\setminus\{0\})^{\mathbb{N}}$, there exists a Borel measure ν_α in $\mathbb{R}^\infty$ which is $S(\alpha)LM$.*

Proof. By Lemma 1.3, the class of measures $(\mu_R)_{R\in(\alpha)\mathcal{SR}}$ is consistent. Since the $(\alpha)\mathcal{SR}$ is translation-invariant and the condition (iii) in Lemma 1.1 has been satisfied with respect to the group of all translations of $\mathbb{R}^\infty$, by Lemma 1.1 we claim that $\nu_\alpha := \lambda_{(\alpha)\mathcal{SR}}$ is S(α)LM. □

Let μ_1 and μ_2 be two measures defined on the measurable space $(\mathbb{E}, \mathbb{S})$.

Definition 1.8 ([46], p. 124)**.** We say that the μ_1 is absolutely continuous with respect to the μ_2, in symbols $\mu_1 \ll \mu_2$, if

$$(\forall X)(X \in \mathbb{S} \,\&\, \mu_2(X) = 0 \rightarrow \mu_1(X) = 0).$$

Definition 1.9 ([46], p. 126)**.** Two measures μ_1 and μ_2 for which both $\mu_1 \ll \mu_2$ and $\mu_2 \ll \mu_1$ are called equivalent, in symbols $\mu_1 \equiv \mu_2$.

We have the following assertion.

Theorem 1.3. *For every $\alpha = (n_i)_{i\in\mathbb{N}} \in (\mathbb{N}\setminus\{0\})^{\mathbb{N}}$, we have $\nu_\alpha \ll \mu_\alpha$ and the measures ν_α and μ_α are not equivalent.*

Proof. Suppose that $\mu_\alpha(D) = 0$ for $D \in \mathcal{B}(\mathbb{R}^\infty)$. This means that D is covered by a countable family $(D_k)_{k\in\mathbb{N}}$ of elements of $(\alpha)\mathcal{OR}$ such that $D_k = \prod_{i\in\mathbb{N}} D_i^{(k)}$, $D_i^{(k)} \in \mathcal{B}(\mathbb{R}^{n_i})(k, i \in \mathbb{N})$ and $\mu_{D_k}(D \cap D_k) = 0$ for each k.

We have to show that for an arbitrary $\varepsilon > 0$ the inequality $\nu_\alpha(D) < \varepsilon$ holds.

If $\mu_{D_k}(D_k) = 0$, then it is obvious that $\mu_{D_k}(D \cap D_k) = 0 < \frac{\varepsilon}{2^{k+1}}$.

Now assume that $\mu_{D_k}(D_k) > 0$. We have $\mu_{D_k}(D \cap D_k) = 0$. By Charatheodory well known theorem there exists a sequence $(A_s^{(k,\varepsilon)})_{s\in\mathbb{N}} = (\prod_{i\in\mathbb{N}} A_i^{(s)})_{s\in\mathbb{N}}$ of elementary measurable rectangles in D_k for which $A_i^{(s)} \in R^{n_i}$ for $s, i \in \mathbb{N}$, $D \cap D_k \subseteq \cup_{s\in\mathbb{N}} A_s^{(k,\varepsilon)}$ and

$$\sum_{s\in\mathbb{N}} \mu_{D_k}\Big(\prod_{i\in\mathbb{N}} A_i^{(s)}\Big) < \frac{\varepsilon}{2^{k+1}}.$$

We set

$$A = \Big\{s : \sum_{i\in\mathbb{N}} ln(m^{n_i}(A_i^{(s)})) \text{ is not absolutely convergent}\Big\}.$$

Then we get

$$\nu_\alpha(D \cap D_k) \le \nu_\alpha(\cup_{s\in\mathbb{N}} A_s^{(k,\varepsilon)}) \le \sum_{s\in\mathbb{N}} \nu_\alpha(A_s^{(k,\varepsilon)}) = \sum_{s\in A} \nu_\alpha(A_s^{(k,\varepsilon)}) +$$

$$+ \sum_{s\in\mathbb{N}\setminus A} \nu_\alpha(A_s^{(k,\varepsilon)}) = \sum_{s\in A} (\mathbf{S}) \prod_{i\in\mathbb{N}} m^{n_i}(A_i^{(s)}) + \sum_{s\in\mathbb{N}\setminus A} (\mathbf{S}) \prod_{i\in\mathbb{N}} m^{n_i}(A_i^{(s)}) =$$

$$= 0 + \sum_{s\in\mathbb{N}\setminus A} (\mathbf{0}) \prod_{i\in\mathbb{N}} m^{n_i}(A_i^{(s)}) = \sum_{s\in\mathbb{N}\setminus A} \mu_\alpha(A_s^{(k,\varepsilon)}) \le \sum_{s\in\mathbb{N}} \mu_\alpha(A_s^{(k,\varepsilon)}) \le \frac{\varepsilon}{2^{k+1}}.$$

Finally, we get

$$\nu_\alpha(D) \le \sum_{k\in\mathbb{N}} \nu_\alpha(D \cap D_k) \le \sum_{k\in\mathbb{N}} \frac{\varepsilon}{2^{k+1}} = \varepsilon.$$

The proof of the fact that the measures ν_α and μ_α are not equivalent can be obtained as follows: Let $D = \prod_{i\in\mathbb{N}} D_i$, $D_i \in \mathcal{B}(\mathbb{R}^{n_i})(i \in \mathbb{N})$ such that $\mu^{n_0}(D_0) = 1$ and $\mu^{n_i}(D_i) = e^{\frac{(-1)^i}{i}}$ for $i \ge 1$. Then we get

$$\mu_\alpha(D) = (\mathbf{O}) \prod_{i\in\mathbb{N}} m^{n_i}(D_i) = 2$$

and

$$\nu_\alpha(D) = (\mathbf{S}) \prod_{i\in\mathbb{N}} m^{n_i}(D_i) = 0. \qquad \square$$

Remark 1.1. Note that the μ_α coincides with Baker's measure [3] for $\alpha = (1,1,\dots)$. By Lemmas 1.1 and 1.2 we can get the construction of Baker's measure [2]. In this direction we must consider a class $\mathcal{R}_B$ of all measurable rectangles $\prod_{i=1}^{\infty}(a_i,b_i)$, $-\infty < a_i < b_i < +\infty$ for which $0 \le (\mathbf{O})\prod_{i\in\mathbb{N}}(b_i - a_i) < +\infty$. Since $\mathcal{R}_B$ is translation-invariant and the family of measures $(\mu_R)_{R\in\mathcal{R}_B}$ is consistent as a subfamily of the consistent family of measures constructed in Lemma 1.2, we claim that Baker's measure [2] coincides with the measure $\lambda_{\mathcal{R}_B}$.

Definition 1.10. Let $\alpha = (n_i)_{i\in\mathbb{N}} \in (\mathbb{N}\setminus\{0\})^{\mathbb{N}}$ such that $n_i = n_j$ for every $i,j\in\mathbb{N}$. We set $F_i = (a_1^{(i)},\dots,a_{n_0}^{(i)})$ for every $i\in\mathbb{N}$(see, notations introduced before Definition 21.3. Let f be any permutation of $\mathbb{N}$ such that for every $i\in\mathbb{N}$ there exists $j\in\mathbb{N}$ such that $f(a_k^{(i)}) = a_k^{(j)}$ for $1\le k\le n_0$. Then a map $A_f : \mathbb{R}^\infty \to \mathbb{R}^\infty$ defined by $A_f((z_k)_{k\in\mathbb{N}}) = (z_{f(k)})_{k\in\mathbb{N}}$ for $(z_k)_{k\in\mathbb{N}} \in \mathbb{R}^\infty$, is called a canonical α-permutations of $\mathbb{R}^\infty$.

A group of transformations generated by all α-permutations and shifts of $\mathbb{R}^\infty$, is denoted by $\mathcal{G}_\alpha$.

Corollary 1.1. *For every $\alpha = (n_i)_{i\in\mathbb{N}} \in (\mathbb{N}\setminus\{0\})^{\mathbb{N}}$ for which $n_i = n_j (i,j\in\mathbb{N})$, the measure ν_α is $\mathcal{G}_\alpha$-invariant.*

One can easily get the validity of the following propositions.

Proposition 1.2. *For every $\alpha = (n_i)_{i\in\mathbb{N}} \in (\mathbb{N}\setminus\{0\})^{\mathbb{N}}$ there exists $\beta \in (\mathbb{N}\setminus\{0\})^{\mathbb{N}}$ such that μ_α and μ_β are different.*

Proposition 1.3. *For every $\alpha = (n_i)_{i\in\mathbb{N}} \in (\mathbb{N}\setminus\{0\})^{\mathbb{N}}$ there exists $\beta \in (\mathbb{N}\setminus\{0\})^{\mathbb{N}}$ such that ν_α and ν_β are different.*

As corollary of Propositions 1.2-1.3 we get

Corollary 1.2. *There does not exist a translation-invariant Borel measure λ in $\mathbb{R}^\infty$ such that for every $\alpha = (n_i)_{i\in\mathbb{N}} \in (\mathbb{N}\setminus\{0\})^{\mathbb{N}}$, and for every $D\in\mathcal{B}(\mathbb{R}^\infty)$ the following equality $\lambda(D) = \mu_\alpha(D)$ holds.*

Corollary 1.3. *There does not exist a translation-invariant Borel measure λ in $\mathbb{R}^\infty$ such that for every $\alpha = (n_i)_{i\in\mathbb{N}} \in (\mathbb{N}\setminus\{0\})^{\mathbb{N}}$, and for every $D\in\mathcal{B}(\mathbb{R}^\infty)$ the following equality $\lambda(D) = \nu_\alpha(D)$ holds.*

Corollary 1.4. *We set*

$$\mathcal{R} = \{R \,:\, R = [0,1]^{\mathbb{N}} + a \text{ for any } a\in\mathbb{R}^\infty\}$$

and

$$\mu_R(X) = \lambda(X - a \cap [0,1]^{\mathbb{N}})$$

for every $X\in\mathcal{B}(R)$, where $\lambda - \mu^{\mathbb{N}}$ and μ is a linear probability Lebesgue measure in $[0,1]$.

Then the family $(\mu_R)_{R\in\mathbb{R}}$ and the class $\mathcal{R}$, being invariant under the group $\mathcal{G}$, satisfy all conditions of Lemma 1.1. Hence $\mu_{\mathcal{R}}$ is a $\mathcal{G}$-invariant measure in $\mathbb{R}^\infty$.

Corollary 1.5. *Let $(L_i^{(n)})_{i\in I}$ be the family of all different n-dimensional vector subspaces of $\mathbb{R}^\infty$ and let $\ell_n^{(i)}$ be the n-dimensional Lebesgue measure in L_i.*

We set

$$\mathcal{R} = \{L_i^{(n)} + a : a \in \mathbb{R}^\infty, i \in I\}$$

and

$$\mu_{L_i^{(n)}+a}(X) = \ell_n^{(i)}(X - a \cap L_i^{(n)})$$

for every $X \in \mathcal{B}(\mathbb{R}^\infty)$.

Then the class $\mathcal{R}$, the family of measures $(\mu_R)_{R\in\mathbb{R}}$ and the group of all translations of $\mathbb{R}^\infty$ satisfy all conditions of Lemma 1.1. Hence there exists a translation-invariant Borel measure $\mu_{\mathcal{R}}$ such that $\mu_{\mathcal{R}}(X) = \mu_{L_i^{(n)}+a}(X)$ for every Borel subset $X \subset L_i^{(n)} + a$.

Though the next three examples are not the particular realizations of Lemma 1.1, they are of our interest.

Example 1.1. The Mankiewicz generator G_M [96] is the usual completion of the functional μ defined by

$$\mu(X) = \sum_{a\in\ell_1^\perp} \mu_{[0,1]^\mathbb{N}}(X - a \cap B_{[0,1]^\mathbb{N}})$$

for every $X \in \mathcal{B}(\mathbb{R}^\infty)$, where

(i) $\mu_{[0,1]^\mathbb{N}}$ denotes Kharazishvili's quasi-generator of shy sets in $\mathbb{R}^\infty$(see, [96]),

(ii) $B_{[0,1]^\mathbb{N}} = \cup_{n\in\mathbb{N}}\left(\mathbb{R}^n \times [0,1]^{\mathbb{N}\setminus\{1,\dots,n\}}\right)$,

(iii) $\ell_1^\perp$ denotes the linear complement of the vector subspace ℓ_1 in $\mathbb{R}^\infty$.

Setting $\mathcal{G} = \mathcal{G}_{\{1,1,\dots\}}$, we can say that G_M is a $\mathcal{G}$-invariant and has the property that X is a standard cube null set iff X is of G_M-measure zero.

The measure described in Corollary 1.4 is different from the Mankiewicz generator G_M. Indeed, if we consider the set $(2\mathbb{Z})^\mathbb{N}$, then we observe that it is not covered by the union of a countable family of elements of the class $\mathcal{R}$, and hence $\mu_{\mathcal{R}}(2\mathbb{Z}^\mathbb{N}) = +\infty$ whenever $G_M(2\mathbb{Z}^\mathbb{N}) = 0$.

Example 1.2. Let $(L_i)_{i\in I}$ be the family of all different n-dimensional vector subspaces of $\mathbb{R}^\infty$ and let $\ell_n^{(i)}$ be an n-dimensional Lebesgue measure on L_i. For $i \in I$, let us denote by $L_i^\perp$ the linear complement of L_i. Then the functional $G_{P\&T}$ defined by

$$G_{P\&T}^{(n)}(X) = \sum_{i\in I}\sum_{a\in L_i^\perp} \ell_n^{(i)}(X - a \cap L_i)$$

for $X \in \mathcal{B}(\mathbb{R}^\infty)$ is a $\mathcal{G}$-invariant Borel measure and $G_{P\&T}(Y) = 0$ iff Y is n-dimensional null in the sense of [122](see, [96]).

Note that the measure $G_{P\&T}^{(n)}$ and the measure $\mu_{\mathcal{R}}$ described in Corollary 1.5 are different. Indeed, for $n > 1$, let S_n be an n-dimensional sphere lying in any $n+1$-dimensional vector subspace of $\mathbb{R}^\infty$. Then, $G_{P\&T}^{(n)}(S_n) = 0$, whenever $\mu_{\mathcal{R}}(S_n) = +\infty$ because it is not covered by countable family of elements of $\mathcal{R}$.

Remark 1.2. For a set $\prod_{k\in\mathbb{N}} X_k$, where $X_k=[0,\frac{1}{2}]$ for even k and $X_k=[0,k]$ for odd k, we have

$$+\infty=\lambda\Big(\prod_{n\in\mathbb{N}} X_k\Big)\neq(\mathbf{S})\prod_{k\in\mathbb{N}} m(X_k)=0$$

for each Baker's measure λ [2], [3].

For $Y_k=[0,1]$ $(k\in\mathbb{N})$, the validity of the following condition

$$+\infty=\mu_{\mathcal{R}}\Big(\prod_{n\in\mathbb{N}} Y_k\Big)=G^{(n)}_{P\&T}\Big(\prod_{n\in\mathbb{N}} Y_k\Big)>1=(\mathbf{S})\prod_{k\in\mathbb{N}} m(Y_k)$$

implies that the measures described in Corollary 1.5 and Example 1.2 are not α-standard Lebesgue measures for $\alpha=(1,1,\dots)$.

For the Mankiewicz generator G_M described in Example 1.1 we have that

$$G_M\Big(\prod_{k\in\mathbb{N}} X_k\Big)=0,$$

but for the set $\prod_{k\in\mathbb{N}} Z_k=\prod_{k\in\mathbb{N}}\big([0,\frac{1}{2}]\cup[1,\frac{3}{2}]\big)$ we get

$$0=G_M\Big(\prod_{n\in\mathbb{N}} Z_k\Big)\neq(\mathbf{S})\prod_{k\in\mathbb{N}} m(Z_k)=1.$$

Example 1.3 ([91]). Let, for $k\in\mathbb{N}$, S_k be the unit circle in the Euclidean plane $\mathbb{R}^2$. We may identify the circle S_k with a compact group of all rotations of $\mathbb{R}^2$ around its origin. Let $\lambda_{\mathbb{N}}$ be the probability Haar measure defined on the compact group $\prod_{k\in N} S_k$. For $k\in\mathbb{N}$, define the function f_k by $f_k(x)=\exp\{2\pi xi\}$ for every $x\in\mathbb{R}$.

For $E\subset\mathbb{R}^{\mathbb{N}}$ and $g\in\prod_{k\in\mathbb{N}} S_k$, put

$$f_E(g)=\begin{cases}\operatorname{card}\Big(\Big(\prod_{k\in\mathbb{N}} f_k\Big)^{-1}(g)\cap E\Big), & \text{if this is finite;}\\ +\infty, & \text{in all other cases.}\end{cases}$$

In the Solovay model [132], we define the functional $\mu_{\mathbb{N}}$ by

$$(\forall E)\Big(E\subseteq\mathbb{R}^\infty\to\mu_{\mathbb{N}}(E)=\int_{\prod_{k\in\mathbb{N}} S_k} f_E(g)d\lambda_{\mathbb{N}}(g)\Big).$$

It was established in [91] that $\mu_{\mathbb{N}}$ is a translation-invariant Borel measure in $\mathbb{R}^\infty$ which takes the value one on the set $[0,1]^{\mathbb{N}}$.

Let us show that the measure $\mu_{\mathbb{N}}$ is not a α-standard Lebesgue measure on $\mathbb{R}^\infty$ for $\alpha=(1,1,\dots)$. Indeed, let us consider an infinite-dimensional measurable rectangles $R\in\mathcal{B}(\mathbb{R}^\infty)$ of the form

$$R=\prod_{i\in\mathbb{N}} R_i,$$

where $R_i=\cup_{k=1}^{i}[k,k+\frac{1}{i}[$ for every $i\in\mathbb{N}$.

It is obvious that $m(R_i) = 1$ for every $i \in \mathbb{N}$, which implies that

$$0 < 1 = (\mathbf{S})\prod_{k\in\mathbb{N}} m(R_k) < +\infty.$$

Note that $f_{\prod_{i\in\mathbb{N}} R_i}(g) = +\infty$ if $g \in \prod_{k\in\mathbb{N}} f_k([0,\frac{1}{k}[)$ and $= 0$, otherwise. Hence

$$\mu_{\mathbb{N}}\Big(\prod_{i\in\mathbb{N}} R_i\Big) = \int\limits_{\prod_{k\in\mathbb{N}} S_k} f_{\prod_{i\in\mathbb{N}} R_i}(g)d\lambda_{\mathbb{N}}(g)) =$$

$$+\infty\times\lambda_{\mathbb{N}}\Big(\prod_{k\in\mathbb{N}} f_k\Big(\Big[0,\frac{1}{k}\Big[\Big)\Big)+0\times\lambda_{\mathbb{N}}\Big(\prod_{k\in\mathbb{N}} S_k\setminus\prod_{k\in\mathbb{N}} f_k\Big(\Big[0,\frac{1}{k}\Big[\Big)\Big)=0<1=(\mathbf{S})\prod_{k\in\mathbb{N}} m(R_k).$$

Remark 1.3. Example 1.3 shows us that Conjecture 1 (see, [99], p 9) is not valid, i.e. for every $\mu_\alpha \in \mathrm{O}(\alpha)\mathrm{LM}$ $(\alpha \in (\mathbb{N}\setminus\{0\})^{\mathbb{N}})$ and for every $D \in \mathcal{B}(\mathbb{R}^\infty)$ with $0 \leq \mathrm{O}(\alpha)\mathrm{LM}(D) < \infty$ the following equality $\mu_{\mathbb{N}}(D) = \mu_\alpha(D)$ is not valid. Corollary 1.2 contains more strict result, in particular, it answers negatively to the Problem 2 (see, [99], p. 9).

Chapter 2

On Uniformly Distributed Sequences of an Increasing Family of Finite Sets in Infinite-Dimensional Rectangles

2.1 Introduction

Following [64], a sequence $s_1, s_2, s_3, \ldots$ of real numbers from an interval $[a,b]$ is said to be equidistributed, or uniformly distributed in that interval if the proportion of terms falling in a subinterval $[c,d]$ is proportional to the length of that interval. Such sequences are studied in Diophantine approximation theory and have applications to Monte Carlo integration (see, for example, [64], [47], [48]).

Let $\mathcal{R}$ be the class of all infinite dimensional rectangles R of the form

$$R = \prod_{i=1}^{\infty} [a_i, b_i], \quad -\infty < a_i < b_i < +\infty$$

with $0 < \prod_{i=1}^{\infty}(b_i - a_i) < +\infty$, where

$$\prod_{i=1}^{\infty}(b_i - a_i) := \lim_{n\to\infty} \prod_{i=1}^{n}(b_i - a_i).$$

In [2] has been constructed such a translation invariant Borel measure λ on $\mathbb{R}^\infty$ that

$$\lambda(R) = \prod_{i=1}^{\infty}(b_i - a_i)$$

for $R \in \mathcal{R}$.

The purpose of the present chapter is to consider a concept of the uniform distribution in infinite-dimensional rectangles for a calculation of Riemann integrals over such rectangles. Note that similar topics have been discussed in [102].

The chapter is organized as follows.

In Section 2.2 we consider some auxiliary notions and facts belonging to Weyl [144]. In Section 2.3 we give proof of main results. In particular, we prove infinite-dimensional versions of Lebesgue [86] and Weyl [144] famous results. In Section 2.4 is discussed the property of a spatial class of real-valued sequences which are uniformly distributed in infinite-dimensional rectangles.

2.2 Auxiliary Notions and Propositions

Definition 2.2.1. A bounded sequence $s_1, s_2, s_3, \dots$ of real numbers is said to be equidistributed or uniformly distributed on an interval $[a,b]$ if for any subinterval $[c,d]$ of the $[a,b]$ we have

$$\lim_{n \to \infty} \frac{\#(\{s_1, s_2, s_3, \dots, s_n\} \cap [c,d])}{n} = \frac{d-c}{b-a},$$

where # denotes a counting measure.

Remark 2.2.1. For $a \leq c < d \leq b$, let $][c,d][$ denotes a subinterval of the $[a,b]$ which has one of the following forms $[c,d], [c,d[,]c,d[$ or $]c,d]$. Then it is obvious to show that a bounded sequence $s_1, s_2, s_3, \dots$ of real numbers is equidistributed or uniformly distributed on an interval $[a,b]$ iff, for any subinterval $][c,d][$ of the $[a,b]$, we have

$$\lim_{n \to \infty} \frac{\#(\{s_1, s_2, s_3, \dots, s_n\} \cap][c,d][)}{n} = \frac{d-c}{b-a}.$$

Definition 2.2.2. (Weyl [144]) The sequence $s_1, s_2, s_3, \dots$ is said to be equidistributed modulo 1 or uniformly distributed modulo 1 if the sequence $(s_n - [s_n])_{n \in N}$ of the fractional parts of the $(s_n)_{n \in N}$'s, is equidistributed (equivalently, uniformly distributed) in the interval $[0,1]$.

Example 2.2.1. ([64], Exercise 1.12, p. 16) The sequence of all multiples of an irrational α

$$0, \alpha, 2\alpha, 3\alpha, \dots$$

is uniformly distributed modulo 1.

Example 2.2.2. ([64], Exercise 1.13, p. 16) The sequence

$$\frac{0}{1}, \frac{0}{2}, \frac{1}{2}, \frac{0}{3}, \frac{1}{3}, \frac{2}{3}, \dots, \frac{0}{k}, \dots, \frac{k-1}{k}, \dots$$

is uniformly distributed modulo 1.

Example 2.2.3. The sequence of all multiples of an irrational α by successive prime numbers

$$2\alpha, 3\alpha, 5\alpha, 7\alpha, 11\alpha, \dots$$

is equidistributed modulo 1. This is a famous theorem of analytic number theory, proved by I. M. Vinogradov in 1935 (see, [139]).

Agreement. In the sequel, unlike N. Bourbaki well known notion, under N we understand a set $\{1, 2, \dots\}$.

Remark 2.2.2. If $(s_k)_{k\in N}$ is uniformly distributed modulo 1, then $((s_k-[s_k])(b-a)+a)_{k\in N}$ is uniformly distributed in an interval $[a,b)$.

The following assertion contains an interesting application of uniformly distributed sequences for a calculation of Riemann integrals.

Lemma 2.2.1 (Weyl [144]). *The following two conditions are equivalent:*
(i) $(a_n)_{n\in N}$ *is equidistributed modulo* 1*;*
(ii) For every Riemann integrable function f *on* $[0,1]$

$$\lim_{n\to\infty}\frac{1}{n}\sum_{j=1}^{n}f(a_j)=\int_{[0,1]}f(x)dx.$$

2.3 On Uniformly Distributed Sequences of Increasing Family of Finite Sets in Infinite-Dimensional Rectangles

Let $s_1,s_2,s_3,\dots$ be a uniformly distributed in an interval $[a,b]$. Setting $Y_n=\{s_1,s_2,s_3,\dots,s_n\}$ for $n\in N$, the $(Y_n)_{n\in N}$ will be such an increasing sequence of finite subsets of the $[a,b]$ that, for any subinterval $[c,d]$ of the $[a,b]$, the following equality

$$\lim_{n\to\infty}\frac{\#(Y_n\cap[c,d])}{\#(Y_n)}=\frac{d-c}{b-a}$$

will be valid.

This remark raises the following

Definition 2.3.1. An increasing sequence $(Y_n)_{n\in N}$ of finite subsets of the $[a,b]$ is said to be equidistributed or uniformly distributed in an interval $[a,b]$ if, for any subinterval $[c,d]$ of the $[a,b]$, we have

$$\lim_{n\to\infty}\frac{\#(Y_n\cap[c,d])}{\#(Y_n)}=\frac{d-c}{b-a}.$$

Definition 2.3.2. Let $\prod_{k\in N}[a_k,b_k]\in\mathcal{R}$. A set U is called an elementary rectangle in the $\prod_{k\in N}[a_k,b_k]$ if it admits the following representation

$$U=\prod_{k=1}^{m}][c_k,d_k][\times\prod_{k\in N\setminus\{1,\dots,m\}}[a_k,b_k],$$

where $a_k\le c_k<d_k\le b_k$ for $1\le k\le m$.

It is obvious that

$$\lambda(U)=\prod_{k=1}^{m}(d_k-c_k)\times\prod_{k=m+1}^{\infty}(b_k-a_k),$$

for the elementary rectangle U.

Definition 2.3.3. An increasing sequence $(Y_n)_{n\in N}$ of finite subsets of the infinite-dimensional rectangle $\prod_{k\in N}[a_k,b_k] \in \mathcal{R}$ is said to be uniformly distributed in the $\prod_{k\in N}[a_k,b_k]$ if for every elementary rectangle U in the $\prod_{k\in N}[a_k,b_k[$ we have

$$\lim_{n\to\infty}\frac{\#(Y_n\cap U)}{\#(Y_n)} = \frac{\lambda(U)}{\lambda(\prod_{k\in N}[a_k,b_k[)}.$$

Theorem 2.3.1. *Let $\prod_{k\in N}[a_k,b_k] \in \mathcal{R}$. Let $(x_n^{(k)})_{n\in N}$ be uniformly distributed in the interval $[a_k,b_k]$ for $k \in N$. We set*

$$Y_n = \prod_{k=1}^{n}(\cup_{j=1}^{n} x_j^{(k)}) \times \prod_{k\in N\setminus\{1,\dots,n\}}\{a_k\}.$$

Then $(Y_n)_{n\in N}$ is uniformly distributed in the rectangle $\prod_{k\in N}[a_k,b_k]$.

Proof. Let U be an elementary rectangle in the $\prod_{k\in N}[a_k,b_k]$.

Since $(x_n^{(k)})_{n\in N}$ is uniformly distributed in the interval $[a_k,b_k]$ for $k \in N$, by Remark 2.2.1, we have

$$\lim_{n\to\infty}\frac{\#(\{x_1^{(k)},x_2^{(k)},\dots,x_n^{(k)}\}\cap][c_k,d_k][)}{n} = \frac{d_k-c_k}{b_k-a_k}.$$

Hence, we have

$$\lim_{n\to\infty}\frac{\#(Y_n\cap U)}{\#(Y_n)} = \lim_{n\to\infty}\prod_{k=1}^{m}\frac{\#(\{x_1^{(k)},x_2^{(k)},\dots,x_n^{(k)}\}\cap][c_k,d_k][)}{n}$$
$$= \prod_{k=1}^{m}\lim_{n\to\infty}\frac{\#(\{x_1^{(k)},x_2^{(k)},\dots,x_n^{(k)}\}\cap][c_k,d_k][)}{n}$$
$$= \prod_{k=1}^{m}\frac{d_k-c_k}{b_k-a_k} = \frac{\lambda(U)}{\lambda(\prod_{k\in N}[a_k,b_k])}.$$

The theorem is proved. □

Remark 2.3.1. In context with Theorem 2.3.1, it is natural to ask *whether there exists an increasing sequence of finite subsets $(Y_n)_{n\in N}$ such that*

$$\lim_{n\to\infty}\frac{\#(Y_n\cap U)}{\#(Y_n)} = \frac{\lambda(U)}{\lambda(\prod_{k\in N}[a_k,b_k])}$$

for every infinite-dimensional rectangle $U = \prod_{k\in N}X_k \subset \prod_{k\in N}[a_k,b_k]$, where, for each $k \in N$, X_k is a finite sum of pairwise disjoint intervals in $[a_k,b_k]$?

Let us show that the answer to this question is no.

Indeed, assume the contrary and let $(Y_n)_{n\in N}$ be such an increasing sequence of finite subsets in $\prod_{k\in N}[a_k,b_k[$. Then we have

$$\cup_{n\in N}Y_n = \{(x_i^{(k)})_{i\in N} : k \in N\}.$$

For $k \in N$, we set $X_k = [a_k, b_k] \setminus x_k^{(k)}$. Then, it is clear that

$$\lambda\Big(\prod_{k\in N} X_k\Big) = \lambda\Big(\prod_{k\in N}[a_k, b_k]\Big)$$

and

$$\frac{\#(Y_n \cap \prod_{k\in N} X_k)}{\#(Y_n)} = 0$$

for $k \in N$, which follows

$$\lim_{n\to\infty} \frac{\#(Y_n \cap \prod_{k\in N} X_k)}{\#(Y_n)} = 0 < 1 = \frac{\lambda(\prod_{k\in N} X_k)}{\lambda(\prod_{k\in N}[a_k, b_k])}.$$

Definition 2.3.4. Let $\prod_{k\in N}[a_k, b_k] \in \mathcal{R}$. A family of pairwise disjoint elementary rectangles $\tau = (U_k)_{1\le k\le n}$ of the $\prod_{k\in N}[a_k, b_k]$ is called Riemann partition of the $\prod_{k\in N}[a_k, b_k]$ if $\cup_{1\le k\le n} U_k = \prod_{k\in N}[a_k, b_k]$.

Definition 2.3.5. Let $\tau = (U_k)_{1\le k\le n}$ be Riemann partition of the $\prod_{k\in N}[a_k, b_k]$. Let $\ell(Pr_i(U_k))$ be a length of the i-th projection $Pr_i(U_k)$ of the U_k for $i \in N$. We set

$$d(U_k) = \sum_{i\in N} \frac{\ell(Pr_i(U_k))}{2^i(1+\ell(Pr_i(U_k)))}.$$

It is obvious that $d(U_k)$ is a diameter of the elementary rectangle U_k for $k \in N$ with respect to Tikhonov metric ρ defined as follows

$$\rho((x_k)_{k\in N}, (y_k)_{k\in N}) = \sum_{k\in N} \frac{|x_k - y_k|}{2^k(1+|x_k - y_k|)}$$

for $(x_k)_{k\in N}, (y_k)_{k\in N} \in \mathbf{R}^\infty$.

A number $d(\tau)$, defined by

$$d(\tau) = \max\{d(U_k) : 1 \le k \le n\}$$

is called mesh or norm of the Riemann partition τ.

Definition 2.3.6. Let $\tau_1 = (U_i^{(1)})_{1\le i\le n}$ and $\tau_2 = (U_j^{(2)})_{1\le j\le m}$ be Riemann partitions of the $\prod_{k\in N}[a_k, b_k]$. We say that $\tau_2 \le \tau_1$ iff

$$(\forall j)((1 \le j \le m) \to (\exists i_0)(1 \le i_0 \le n \,\&\, U_j^{(2)} \subseteq U_{i_0}^{(1)})).$$

Definition 2.3.7. Let f be a real-valued bounded function defined on the $\prod_{i\in N}[a_i, b_i]$. Let $\tau = (U_k)_{1\le k\le n}$ be Riemann partition of the $\prod_{k\in N}[a_k, b_k]$ and $(t_k)_{1\le k\le n}$ be a sample such that, for each k, $t_k \in U_k$. Then

(i) a sum $\sum_{k=1}^n f(t_k)\lambda(U_k)$ is called Riemann sum of the f with respect to Riemann partition $\tau = (U_k)_{1\le k\le n}$ together with sample $(t_k)_{1\le k\le n}$;

(ii) a sum $S_\tau = \sum_{k=1}^n M_k\lambda(U_k)$ is called the upper Darboux sum with respect to Riemann partition τ, where $M_k = \sup_{x\in U_k} f(x)(1 \le k \le n)$;

(ii) a sum $s_\tau = \sum_{k=1}^n m_k\lambda(U_k)$ is called the lower Darboux sum with respect to Riemann partition τ, where $m_k = \inf_{x\in U_k} f(x)(1 \le k \le n)$.

Definition 2.3.8. Let f be a real-valued bounded function defined on $\prod_{i\in N}[a_i,b_i[$. We say that the f is Riemann-integrable on $\prod_{i\in N}[a_i,b_i]$ if there exists a real number s such that for every positive real number ε there exists a real number $\delta > 0$ such that, for every Riemann partition $(U_k)_{1\le k\le n}$ of the $\prod_{k\in N}[a_k,b_k]$ with $d(\tau) < \delta$ and for every sample $(t_k)_{1\le k\le n}$, we have

$$\Big|\sum_{k=1}^{n} f(t_k)\lambda(U_k) - s\Big| < \varepsilon.$$

The number s is called Riemann integral and is denoted by

$$(R)\int_{\prod_{k\in N}[a_k,b_k]} f(x)d\lambda(x).$$

Definition 2.3.9. A function f is called a step function on $\prod_{k\in N}[a_k,b_k]$ if it can be written as

$$f(x) = \sum_{k=1}^{n} c_k \mathcal{X}_{U_k}(x),$$

where $\tau = (U_k)_{1\le k\le n}$ is any Riemann partition of the $\prod_{k\in N}[a_k,b_k]$, $c_k \in R$ for $1 \le k \le n$ and $\mathcal{X}_A$ is the indicator function of the A

Theorem 2.3.2. *Let f be a continuous function on $\prod_{k\in N}[a_k,b_k]$ with respect to Tikhonov metric ρ. Then the f is Riemann-integrable on $\prod_{k\in N}[a_k,b_k]$.*

Proof. It is obvious that, for every Riemann partition $\tau = (U_k)_{1\le k\le n}$ of the $\prod_{k\in N}[a_k,b_k]$ and for every sample $(t_k)_{1\le k\le n}$ with $t_k \in U_k (1 \le k \le n)$, we have

$$s_\tau \le \sum_{k=1}^{n} f(t_k)\lambda(U_k) \le S_\tau.$$

Note that if τ_1 and τ_2 are two Riemann partitions of the $\prod_{k\in N}[a_k,b_k]$ such that $\tau_2 \le \tau_1$, then

$$s_{\tau_1} \le s_{\tau_2} \le \sum_{k=1}^{n} f(t_k)\lambda(U_k) \le S_{\tau_2} \le S_{\tau_1}.$$

Let us show the validity of the condition

$$(\forall\varepsilon)(\varepsilon > 0 \to (\exists r)(\forall\tau)(d(\tau) < r \to S_\tau - s_\tau < \varepsilon)),$$

which yields that $\inf_\tau S_\tau = \sup_\tau s_\tau$.

Following Tikhonov theorem, the $\prod_{k\in N}[a_k,b_k]$ is compact set in the Polish group $\mathbf{R}^\infty$ equipped with Tikhonov metric ρ.

Following Cantor well known result, the function f is uniformly continuous on the $\prod_{k\in N}[a_k,b_k]$. Hence, for $\varepsilon > 0$, there exists $r > 0$ such that

$$(\forall x,y)\Big(x,y \in \prod_{k\in N}[a_k,b_k] \,\&\, \rho(x,y) < r \to |f(x)-f(y)| \le \frac{\varepsilon}{\lambda(\prod_{k\in N}[a_k,b_k])}\Big).$$

Thus, for every Riemann partition $\tau = (U_k)_{1 \leq k \leq n}$ with $d(\tau) < r$ we get

$$S_\tau - s_\tau \leq \frac{\varepsilon}{\lambda(\prod_{k \in N}[a_k, b_k[)} \times \sum_{1 \leq k \leq n} \lambda(U_k) = \varepsilon.$$

Thus, $\inf_\tau S_\tau = \sup_\tau s_\tau$.

Finally, setting $\delta = r$ and $s = \inf_\tau S_\tau$, we deduce that for every Riemann partition $(U_k)_{1 \leq k \leq n}$ of the $\prod_{k \in N}[a_k, b_k]$ with $d(\tau) < \delta$ and for every sample $(t_k)_{1 \leq k \leq n}$ with $t_k \in U_k (1 \leq k \leq n)$, we have

$$\left| \sum_{k=1}^{n} f(t_k)\lambda(U_k) - s \right| \leq S_\tau - s_\tau \leq \varepsilon.$$

This ends the proof of Theorem 2.3.2. □

We have the following infinite-dimensional version of the Lebesgue theorem (see, [85], Lebesgue Theorem, p. 359).

Theorem 2.3.3. *Let f be a bounded real-valued function on $\prod_{k \in N}[a_k, b_k] \in \mathcal{R}$. Then f is Riemann integrable on $\prod_{k \in N}[a_k, b_k]$ if and only if f is λ-almost continuous on $\prod_{k \in N}[a_k, b_k]$.*

Proof. Necessity. Let f be a Riemann integrable function on $\prod_{k \in N}[a_k, b_k] \in \mathcal{R}$.

Then, for every $\varepsilon > 0$ and $\mu > 0$, there exists a Riemann partition $\tau = (U_k)_{1 \leq k \leq n}$ such that

$$\begin{aligned} \varepsilon \times \mu \geq S_\tau - s_\tau &\geq \sum_{1 \leq k \leq n} (M_k - m_k)\lambda(U_k) \\ &\geq \sum_{k \in I_1} (M_k - m_k)\lambda(U_k) \geq \mu \sum_{k \in I_1} \lambda(U_k), \end{aligned} \tag{2.3.1}$$

where $I_1 = \{k : 1 \leq k \leq n \ \& \ U_k$ contains at least one inner point p belonging to the set $E_\mu\}$, where

$$E_\mu = \left\{ x : x \in \prod_{k \in N} [a_k, b_k] \ \& \ \omega(f, x) \geq \mu \right\}$$

and

$$\omega(f, x) = \lim_{\delta \to 0} \sup_{x', x'' \in V(x, \delta) \cap \prod_{k \in N}[a_k, b_k]} \left| f(x') - f(x'') \right|.$$

Here, for $x \in \mathbf{R}^\infty$ and $\delta > 0$, $V(x, \delta)$ is denoted by

$$V(x, \delta) = \left\{ y : y \in \prod_{k \in N} [a_k, b_k] \ \& \rho(x, y) \leq \delta \right\}.$$

Since, for $k \in I_1$, p is an inner point of the U_k, there exists $V(p, \delta(k, p))$ such that $V(p, \delta(k, p)) \subseteq U_k$. Note that

Since $\omega(f, p) \geq \mu$, we have

$$M_k - m_k \geq M_p - m_p \geq \omega(f, p) \geq \mu,$$

where

$$M_p = \sup_{x \in V(p,\delta(k,p))} f(x), \; m_\delta = \inf_{x \in V(p,\delta(k,p))} f(x).$$

From (2.3.1), we get

$$\varepsilon \geq \sum_{k \in I_1} \lambda(U_k).$$

Another points of the E_μ, which are not inner points of elements of the partition τ, may be placed on the boundary of elements of the τ, whose λ-measure is zero.

Thus, for $\mu > 0$, we have

$$\lambda(E_\mu) \leq \sum_{k \in I_1} \lambda(U_k) + \lambda(\cup_{1 \leq k \leq n} \partial(U_k)) \leq \frac{\varepsilon}{\mu},$$

which yields that $\lambda(E_\mu) = 0$. Since a set E of all points of discontinuity of the f admits the representation $E = \cup_{k=1}^{\infty} E_{\frac{1}{k}}$, we deduce that $\lambda(E) = 0$.

This ends Necessity.

Proof of the sufficiency. Let, for $K \in \mathbf{R}^+$, we have $|f(x)| \leq K$ whenever $x \in \prod_{k \in N}[a_k, b_k]$.

Suppose that f is λ-almost continuous on $\prod_{k \in N}[a_k, b_k]$.

For $\varepsilon > 0$, let μ be such a positive number that

$$4\mu\lambda\Big(\prod_{k \in N}[a_k, b_k]\Big) < \varepsilon.$$

Since, for a set E of all points of discontinuity of the f on $\prod_{k \in N}[a_k, b_k]$ we have $\lambda(E) = 0$, we easily claim that $\lambda(E_\mu) = 0$. Since E_μ is closed in $\prod_{k \in N}[a_k, b_k]$, we claim that E_μ is compact. Hence, for $\varepsilon > 0$, there exists a finite family of open elementary rectangles in $\prod_{k \in N}[a_k, b_k]$ whose union covers E_μ such that

$$\lambda(\cup_{1 \leq k \leq n} U_k) < \frac{\varepsilon}{4K}.$$

Finally, we have

$$\prod_{k \in N}[a_k, b_k] = \cup_{1 \leq k \leq n} U_k \cup F,$$

where F is a compact subset in $\prod_{k \in N}[a_k, b_k]$.

It is obvious that, for every point $x \in F$, we have $\omega(f, x) < \mu$. Since F is compact, we can choose $\delta > 0$ such that for every $x, x' \in F$ the condition $\rho(x, x') < \delta$ yields $|f(x) - f(x')| < 2\lambda$.

Since F is a finite union of elementary rectangles in $\prod_{k \in N}[a_k, b_k]$ (this follows from the fact that the class of all elementary rectangles in $\prod_{k \in N}[a_k, b_k]$ is ring), there exists a partition $\tau_1 = (F_i)_{2 \leq i \leq m}$ of the F such that, for i with $2 \leq i \leq m$, F_i is an elementary rectangle in $\prod_{k\in N}[a_k,b_k]$ with $d(F_i)<\delta$. Then $\tau=\{\cup_{1\le k\le n}U_k,F_2,\dots,F_m\}$ will be such a Riemann partition of the $\prod_{k\in N}[a_k,b_k]$ that

$$S_\tau - s_\tau = (M_1-m_1)\lambda(\cup_{1\le k\le n}U_k)+\sum_{1\le i\le m}(M_i-m_i)\lambda(F_k)$$
$$\le \frac{\varepsilon}{2}+2\mu\lambda(\prod_{k\in N}[a_k,b_k]))\le\frac{\varepsilon}{2}+\frac{\varepsilon}{2}=\varepsilon.$$

The theorem is proved. □

Remark 2.3.2. Note that Theorem 2.3.2 can be considered as a simple consequence of Theorem 2.3.3. Now, by using Theorem 2.3.3, one can extend the concept of Riemann integrability theory for functions defined in the topological vector space $\mathbf{R}^\infty$ of all real-valued sequences equipped with Tikhonov topology.

In the sequel we need some important notions and well-known results from general topology and measure theory.

Definition 2.3.10. A topological Hausdorff space X is called normal if given any disjoint closed sets E and F, there are neighborhoods U of E and V of F that are also disjoint.

Lemma 2.3.1 (Urysohn [137]). *A topological space X is normal if and only if any two disjoint closed sets can be separated by a function. That is, given disjoint closed sets E and F, there is a continuous function f from X to $[0,1]$ such that the preimages of 0 and 1 under f are E and F respectively.*

Remark 2.3.3. Since all compact Hausdorff spaces are normal, we deduce that $\prod_{k\in N}[a_k,b_k]$ equipped with Tikhonov topology, is normal. By Urysohn's lemma we deduce that any two disjoint closed sets in $\prod_{k\in N}[a_k,b_k]$ can be separated by a function.

Definition 2.3.11. A Borel measure μ, defined on a Hausdorff topological space X, is called Radon if

$$(\forall Y)(Y\in\mathcal{B}(X)\ \&\ 0\le\mu(Y)<+\infty\rightarrow\mu(Y)=\sup_{\substack{K\subseteq Y\\ \text{K is compact in X}}}\mu(K)).$$

Lemma 2.3.2 (Ulam [136]). *Every probability Borel measure defined on Polish metric space is Radon.*

In the sequel we denote by $C(\prod_{k\in N}[a_k,b_k])$ a class of all continuous (with respect to Tikhonov topology) real-valued functions on $\prod_{k\in N}[a_k,b_k]$.

Theorem 2.3.4. *For $\prod_{i\in N}[a_i,b_i]\in\mathcal{R}$, let $(Y_n)_{n\in N}$ be an increasing family its finite subsets. Then $(Y_n)_{n\in N}$ is uniformly distributed in the $\prod_{k\in N}[a_k,b_k]$ if and only if for every $f\in C(\prod_{k\in N}[a_k,b_k])$ the following equality*

$$\lim_{n\to\infty}\frac{\sum_{y\in Y_n}f(y)}{\#(Y_n)}=\frac{(R)\int_{\prod_{k\in N}[a_k,b_k]}f(x)d\lambda(x)}{\lambda(\prod_{i\in N}[a_i,b_i])}$$

holds.

Proof. Necessity. Let $(Y_n)_{n\in N}$ be a uniformly distributed in the $\prod_{k\in N}[a_k,b_k]$ and let $f(x)=\sum_{k=1}^m c_k\mathcal{X}_{U_k}(x)$ be a step function. Then we have

$$\lim_{n\to\infty}\frac{\sum_{y\in Y_n}f(y)}{\#(Y_n)}=\lim_{n\to\infty}\frac{\sum_{y\in Y_n}\sum_{k=1}^m c_k\mathcal{X}_{U_k}(y)}{\#(Y_n)}$$

$$=\lim_{n\to\infty}\frac{\sum_{k=1}^m c_k\#(U_k\cap Y_n)}{\#(Y_n)}=\sum_{k=1}^m c_k\lim_{n\to\infty}\frac{\#(U_k\cap Y_n)}{\#(Y_n)}$$

$$=\sum_{k=1}^m c_k\frac{\lambda(U_k)}{\lambda(\prod_{i\in N}[a_i,b_i])}=\frac{(R)\int_{\prod_{k\in N}[a_k,b_k]}f(x)d\lambda(x)}{\lambda(\prod_{i\in N}[a_i,b_i])}.$$

Now, let $f\in C(\prod_{k\in N}[a_k,b_k])$. By Theorem 2.3.2 we deduce that f is Riemann-integrable. From the definition of the Reimann integral we deduce that, for every positive ε, there exists two step functions f_1 and f_2 on $\prod_{i\in N}[a_i,b_i]$ such that

$$f_1(x)\le f(x)\le f_2(x)$$

and

$$(R)\int_{\prod_{i\in N}[a_i,b_i]}(f_1(x)-f_2(x))d\lambda(x)<\varepsilon.$$

Then we have

$$(R)\int_{\prod_{i\in N}[a_i,b_i]}f(x)d\lambda(x)-\varepsilon\le(R)\int_{\prod_{i\in N}[a_i,b_i]}f_1(x)d\lambda(x)$$

$$=\lambda\Big(\prod_{i\in N}[a_i,b_i]\Big)\times\lim_{n\to\infty}\frac{\sum_{y\in Y_n}f_1(y)}{\#(Y_n)}\le\lambda\Big(\prod_{i\in N}[a_i,b_i]\Big)\times\underline{\lim}_{n\to\infty}\frac{\sum_{y\in Y_n}f(y)}{\#(Y_n)}$$

$$\le\lambda\Big(\prod_{i\in N}[a_i,b_i]\Big)\times\overline{\lim}_{n\to\infty}\frac{\sum_{y\in Y_n}f(y)}{\#(Y_n)}\le\lim_{n\to\infty}\frac{\sum_{y\in Y_n}f_2(y)}{\#(Y_n)}$$

$$\le\lambda\Big(\prod_{i\in N}[a_i,b_i]\Big)\times(R)\int_{\prod_{i\in N}[a_i,b_i]}f_2(x)d\lambda(x)\le(R)\int_{\prod_{i\in N}[a_i,b_i]}f(x)d\lambda(x)+\varepsilon.$$

The latter relation yields an existence of the limit $\lim_{n\to\infty}\frac{\sum_{y\in Y_n}f(y)}{\#(Y_n)}$ such that

$$\lim_{n\to\infty}\frac{\sum_{y\in Y_n}f(y)}{\#(Y_n)}=\frac{(R)\int_{\prod_{k\in N}[a_k,b_k]}f(x)d\lambda(x)}{\lambda(\prod_{i\in N}[a_i,b_i])}.$$

This ends the proof of Necessity.

Sufficiency. Assume that $(Y_n)_{n\in N}$ is an increasing sequence of subsets of the $\prod_{k\in N}[a_k,b_k]$ such that for every $f\in C(\prod_{k\in N}[a_k,b_k])$ the following equality

$$\lim_{n\to\infty}\frac{\sum_{y\in Y_n}f(y)}{\#(Y_n)}=\frac{(R)\int_{\prod_{k\in N}[a_k,b_k[}f(x)d\lambda(x)}{\lambda(\prod_{i\in N}[a_i,b_i])}$$

holds.

Let U be any elementary rectangle in $\prod_{i \in N}[a_i, b_i]$.

For $\varepsilon > 0$, by Ulam's lemma we can choose such a compact set

$$F \subset \prod_{k \in N}[a_k, b_k] \setminus [U]_T,$$

that $\lambda((\prod_{k \in N}[a_k, b_k] \setminus [U]_T) \setminus F) < \frac{\varepsilon}{2}$, where $[U]_T$ denotes completion of the set U by Tikhonov topology in $\prod_{k \in N}[a_k, b_k]$. Then, by Urysohn's lemma we deduce that there is a continuous function g_2 from $\prod_{k \in N}[a_k, b_k]$ to $[0, 1]$ such that the preimages of 0 and 1 under g_2 are F and $[U]_T$ respectively. Then, for $x \in \prod_{k \in N}[a_k, b_k]$, we have

$$\mathcal{X}_U(x) \leq g_2(x)$$

and

$$(R) \int_{\prod_{k \in N}[a_k, b_k]} (g_2(x) - \mathcal{X}_U(x)) d\lambda(x) \leq \frac{\varepsilon}{2},$$

where $\mathcal{X}_U$ is indicator of the U defined on the $\prod_{k \in N}[a_k, b_k]$.

Now let consider a set $[\prod_{k \in N}[a_k, b_k] \setminus U]_T$. Using Ulam's lemma, we can choose such a compact set

$$F_1 \subset \prod_{k \in N}[a_k, b_k] \setminus \Big[\prod_{k \in N}[a_k, b_k] \setminus U\Big]_T$$

that

$$\lambda\Big(\Big(\prod_{k \in N}[a_k, b_k] \setminus \Big[\prod_{k \in N}[a_k, b_k] \setminus U\Big]_T\Big) \setminus F_1\Big) < \frac{\varepsilon}{2}.$$

Then, by Urysohn's lemma we deduce that there is a continuous function g_1 from $\prod_{k \in N}[a_k, b_k]$ to $[0, 1]$ such the preimages of 0 and 1 under g_1 are $[\prod_{k \in N}[a_k, b_k] \setminus U]_T$ and F_1, respectively. Then, for $x \in \prod_{k \in N}[a_k, b_k]$, we have

$$g_1(x) \leq \mathcal{X}_U(x)$$

and

$$(R) \int_{\prod_{k \in N}[a_k, b_k]} (\mathcal{X}_U(x) - g_1(x)) d\lambda(x) \leq \frac{\varepsilon}{2}.$$

Now, we deduce that for every elementary rectangle U in $\prod_{i \in N}[a_i, b_i]$ there exists two continuous functions g_1 and g_2 on the $\prod_{i \in N}[a_i, b_i]$ such that

$$g_1(x) \leq \mathcal{X}_U(x) \leq g_2(x)$$

and

$$(R) \int_{\prod_{i \in N}[a_i, b_i]} (g_2(x) - g_1(x)) d\lambda(x) \leq \varepsilon.$$

Then we have

$$\lambda(U)-\varepsilon \le (R)\int_{\prod_{i\in N}[a_i,b_i]} g_2(x)d\lambda(x)-\varepsilon \le (R)\int_{\prod_{i\in N}[a_i,b_i]} g_1(x)d\lambda(x)$$

$$=\lambda\Big(\prod_{i\in N}[a_i,b_i]\Big)\times \lim_{n\to\infty}\frac{\sum_{y\in Y_n} g_1(y)}{\#(Y_n)} \le \lambda\Big(\prod_{i\in N}[a_i,b_i]\Big)\times \underline{\lim}_{n\to\infty}\frac{\#(Y_n\cap U)}{\#(Y_n)}$$

$$\le \lambda\Big(\prod_{i\in N}[a_i,b_i]\Big)\times \overline{\lim}_{n\to\infty}\frac{\#(Y_n\cap U)}{\#(Y_n)} \le \lambda\Big(\prod_{i\in N}[a_i,b_i]\Big)\times \lim_{n\to\infty}\frac{\sum_{y\in Y_n} g_2(y)}{\#(Y_n)}$$

$$=(R)\int_{\prod_{i\in N}[a_i,b_i]} g_2(x)d\lambda(x) \le (R)\int_{\prod_{i\in N}[a_i,b_i]} g_1(x)d\lambda(x)+\varepsilon \le \lambda(U)+\varepsilon.$$

Since ε was taken arbitrary, we deduce that

$$\lambda\Big(\prod_{i\in N}[a_i,b_i]\Big)\times \lim_{n\to\infty}\frac{\#(Y_n\cap U)}{\#(Y_n)}=\lambda(U).$$

This ends the proof of Theorem 2.3.4. □

Now by the scheme used in the proof of Theorem 2.3.4, one can get the validity of an infinite-dimensional analog of Lemma 2.3.1. In particular, the following assertion is valid

Theorem 2.3.5. *For $\prod_{i\in N}[a_i,b_i]\in\mathcal{R}$, let $(Y_n)_{n\in N}$ be an increasing family its finite subsets. Then $(Y_n)_{n\in N}$ is uniformly distributed in the $\prod_{k\in N}[a_k,b_k]$ if and only if for every Riemann integrable function f on $\prod_{k\in N}[a_k,b_k])$ the following equality*

$$\lim_{n\to\infty}\frac{\sum_{y\in Y_n} f(y)}{\#(Y_n)}=\frac{(R)\int_{\prod_{k\in N}[a_k,b_k]} f(x)d\lambda(x)}{\lambda(\prod_{i\in N}[a_i,b_i])}$$

holds.

2.4 On Uniformly Distribution in Infinite-Dimensional Rectangles

Definition 2.4.1. (Weyl [144]) A sequence $s_1,s_2,s_3,\ldots$ is said to be equidistributed modulo 1 or uniformly distributed modulo 1 if the sequence $<s_n>)_{n\in N}$ of the fractional parts of $(s_n)_{n\in N}$ is equidistributed (equivalently, uniformly distributed) on the interval $[0,1]$.

Lemma 2.4.1 ([64], Theorem 4.1, p. 42)**.** *Let $(\alpha_n)_{n\in N}$ be a sequence of different integer numbers. Then a sequence of real numbers $(\alpha_n x)_{n\in N}$ is u.d. mod 1 for l_1-almost points of R, where l_1 denotes a linear Lebesgue measure in $\mathbb{R}$.*

Definition 2.4.2. Following Brian R. Hunt, Tim Sauer and James A. Yorke (cf. [50]), a set X is called shy if it is a subset of a Borel set X' for which $\mu(X'+v)=0$ for every $v\in B$ and some Borel probability measure μ such that $\mu(K)=\mu(B)$ for some compact K.

Definition 2.4.3. A Borel measure μ in V is called a generator (of shy sets) in V, if

$$(\forall X)(\overline{\mu}(X)=0 \rightarrow X \in S(V)),$$

where $\overline{\mu}$ denotes a usual completion of the Borel measure μ.

Lemma 2.4.2 ([96], Example 2.1, p. 242). *"Lebesgue measure"* [2] *is a generator of shy sets.*

Theorem 2.4.1. *Let $(\alpha_n^{(k)})_{n\in N}$ be an infinite sequence of different integer numbers for every $k \in N$. Then a set of all sequences $(x_k)_{k\in N}$ in R^∞ for which a sequence of increasing sets $(Y_n((x_k)_{k\in N}))_{n\in N}$ is not λ-u.d. on the $\prod_{k\in N}[a_k,b_k]$), where*

$$Y_n((x_k)_{k\in N}) = \prod_{k=1}^{n}(\cup_{j=1}^{n}\{<\alpha_j^{(k)}x_k>(b_k-a_k)\}) \times \prod_{k\in N\setminus\{1,\dots,n\}}\{a_k\},$$

is of λ-measure zero.

Proof. For $k \in N$, we denote D_k by

$D_k = \{x_k : x_k \in R \;\&\; (<\alpha_j^{(k)}x_k>(b_k-a_k))_{j\in N}$ is l_1-u.d. on $[a_k,b_k]\;\}$.

By Lemma 2.4.1, we have that $l_1(R\setminus D_k)=0$ for $k\in N$.

We set $D=\prod_{k\in N}D_k$. It is clear that $\lambda(R^\infty\setminus D)=0$. For $(x_k)_{k\in N}\in D$, we have that $(<\alpha_j^{(k)}x_k>(b_k-a_k))_{j\in N}$ is l_1-u.d. on $[a_k,b_k]$ for every $k\in N$. By Theorem 2.3.1 we claim that $(Y_n((x_k)_{k\in N}))_{n\in N}$ is λ-u.d. on the $\prod_{k\in N}[a_k,b_k]$ for $(x_k)_{k\in N}\in D$. Since $\lambda(R^\infty\setminus D)=0$, Theorem 2.4.1 is proved. □

Remark 2.4.1. Following Lemma 2.4.2, λ is generator of shy sets. The later relation means that every set of λ-measure zero is shy in R^∞. Following Theorem 2.4.1, we claim that a set of all sequences $(x_j)_{j\in N}$ in R^∞ for which a sequence of increasing sets $(Y_n((x_j)_{j\in N}))_{n\in N}$ is λ-u.d. on the $\prod_{k\in N}[a_k,b_k]$) is shy.

Chapter 3

Change of Variable Formula for the α-Ordinary Lebesgue Measure in $\mathbb{R}^{\mathbb{N}}$

Let $\mathbb{R}^n(n > 1)$ be an n-dimensional Euclidean space and let μ_n an n-dimensional standard Lebesgue measure on $\mathbb{R}^n$. Further, let T be a linear μ_n-measurable transformation of $\mathbb{R}^n$.

It is obvious that $\mu_n T^{-1}$ is absolutely continuous with respect to μ_n, and there exists a non-negative μ_n-measurable function Φ on $\mathbb{R}^n$ such that

$$\mu_n(T^{-1}(X)) = \int_X \Phi(y) d\mu_n(y)$$

for every μ_n-measurable subset X of $\mathbb{R}^n$.

The function Φ plays the role of the Jacobian $J(T^{-1})$ of the transformation T^{-1}(or, rather the absolute value of the Jacobian)(see, e.g., [46]) in the theory of transformations of multiple integrals. It is clear that $J(T^{-1})$ coincides with a Radon-Nikodym derivative $\frac{d\mu_n T^{-1}}{d\mu_n}$, which is unique a.e. with respect to μ_n.

It is clear that

$$\frac{d\mu_n T^{-1}}{d\mu_n}(x) = \lim_{k\to\infty} \frac{\mu_n(T^{-1}(U_k(x)))}{\mu_n(U_k(x))} (\mu_n - \text{a.e.}),$$

where $U_k(x)$ is a spherical neighborhood with the center in $x \in \mathbb{R}^n$ and radius $\mathbf{r}_k > 0$ so that $\lim_{k\to\infty} r_k = 0$. The class of such spherical neighborhoods generate so-called Vitali differentiability class of subsets which allows us to calculate the Jacobian $J(T^{-1})$ of the transformation T^{-1}.

If we consider a vector space of all real-valued sequences $\mathbb{R}^{\mathbb{N}}$(equipped with Tychonoff topology), then we observe that for the infinite-dimensional Lebesgue measure [2] (or [3]) defined in $\mathbb{R}^{\mathbb{N}}$ there does not exist any Vitali system of differentiability, but in spite of non-existence of such a system the inner structure of this measure allows us to define a form of the Radon-Nikodym derivative defined by any linear transformation of $\mathbb{R}^{\mathbb{N}}$. In order to show it, let consider the following

Example 3.1. Let $\mathcal{R}_1$ be the class of all infinite dimensional rectangles $R \in B(\mathbb{R}^{\mathbb{N}})$ of the form

$$R=\prod_{i=1}^{\infty}R_i,\ \ R_i=(a_i,b_i),-\infty<a_i\leq b_i<+\infty,$$

such that

$$0\leq\prod_{i=1}^{\infty}(b_i-a_i)<\infty.$$

Let τ_1 be the set function on $\mathcal{R}_1$ defined by

$$\tau_1(R)=\prod_{i=1}^{\infty}(b_i-a_i).$$

R. Baker [2] proved that the functional λ_1 defined by

$$(\forall X)\Big(X\in B(\mathbb{R}^{\mathbb{N}})\rightarrow\lambda_1(X)=\inf\Big\{\sum_{j=1}^{\infty}\tau_1(R_j):R_j\in\mathcal{R}_1\ \&\ X\subseteq\cup_{j=1}^{\infty}R_j\Big\}\Big)$$

is a quasi-finite translation-invariant Borel measure in $\mathbb{R}^{\mathbb{N}}$.

The following change of variable formula has been established in [2] (cf. p. 1029): *Let $T^n:\mathbb{R}^n\rightarrow\mathbb{R}^n$, $n>1$, be a linear transformation with Jacobian $\Delta\neq0$, and let $T^{\mathbb{N}}:\mathbb{R}^{\mathbb{N}}\rightarrow\mathbb{R}^{\mathbb{N}}$ be the map defined by*

$$T^{\mathbb{N}}(x)=(T^n(x_1,\dots,x_n),x_{n+1},x_{n+2},\dots),\ x=(x_i)_{i\in N}\in\mathbb{R}^{\mathbb{N}}.$$

Then for each $E\in\mathcal{B}(\mathbb{R}^{\mathbb{N}})$, we have

$$\lambda_1(T^{\mathbb{N}}(E))=|\Delta|\lambda_1(E).$$

The main goal of the present chapter is to extend change of variable formula presented in Example 3.1 for the α-ordinary Lebesgue measure and wider class of linear transformations of $\mathbb{R}^{\mathbb{N}}$.

Now, let (X_i,M_i,μ_i) be a sequence of measure spaces and for each i let ρ_i be a metric on the set X_i. Assume that the following conditions are satisfied.

(i) Each (X_i,ρ_i) is a locally compact metric space.

(ii) Each M_i contains the family, $B(X_i)$, of Borel subsets of X_i and μ_i is a regular Borel measure on X_i (cf. [125], Def. 2.15).

(iii) For all i, and for every $\delta>0$, there exists a sequence (A_j) of Borel subsets of X_i such that $d_i(A_j)<\delta$ and $X_i=\cup_{j=1}^{\infty}A_j$, where $d_i(A_j)$ is the diameter of A_j in X_i.

Define $X=\prod_{i=1}^{\infty}X_i$ and let equip X with the product topology. Let denote by $\mathcal{R}$ the family of all rectangles $R\subseteq X$ of the form $R=\prod_{i=1}^{\infty}R_i$, $R_i\in\mathcal{B}(X_i)$, and

$$0\leq\prod_{i=1}^{\infty}\mu_i(R_i):=\lim_{n\rightarrow\infty}\prod_{i=1}^{n}\mu_i(R_i)<+\infty.$$

For $R\in\mathcal{R}$, we define

$$\tau(R)=\prod_{i=1}^{\infty}\mu_i(R_i).$$

Let τ^* be the set function on the powerset $\mathcal{P}(X)$ defined by

$$\tau^*(E) = \inf\left\{\sum \tau(R_j) : R_j \in \mathbb{R} \,\&\, E \subseteq \cup R_j\right\}$$

for $E \subseteq X$.

Also, let use the convention that $0 \cdot +\infty = +\infty \cdot 0 = 0$, $+\infty \cdot +\infty = +\infty$, and that the infimum taken over the empty set has the value $+\infty$.

Theorem 3.1 ([3], Theorem I, p. 2579). *The set function τ^* is an outer measure on X. Let M be the σ-algebra of subsets of X which are measurable with respect to τ^*, and let μ be the measure on M obtained by restricting τ^* to M. Then $B(X) \subseteq M$, and for all $R = \prod_{i=1}^{\infty} R_i \in \mathcal{R}$, we have $\mu(R) = \prod_{i=1}^{\infty} \mu_i(R_i)$. If each space X_i contains disjoint subsets A_i, B_i such that $\mu_i(A_i) = \mu_i(B_i) = 1$, then the measure μ is not σ-finite. Finally, assume that each (X_i, ρ_i) is an M_i-measurable group. If each μ_i is left-invariant measure on M_i, then μ is a left-invariant measure on M. Similarly, if each μ_i is right-invariant measure on M_i, then μ is right-invariant measure on M.*

Let $\mathcal{R}_2$ be the class of all infinite dimensional rectangles $R \in \mathbb{B}(\mathbb{R}^{\mathbb{N}})$ of the form

$$R = \prod_{i=1}^{\infty} R_i, \quad R_i \in B(\mathbb{R})$$

such that

$$0 \le \prod_{i=1}^{\infty} m(R_i) := \lim_{n\to\infty} \prod_{i=1}^{n} m(R_i) < \infty,$$

where m denotes a one-dimensional classical Borel measure on $\mathbb{R}$.

Let τ_2 be a set function on $\mathcal{R}_2$, defined by

$$\tau_2(R) = \prod_{i=1}^{\infty} m(R_i).$$

As a simple consequence of Theorem 3.1, R.Baker [3] had obtained that the functional λ_2 defined by

$$(\forall X)\left(X \in \mathcal{B}(\mathbb{R}^{\mathbb{N}}) \to \lambda_2(X) = \inf\left\{\sum_{j=1}^{\infty} \tau_2(R_j) : R_j \in \mathcal{R}_2 \,\&\, X \subseteq \cup_{j=1}^{\infty} R_j\right\}\right)$$

is a quasi-finite translation-invariant Borel measure in $\mathbb{R}^{\mathbb{N}}$.

Below we consider a certain example of infinite-dimensional Lebesgue measure defined by Theorem 3.1, which can be considered as a generalization of the construction [3]. In order to do it we need the following simple auxiliary proposition.

Lemma 3.1. *Let $(n_i)_{i\in\mathbb{N}}$ be the sequence of non-zero natural numbers. Then the following formula is valid*

$$\prod_{i\in\mathbb{N}} \mathcal{B}(\mathbb{R}^{n_i}) = \mathcal{B}(\mathbb{R}^{\mathbb{N}}).$$

Theorem 3.2. *Let $\alpha = (n_i)_{i \in N}$ be the sequence of non-zero natural numbers and let $\mathcal{R}_\alpha$ be the class of all infinite dimensional rectangles $R \in B(\mathbb{R}^{\mathbb{N}})$ of the form*

$$R = \prod_{i=1}^{\infty} R_i, \quad R_i \in B(\mathbb{R}^{n_i})$$

such that

$$0 \leq \prod_{i=1}^{\infty} m_{n_i}(R_i) := \lim_{n \to \infty} \prod_{i=1}^{n} m_{n_i}(R_i) < \infty,$$

where m_{n_i} denotes an n_i-dimensional classical Borel measure in $\mathbb{R}^{n_i}$.

Let τ_α be a set function on $\mathcal{R}_\alpha$, defined by

$$\tau_\alpha(R) = \prod_{i=1}^{\infty} m_{n_i}(R_i).$$

Then the functional λ_α defined by

$$(\forall X)\Big(X \in B(\mathbb{R}^{\mathbb{N}}) \to \lambda_\alpha(X) = \inf\Big\{\sum_{j=1}^{\infty} \tau_\alpha(R_j) : R_j \in \mathcal{R}_\alpha \ \& \ X \subseteq \cup_{j=1}^{\infty} R_j\Big\}\Big)$$

is α-ordinary Lebesgue measure in $\mathbb{R}^{\mathbb{N}}$.

Theorem 3.3. *Let $\alpha = (n_i)_{i \in N}$ be the sequence of non-zero natural numbers, and let $T^{n_i} : \mathbb{R}^{n_i} \to \mathbb{R}^{n_i}, i > 1$, be a family of linear transformation with Jacobians $\Delta_i \neq 0$ and $0 < \prod_{i=1}^{\infty} \Delta_i < \infty$. Let $T^N : \mathbb{R}^{\mathbb{N}} \to \mathbb{R}^{\mathbb{N}}$ be the map defined by*

$$T^{\mathbb{N}}(x) = (T^{n_1}(x_1, \dots, x_{n_1}), T^{n_2}(x_{n_1+1}, \dots, x_{n_1+n_2}), \dots),$$

where $x = (x_i)_{i \in N} \in \mathbb{R}^{\mathbb{N}}$. Then for each $E \in \mathcal{B}(\mathbb{R}^{\mathbb{N}})$, we have

$$\lambda_\alpha(T^N(E)) = \Big(\prod_{i=1}^{\infty} \Delta_i\Big)\lambda_\alpha(E).$$

Proof. Taking into account an equality

$$T^{\mathbb{N}}(\mathcal{R}_\alpha) = \mathcal{R}_\alpha,$$

Lemmas 3.1 and Theorem 3.2, for $E \in \mathcal{B}(\mathbb{R}^{\mathbb{N}})$, we have

$$\lambda_\alpha(T^{\mathbb{N}}(E)) = \inf\Big\{\sum_{j=1}^{\infty} \tau_\alpha(R_j) : R_j \in \mathcal{R}_\alpha \ \& \ T^{\mathbb{N}}(E) \subseteq \cup_{j=1}^{\infty} R_j\Big\} =$$

$$= \inf\Big\{\sum_{j=1}^{\infty} \tau_\alpha(R_j) : R_j = \prod_{i=1}^{\infty} R_i^{(j)} \in \mathcal{R}_\alpha \ \& \ R_i^{(j)} \in \mathcal{B}(\mathbb{R}^{n_i}) \ \& \ T^{\mathbb{N}}(E) \subseteq \cup_{j=1}^{\infty} R_j\Big\} =$$

$$= \inf\Big\{\sum_{j=1}^{\infty} \tau_\alpha(R_j) : R_j = \prod_{i=1}^{\infty} R_i^{(j)} \in \mathcal{R}_\alpha \& \ R_i^{(j)} \in \mathcal{B}(\mathbb{R}^{n_i}) \ \& \ E \subseteq (T^{\mathbb{N}})^{-1}(\cup_{j=1}^{\infty} R_j)\Big\} =$$

$$= \inf\Big\{\sum_{j=1}^{\infty} \tau_\alpha(R_j) : R_j = \prod_{i=1}^{\infty} R_i^{(j)} \in \mathcal{R}_\alpha \ \&\ R_i^{(j)} \in \mathcal{B}(\mathbb{R}^{n_i}) \ \&\ E \subseteq \cup_{j=1}^{\infty}(T^{\mathbb{N}})^{-1}(R_j)\Big\} =$$

$$= \inf\Big\{\sum_{j=1}^{\infty} \tau_\alpha\Big(\prod_{i=1}^{\infty} R_i^{(j)}\Big) : R_j = \prod_{i=1}^{\infty} R_i^{(j)} \in \mathcal{R}_\alpha \ \&\ R_i^{(j)} \in \mathcal{B}(\mathbb{R}^{n_i}) \ \&$$

$$E \subseteq \cup_{j=1}^{\infty}\Big(\prod_{i=1}^{\infty}(T^{n_i})^{-1}R_i^{(j)}\Big)\Big\} =$$

$$= \Big(\prod_{i=1}^{\infty}\Delta_i\Big)\inf\Big\{\sum_{j=1}^{\infty} \tau_\alpha\Big(\prod_{i=1}^{\infty}(T^{n_i})^{-1}R_i^{(j)}\Big) : R_j = \prod_{i=1}^{\infty} R_i^{(j)} \in \mathcal{R}_\alpha \ \&\ R_i^{(j)} \in \mathcal{B}(\mathbb{R}^{n_i}) \ \&$$

$$E \subseteq \cup_{j=1}^{\infty}\prod_{i=1}^{\infty}(T^{n_i})^{-1}(R_i^{(j)})\Big\} = \Big(\prod_{i=1}^{\infty}\Delta_i\Big)\inf\Big\{\sum_{j=1}^{\infty}\tau_\alpha(S_j) : S_j = \prod_{i=1}^{\infty} S_i^{(j)} \in \mathcal{R}_\alpha \ \&$$

$$S_i^{(j)} = (T^{n_i})^{-1}(R_i^{(j)}) \in \mathcal{B}(\mathbb{R}^{n_i}) \ \&\ E \subseteq \cup_{j=1}^{\infty}\prod_{i=1}^{\infty}(T^{n_i})^{-1}(R_i^{(j)})\Big\} =$$

$$= \Big(\prod_{i=1}^{\infty}\Delta_i\Big)\inf\Big\{\sum_{j=1}^{\infty}\tau_\alpha(S_j) : S_j \in \mathcal{R}_\alpha \ \&\ E \subseteq \cup_{j=1}^{\infty} S_j\Big\} = \Big(\prod_{i=1}^{\infty}\Delta_i\Big)\lambda_\alpha(E). \qquad \square$$

Remark 3.1. Theorem 3.3 is change of variable formula for the α-ordinary Lebesgue measure. It extends change of variable formula for Baker's measure considered in Example 3.1. Indeed, let $T^n : \mathbb{R}^n \to \mathbb{R}^n, n > 1$, be a linear transformation with Jacobian $\Delta \neq 0$. Let $n_1 = n$ and $n_i = 1$ for $i > 1$. Further, we set $T^{n_1} = T^n$ and $T^{n_k} = I$, where $I : \mathbb{R} \to \mathbb{R}$ is an identity transformation of $\mathbb{R}$ defined by $I(x) = x$ for $x \in \mathbb{R}$.

Let a map $T^{\mathbb{N}} : \mathbb{R}^{\mathbb{N}} \to \mathbb{R}^{\mathbb{N}}$ be defined by

$$T^{\mathbb{N}}(x) = (T^n(x_1,\dots,x_n), x_{n+1}, x_{n+2},\dots),\ x = (x_i)_{i\in\mathbb{N}} \in \mathbb{R}^{\mathbb{N}}.$$

Then, by Theorems 3.1–3.2, for $T^{\mathbb{N}}$ and for each $E \in \mathcal{B}(\mathbb{R}^{\mathbb{N}})$, we have

$$\lambda_2(T^{\mathbb{N}}(E)) = \lambda_\alpha(T^{\mathbb{N}}(E)) = |\Delta|\lambda_\alpha(E) = |\Delta|\lambda_\alpha(E).$$

Chapter 4

On Existence and Uniqueness of Generators of Shy Sets in Polish Groups

4.1 Introduction

In [96], for a Polish topological vector space $\mathbb{V}$ has been introduced the notion of a generator of shy sets μ, which is such a Borel measure in $\mathbb{V}$ that a condition $\overline{\mu}(X) = 0$ implies that X is Haar null(or shy) in the sense [50], where $\overline{\mu}$ denotes a usual completion of the μ. Here has been shown that the class of generators of shy sets in $\mathbb{V}$ contains specific Borel measures which generate implicitly introduced subclasses of shy sets (see, for example, [19], [50], [24], [76], [115], [122], [129]). Moreover, such measures possess many interesting properties (see, for example, [96], [92]). Some applications of generators of shy sets in infinite-dimensional analysis can be found in [96], [93], [95], [94].

In [104], it has been shown that the class of generators of shy sets in an abelian Polish group G is non-empty if G contains any uncountable locally compact Hausdorff topological subgroup. This result (see, [104], Theorem 2.1) extends a certain result early obtained for Polish topological vector spaces in [96](see, Theorem 2.1, p. 238). For a Borel probability measure μ in a Polish topological vector space $\mathbb{V}$, a generator of shy sets $\mathbf{G}_\mu$ has been constructed such that a subclass of shy sets generated by the measure μ coincides with the class of $\overline{\mathbf{G}}_\mu$-zero sets, where $\overline{\mathbf{G}}_\mu$ denotes a usual completion of the generator $\mathbf{G}_\mu$ (see, [96], Theorem 3.1, p.245).

A notion of Haar measure zero (or shy) in abelian Polish groups to all non-abelian Polish groups has been given in Mycielski [83]. Like this approach the notion of generators of shy sets introduced for Polish topological vector spaces [96] has been extended to all Polish groups in [104], where an example of a two sided invariant generator of two-sided-shy sets has been constructed in the product of unimodular Polish groups and Dougherty's criterion of shyness has been extended. In the same paper have been posed the following unsolving problems:

Problem 4.1.1. ([104],Question 2.1, p.5) Let G be a Polish group. Whether there exists a generator of two-sided (left or right)-shy sets in G?

Problem 4.1.2. ([104],Question 2.2, p.5) Let G be a Polish group and let a class of generators of two-sided(left or right)-shy sets in G is non-empty. Whether there exists a generator of two-sided(left or right)-shy sets with the property of uniqueness in the entire class of generators?

So each a quasi-finite two sided (left or right) invariant Borel measure in Polish group same times is a generator of two-sided(left or right)-shy sets in the entire group, we show that a solution of Problems 4.1.1-4.1.2 is directly connected with the problem of a measure in groups which is the problem of defining a two sided (left or right) invariant Borel measure, with specified properties, in a given group. In this direction can be mentioned especially the paper [88] which contains a solution of the problem of an existence and uniqueness of a quasi-finite left-invariant Borel measure in a complete separable metric group which is dense in itself. In this paper has been shown that there exists a quasi-finite left-invariant Borel measure in any Polish group which is dense in itself, but a locally finite Borel measure is possible only if that group is locally compact. Take into account the validity of these facts, Oxtoby stated Haar's theorem in the following generalized form: *In a complete separable metric group there exists a locally finite left-invariant Borel measure if and only if the group is locally compact and dense in itself.*

The generalized form of Haar's theorem stated above is an immediate consequence of the following theorem, the proof of which is due to S. Ulam [88](see, p. 561, footnote 3).

Proposition 4.1.1. *Let G be any complete separable metric group which is not locally compact, and let m be any left-invariant Borel measure in G. Then every neighborhood contains non-denumerably many disjoint mutually congruent sets of equal finite positive measure.*

By a refinement of the foregoing reasoning Oxtoby has proved the following stronger theorem.

Proposition 4.1.2. *([88], Theorem 2, p. 217) Let G be any complete separable metric group which is not locally compact, and let m be any left-invariant Borel measure in G. Then any neighborhood contains a compact set which is the union of non-denumerably many disjoint mutually congruent compact sets of equal finite positive measure.*

Remark 4.1.1. Concerning with Proposition 4.1.2, it can be noted that there exists a complete separable metric group G which is locally compact and a left-invariant non-locally finite Borel measure μ in the entire group such that every neighborhood contains a compact set which is the union of non-denumerably many disjoint mutually congruent compact sets of equal finite positive measure. Indeed, in order to give such an example it is sufficient to set $G = R^2$(with a usual Euclidian topology) and to consider under μ a one-dimensional Preiss-Tišer generator of shy sets in G (see, [96], Section 7, p. 255).

The purpose of the present chapter is to answer to Problems 4.1.1-4.1.2 by a technique of invariant measure theory developed by Oxtoby in [88].

The present chapter is organized as follows.

Section 4.2 contains some auxiliary notions and propositions from the theory of invariant measures in Polish groups. In Section 4.3 we give solutions to Problems 4.1.1-4.1.2 for generators of left(or right)-shy sets in a Polish group which is dense in itself. In addition, we give solutions to the same problems for generators of two-sided-shy sets in a Polish group which is dense in itself and is equipped with two-sided invariant metric.

4.2 Some Auxiliary Propositions from the Theory of Invariant Measures in Polish Groups

Let $\mathbf{G}$ be a Polish group, by which we mean a separable group with a complete metric for which the transformation (from $\mathbf{G} \times \mathbf{G}$ onto $\mathbf{G}$) which sends $(\mathbf{x}, \mathbf{y})$ into $\mathbf{x}^{-1}\mathbf{y}$ is continuous. Let $\mathcal{B}(\mathbf{G})$ denotes the σ-algebra of Borel subsets of $\mathbf{G}$.

Definition 4.2.1. A Borel set $X \subseteq G$ is called two-sided-shy if there exists a Borel probability measure μ over G such that $\mu(fXg) = 0$ for all $f, g \in G$. A subset of a Borel two-sided-shy set is called also two-sided-shy.

Definition 4.2.2. A Borel set $X \subseteq G$ is called left (or right)-shy if there exists a Borel probability measure μ over G such that $\mu(fX) = 0$(or $\mu(Xf) = 0$) for all $f \in G$. A subset of a Borel left (or right)-shy set is called also left (or right)-shy.

Definition 4.2.3. A Borel measure μ in G is called a generator of two-sided-shy sets in G, if

$$(\forall X)(\overline{\mu}(X) = 0 \rightarrow X \in S(G)),$$

where $\overline{\mu}$ denotes a usual completion of the Borel measure μ and $S(G)$ denotes a class of all two-sided-shy sets.

Definition 4.2.4. A Borel measure μ in G is called a generator of left (or right) shy sets in G, if

$$(\forall X)(\overline{\mu}(X) = 0 \rightarrow X \in \mathcal{LS}(G) \text{ (or } \mathcal{RS}(G)),$$

where $\mathcal{LS}(G)$ and $\mathcal{RS}(G)$ denote classes of all left-shy and right-shy sets in G, respectively.

Definition 4.2.5. A Borel measure μ in G is called quasi-finite if there exists a compact set $U \subseteq G$ for which $0 < \mu(U) < \infty$.

Definition 4.2.6. A Borel measure μ in G is called semi-finite if for X with $\mu(X) > 0$ there exists a compact subset $F \subseteq X$ for which $0 < \mu(F) < \infty$.

Definition 4.2.7. A Borel measure μ in G is called left invariant if

$$(\forall X)(\forall g)(X \in B(G) \;\&\; g \in G \rightarrow \mu(gX) = \mu(X)).$$

Definition 4.2.8. A Borel measure μ in G is called right invariant if

$$(\forall X)(\forall g)(X \in B(G) \;\&\; g \in G \rightarrow \mu(Xg) = \mu(X)).$$

Definition 4.2.9. A Borel measure μ in G is called two sided invariant if

$$(\forall X)(\forall g, f)(X \in B(G) \;\&\; g, f \in G \rightarrow \mu(gXf) = \mu(X)).$$

Definition 4.2.10. A Borel measure μ in G is called locally finite if there is a neighborhood U of unity such that $0 < \mu(U) < +\infty$.

Definition 4.2.11. Let K be the class of measures in G. We say that a measure $\mu \in K$ has the property of uniqueness in the class K if μ and λ are equivalent for every $\lambda \in K$.

Definition 4.2.12. Let G be equipped with a left invariant metric. Two sets A and B are said to be congruent if there exists an element $a \in G$ such that $B = aA$.

Definition 4.2.13. Let G be equipped with a two sided invariant metric. Two sets A and B are said to be congruent if there exists elements $a, b \in G$ such that $B = aAb$.

The following lemma plays a key role in our further considerations.

Lemma 4.2.1. *([88], Theorem 3, p. 220) In any complete separable metric group which is dense in itself*[1] *there exists a left-invariant quasi-finite Borel measure.*

Remark 4.2.1. In [88] has been demonstrated, that there always exist infinitely many left-invariant quasi-finite Borel measures and there is certainly no uniqueness theorem for the problem of measure. In case the group is locally compact, the construction may in some cases generate Haar's measure. In the additive group of real numbers, that construction gives such measures which are not locally finite(see, also, Remark 4.1.1.).

In the sequel we need the following auxiliary proposition.

Lemma 4.2.2. *Every two-sided (left or right) invariant quasi-finite Borel measure μ defined on a Polish group G is a generator of two-sided(left or right)-shy sets in G.*

Proof. We present the proof of Lemma 4.2.2 for a two sided invariant quasi-finite Borel measure μ. One can get the validity of Lemma 4.2.2 similarly for left (or right) invariant quasi-finite Borel measures.

Let $\overline{\mu}(S) = 0$ for $S \subseteq G$. Since $\overline{\mu}(S) = 0$, there exists a Borel set S' for which $S \subseteq S'$ and $\mu(S') = 0$. By using a two sided invariance of the Borel measure μ, we have

$$(\forall f, g)(f, g \in G \rightarrow \mu(fX'g) = 0).$$

Since μ is quasi-finite there is a Borel set F with $0 < \mu(F) < +\infty$. We set

$$(\forall X)(X \in \mathcal{B}(G) \rightarrow \lambda(X) = \frac{\mu(X \cap F)}{\mu(F)}).$$

Let us show that λ is a two sided transverse to the Borel set S'. Indeed, we have:

$$(\forall f, g)(f, g \in G \rightarrow \lambda(fX'g) = \frac{\mu((fX'g) \cap F)}{\mu(F)} \leq \frac{\mu(fX'g)}{\mu(F)} = 0).$$

The latter relation means that S' is a Borel two-sided–shy set. S being a subset of S' also is two-sided-shy set. This ends the proof of Lemma 4.2.2. □

[1] A topological space is called dense in itself if it has no an isolated point.

Here we a certain modification of Theorem 3 presented in [88](see, p.200) with a small change.

Lemma 4.2.3. *(Oxtoby) In any complete separable metric group with two-sided invariant metric which is dense in itself there exists a two-sided-invariant quasi-finite Borel measure and no any two-sided-invariant quasi-finite Borel measure possesses a uniqueness property.*

Proof. Recall that a metric ρ is two-sided invariant if $\rho(zxh, zyh) = \rho(x, y)$ for all x, y, z and h in G. Let $(U_k)_{k \in N}$ be a sequence of non-empty open sets in G whose diameters tend to zero. Let $(W_k)_{k \in N}$ be a sequence of finite positive numbers tending to zero. Let U denote the family of sets in G congruent to one of the sets $\{U_k : k \in N\}$, i.e.,

$$U = \{bU_k a : a, b \in G \; \& \; k \in N\}.$$

Consider any set $A \subseteq G$, and any covering of A by a finite or infinite sequence of sets of the family U, say $A \subseteq \cup_{i \in I} b_i U_{n_i} a_i$. Form the sum $\sum_{i \in I} W_{n_i}$ and let $L_r(A)$ denote the lower bound of these sums for all covering sequences of sets U with diameters less than r.(Such coverings always exist, indeed A can be covered by a sequence of sets congruent to U_k, for any fixed k, since the space is separable).It is evident that as r decreases, $L_r(A)$ will not decrease. Hence the limit $m^*(A) = lim_{r \to 0} L_r(A)$ exists or is equal to $+\infty$. In other words m^* is defined as follows

$$m^*(A) = lim_{r \to 0} \inf\{\sum_{i \in I} W_{n_i} : A \subseteq \cup_{i \in I} b_i U_{n_i} a_i \; \& \mathrm{diam}(U_{n_i}) \leq r\}$$

for all $A \subseteq G$.

From this definition it is an easy to verify that the following conditions are satisfied:

(a1) $0 \leq m^*(A) \leq +\infty$;

(a2) $m^*(A) \leq m^*(B)$ if $A \subseteq B$;

(a3) $m^*(\cup_{n \in N} A_n) \leq \sum_{n \in N} m^*(A_n)$ for any sequence $(A_n)_{n \in N}$ of subsets of G;

(a4) $m^*(A \cup B) = m^* A + m^* B$ if A and B are separated by a positive distance;

(a5) $m^*(xAy) = m^*(A)$ for arbitrary $x, y \in G$ and $A \subseteq G$;

(a6) $m^*(A) = 0$ if A contains only one point;

(a7) $m^*(A) = \inf m^*(B)$ for all G_δ sets B which contain A.

Condition (a7) follows from the fact that any set is contained in a G_δ set having the same outer measure, since only open coverings are employed.

It follows that m^* is a regular outer measure in the sense of Caratheodory [16](see, p. 238) provided that it is finite and positive for at least one set and therefore defines a left-invariant measure in G. In view of (a7), this measure is the complete extension of a Borel measure m. We shall say that the Borel measure m is the measure generated by the measuring system $\{U_n, W_n : n \in N\}$. Now we need to show that it is always possible to find a measuring system $\{U_n, W_n : n \in N\}$ which generates a measure m such that $0 < m(A) < +\infty$ for at least one set. It is clear that not every measuring system will do this. For example, in the additive group of real numbers, if U_n is an interval of length $1/n$, and we take $W_n = n^{-2}$, every set will have measure zero. To secure a non-trivial measure it is therefore necessary to choose the measuring system with some care.

The sets $C(n;i_l,\cdots,i_n)$ with primary index n will be said to be of rank n. To this end, we shall construct in G a family of sets $C(n;i_1,\cdots,i_n)(i_1,\cdots,i_n=0,1;n=1,2,\cdots)$ with the following properties:

(i) $C(n;i_l,\cdots,i_n)$ is compact;

(ii) $C(n;i_l,\cdots,i_n)=C(n+1;i_l,\cdots,i_n,0)\cup C(n+1;i_l,\cdots,i_n,1)$;

(iii) The 2^n sets of rank n are mutually congruent;

(iv) The diameter d_n of the sets of rank n tends to zero as $n\to\infty$;

(v) Any two sets of rank n are separated by a distance greater than d_n.

Define $C=C(1;0)\cup C(1;1)$. Then $C=\cup_{i_1\cdots i_n}C(n;i_l,\cdots,i_n)$ for each n. It may be remarked that the family of sets $C(n;i_l,\cdots,i_n)$ evidently represent C as a dyadic discontinuum (see [49], p. 134), and therefore C is homeomorphic to the Cantor set.

To construct such a family of sets, note first that since G is dense in itself we can choose a sequence of points x_n tending to the unity element e, such that

$$0<(e,x_{n+1})<(e,x_n)/9\ (n=1,2,\cdots). \tag{4.2.1}$$

Define

$$a_{i_1\cdots i_n}=x_1^{i_1}x_2^{i_2}\cdots x_n^{i_n}\ (i_l,\cdots,i_n=0,1).$$

Let $A(n;i_l,\cdots,i_n)$ denote the set of all points $a_{i_1\cdots i_{n+p}}$ $(p\geq 0)$, whose first n indices are equal to $i_l,i_2,\cdots,i_n$ and define $C(n;i_l,\cdots,i_n)$ as the closure of the countable set $A(n;i_l,\cdots,i_n)$. Evidently we have

$$A(n;i_l,\cdots,i_n)=A(n+1;i_l,\cdots,i_n,0)\cup A(n+1;i_l,\cdots,i_n,1).$$

Properties (ii) and (iii) follow at once by taking closures. Furthermore, the diameter d_n of the sets $C(n;i_i,\cdots,i_n)$ of rank n is given by $d_n=diamA(n;0,\cdots,0)$. We proceed to obtain an estimate of d_n.

Note that the left-invariance of the metric implies that for arbitrary elements y,y_1, and y_2 in G we have

$$(e,y_ly_2)\leq(e,y_1)+(y_1,y_1y_2)=(e,y_1)+(e,y_2).$$

Also

$$(e,y)=(y^{-1},e)=(e,y^{-1}),$$

so that $(e,y^k)\leq(e,y)$ if $k=-1,0,1$. A simple induction then shows that

$$(e,y_1^{k_1}\cdots y_n^{k_n})<(e,y_1)+(e,y_2)+\cdots+(e,y_n)$$

for arbitrary $y_1,\cdots,y_n$ in G and exponents $k_1,\cdots,k_n$ equal to $-1,0$ or 1.

Now consider any two points of $A(n;0,\cdots,0)$. These may be written $x_{n+1}^{i_1}\cdots x_{n+p}^{i_p}$ and $x_{n+1}^{j_1}\cdots x_{n+p}^{j_p}$ where the exponents i_k and j_k are either 0 or 1. The distance between these points can be estimated as follows

$$(x_{n+1}^{i_1}\cdots x_{n+p}^{i_p},x_{n+1}^{j_1}\cdots x_{n+p}^{j_p})=$$

$$(e,x_{n+p}^{i_p}\cdots x_{n+1}^{-i_1}x_{n+1}^{j_1}\cdots x_{n+p}^{j_p})\leq 2[(e,x_{n+1})+(e,x_{n+2}+\cdots+(e,x_{n+p})]$$

$$\leq \frac{2}{9}(e,x_n)+\frac{2}{9^2}(e,x_n)+\cdots+\frac{2}{9^p}(e,x_n)<(e,x_n)/4.$$

This being true for any two points of $A(n;0,\cdots,0)$ it follows that $d_n<(e,x_n)/4$. On the other hand, e and x_{n+1} are both contained in $A(n;0,\cdots,0)$, and so $(e,x_{n+1})<d_n$. Consequently

$$d_{n+1}\leq (e,x_{n+l})/4<d_n/4,$$

and so $d_{n+1}\leq d_n/4, n=1,2,\cdots$.. It follows, in particular, that (iv) is satisfied.

Next consider any two distinct sets $A(n;i_1,\cdots,i_n)$ and $A(n;j_1,\cdots,j_n)$ and any two representative points of these sets. These may be written as $a_{i_1\cdots i_p}$ and $a_{j_1\cdots j_p}$, where $p\geq n$. Suppose their indices differ first in the kth place. Then $k\leq n$ and we may assume $i_k=0$ and $j_k=1$. Let $u=x_{k+1}^{i_{k+1}}\cdots x_p^{i_p}$ and $v=x_kx_{k+1}^{j_{k+1}}\cdots x_p^{j_p}$. Then $(a_{i_1\cdots i_p},a_{j_1\cdots j_p})=(u,v)$ and

$$(e,x_k)\leq (e,u)+(u,v)+(v,x_k)$$

$$(e,x_{k+1}^{i_{k+1}}\cdots x_p^{i_p})+(x_{k+1}^{i_{k+1}}\cdots x_p^{i_p},x_kx_{k+1}^{j_{k+1}}\cdots x_p^{j_p})+(x_kx_{k+1}^{j_{k+1}}\cdots x_p^{j_p},x_k)\leq$$

$$(u,v)+2[(e,x_{k+1})+(e,x_{k+2})+\cdots+(e,x_p)]\leq (u,v)+2(e,x_k)[\frac{1}{9}+\frac{1}{9^2}+\cdots+\frac{1}{9^{n-k}}]\leq$$

$$(u,v)+(e,x_k)/4.$$

Therefore $(u,v)\geq 3(e,x_k)/4>3(e,x_n)/4\geq 3d_n$. Consequently any two sets $C(n;i_l,\cdots,i_n)$ of rank n are separated by a distance at least equal to $3d_n$, and condition (v) is satisfied.

Finally, $C(n;i_1,\cdots,i_n)$ is totally bounded, since it is the union of a finite number 2^{p-n} of sets $C(n;i_1,\cdots,i_p)$ having arbitrarily small diameter d_p. Any totally bounded closed subset of a complete space is compact, and therefore (i) is satisfied, provided G is complete with respect to the metric (x,y). If this is not the case, G can be embedded isometrically in a complete space $\overline{G}$ and the same reasoning then shows that the closure of $C(n;i_1,\cdots,i_n)$ in G is compact. Consequently, the sets $C(n;i_1,\cdots,i_n)$ will be compact if they are all closed in $\overline{G}$. To ensure this, the sequence $\{x_n:n\in N\}$, on which the construction rests, must be subjected to a further condition. Introduce a topologically equivalent metric $\rho(x,y)$ with respect to which G is complete, and let the sequence $\{x_n:n\in N\}$ be so chosen that it satisfies not only condition (4.2.1) but also the condition

$$\rho(a_{i_1\cdots i_n},a_{i_1\cdots i_n}x_{n+1})\leq 2^{-n}(i_1,\cdots,i_n=0,1). \quad (4.2.2)$$

Both conditions are satisfied by any sequence that converges to e sufficiently rapidly. Condition (4.2.2) implies that for any sequence of subscripts $i_i,i_2,\cdots$ the sequence $a_{i_1\cdots i_n}(n=1,2,\cdots)$ is a Cauchy sequence with respect to the metric ρ, and therefore converges to an element of G. But any point of accumulation of C in $\overline{G}$ is the limit of such a sequence. Hence C is closed in V, and likewise the sets $C(n;i_1,\cdots,i_n)$. Thus in all cases we have constructed in G a family of sets having properties (i) to (v).

Now consider any family of sets $C(n;i_1,\cdots,i_n)$ satisfying conditions (i) to (v). We proceed to define a measure such that $m(C)=1$. From condition (v) it follows that there exist positive numbers $\{\varepsilon_n:n\in N\}$, $\lim_{r\to\infty}\varepsilon_r=0$, such that any two sets $C(n;i_1,\cdots,i_n)$ of rank n are separated by a distance at least equal to $d_n+2\varepsilon_n$. (In the family constructed above it suffices to take ε_n equal to d_n.) Consider the measuring system with $\{U_n:n\in N\}$

equal to the ε_n-neighborhood of $C(n;0,\cdots,0)$, and $W_n = 2^{-n}$. Then $diam(U_n) < d_n + 2\varepsilon_n$, hence $diam(U_n) \to 0$, and any set congruent to U_n can overlap at most one set $C(n;i_i,\cdots,i_n)$ of rank n.

To show that $m(C) = 1$, consider any $r > 0$ and choose n such that $diam(U_n) < r$. From (iii) it follows that C can be covered by 2^n sets congruent to U_n.Hence $L_r(C) \leq 2^n W_n = 1$. To establish the opposite inequality, consider any covering of C by sets of the form $b_i U_{n_i} a_i$. Since C is compact, only finite coverings need be considered. Suppose $C \subseteq \cap_{s=1}^{k} b_s U_{n_s} a_s$. Let p be the largest of the indices $n_1, n_2, \cdots, n_k$ and let q_i be the number of sets $C(p;i_1,\cdots,i_p)$ of rank p overlapped by $b_i U_{n_i} a_i$. Since the sets $b_i U_{n_i} a_i$ cover C we must have $q_1 + q_2 + \cdots + q_k \geq 2^p$. On the other hand, each set $b_i U_{n_i} a_i$ can overlap at most one set of rank n_i, and therefore at most 2^{p-n_i} sets of rank p. Hence $2^{p-n_i} \geq q_i$. Therefore $W_{n_i} = 2^{-n_i} = 2^{p-n_i}/2^p \geq q_i/2^p$, and so $\sum_{i=1}^{k} W_{n_i} \geq 2^{-p} \sum_{i=1}^{k} q_i \geq 1$, which shows that $L_r(C) \geq 1$. Thus $L_r(C) = 1$ for every positive r. Therefore $m(C) = 1$ and we have established the existence of a two-sided-invariant Borel measure in G.

The structure of the measure m within the unit set C is easy to recognize. To each point $x \in C$ corresponds uniquely a sequence of indices $i_i, i_2, \cdots$ such that $x = \cap_n C(n;i_i,\cdots,i_n)$. If we regard this sequence $\{i_n : n \in N\}$ as an element of the infinite product group C^* of the group of order two with itself, with the usual topology [120](see, p. 51), it is not hard to see that this correspondence is a homeomorphism between C and C^*, and that the measure m in C corresponds to the normalized Haar measure in C^*. Hence the measure m in G may be regarded as an extension of the measure in C defined by this correspondence and the Haar measure in C^*.

The foregoing construction evidently contains a large degree of arbitrariness. In fact, it associates an invariant measure with any sequence $\{x_n : n \in N\}$ that satisfies conditions (4.2.1) and (4.2.2). If a sequence satisfies these conditions, any subsequence will also satisfy them. In particular, the subsequence $\{x_{2n} : n \in N\}$ defines a unit set $C_1 \subseteq C$ such that $m(C_1) = 0$, where m is the measure associated with the sequence $\{x_n : n \in N\}$.Repeated application of this result gives rise to a sequence of measures $m_1, m_2, \cdots$, where m_k is the measure associated with the subsequence $\{x_{k(n)} : k(n) = n2^k\}$. These measures are all distinct and form an increasing sequence, $m_1(A) \leq m_2(A) \leq \cdots$ for every Borel set A. It therefore appears that infinitely many invariant measures always exist, and there is certainly no uniqueness theorem for the problem of measure we have considered. □

Remark 4.2.2. Every Polish group admits a compatible left-invariant metric. Such a metric need not be complete. A compatible two-sided invariant metric on a Polish group is necessarily complete (c.f. [8]). Each abelian Polish group admit a compatible two-sided invariant metric; it is also known that all compact Polish groups have an analogous property. There are locally compact Polish group which do not admit a compatible two-sided invariant metric (an example is $SL_2(\mathbb{R})$). For a diagram of the relationship among classes of Polish groups which admit compatible two-sided invariant metric, see [9].

4.3 Main Results

The following two assertions answers to Problems 4.2.1-4.2.2 for generators of left(or right)-shy sets in a Polish groups which is dense in itself.

Theorem 4.3.1. *In any Polish group G which is dense in itself there exists a left(or right)-invariant quasi-finite generator of left(or right)-shy sets.*

Proof. By Lemma 4.2.1, there exists a left-invariant quasi-finite Borel measure μ in G. By Lemma 4.2.2, the measure μ is a generator of left-shy sets in G. □

By Remark 4.2.1, we get the following

Theorem 4.3.2. *In any Polish group G which is dense in itself there exists no a generator of left(or right)-shy sets with the property of uniqueness in the class of quasi-finite generators of left(or right)-shy sets.*

Theorems 4.3.1–4.3.2 extend the following early obtained propositions:

Proposition 4.3.1. *([104], Theorem 2.1, p.5) Let $(G,+)$ be an abelian Polish group which admits the following representation $G = G_0 + G_1$, where G_0 is an uncountable locally compact Hausdorff topological group and G_1 is such a group that $G_0 \cap G_1 = \{\mathbf{0}\}$, where $\{\mathbf{0}\}$ denotes the zero of the group $(G,+)$. Then there exists a semi-finite inner regular invariant generator λ of left shy sets in G. The generator λ is non-σ-finite iff the G_1 is uncountable.*

Proposition 4.3.2. *([104], Theorem 2.2, p.8) Let G be a non-locally compact abelian Polish group with an invariant metric which contains any uncountable locally compact Hausdorff topological group. Then the class of generators of shy sets in G is non-empty and each its element is non-σ-finite.*

Proposition 4.3.3. *([104], Corollary 2.1., p.8) There always exists a semi-finite inner regular generator of shy sets in the Polish topological vector space G and no any generator has a property of uniqueness if $dim(G) > 1$. In addition, if G is infinite-dimensional then every generator of shy sets in G is non-σ-finite.*

Remark 4.3.1. Let G be a Polish group. Then there does not exist a family of mutually singular generators of left(or right) shy sets in G. This is a simple consequence of Theorem 3.2 established in [105] (see, p.27).

Remark 4.3.2. Let G be a non-locally compact Polish group. Then each generator m of left-shy sets is non-σ-finite. Indeed, following Kakutani [53], any topological group that satisfies the first countability axiom is metrizable, and that the metric can be taken to be left-invariant. Following Dougherty [31], any compact set in non-locally compact Polish group equipped with left invariant metric, is left-shy set. If we assume that m is σ-finite, then by using a certain result of Oxtoby (see [88], Lemma 1, p. 216), there will be a countable family of compact sets $\{F_k : k \in N\}$ such that $m(G \setminus \cup_{k \in N} F_k) = 0$. Following [83](see Theorem 3, p.32), $\cup_{k \in N} F_k$ is left- shy set. The previous relation implies that $G \setminus \cup_{k \in N} F_k$ is left-shy since m is generator of left-shy sets in G. Finally we get that G is left-shy set as a union of the following two left-shy sets $\cup_{k \in N} F_k$ and $G \setminus \cup_{k \in N} F_k$. This is a required contradiction.

Theorem 4.3.3. *In any Polish group G which is dense in itself and is equipped with a two sided invariant metric there exists a two-sided invariant quasi-finite generator of two-sided-shy sets and there does not exist a generator of two-sided-shy sets with the property of uniqueness in the entire class of generators.*

Proof. By Lemma 4.2.3, there exists a two sided invariant quasi-finite Borel measure μ in G. By Lemma 4.2.2, the measure μ is a generator of two-sided-shy sets in G. Using again Lemma 4.2.3, we deduce that there exists no a generator of two-sided-shy sets with the property of uniqueness in the class of quasi-finite generators of two-sided-shy sets. □

Chapter 5

On a Certain Criterion of Shyness for Subsets in the Product of Unimodular Polish Groups that are not Compact

5.1 Introduction

A suitable extension of the property of being of Haar measure zero in abelian Polish groups [19] to all non-abelian Polish groups is given in Topsøe and Hoffmann-Jørgens [135] and Mycielski [83] as follows: A Borel set A in a Polish group G is called *shy* if there exists a Borel measure μ over G such that $\mu(F) > 0$ for any compact $F \subset G$ and every two sided translation of A is of μ-measure zero; Similarly, a Borel set B in a Polish topological group G is called *left* (or *right*)-shy iff there exists a Borel measure λ over G such that every left (or right) translation of B is of λ-measure zero and $\lambda(F) > 0$ for any compact $F \subset G$. The measure μ is said to be a transverse to A, and a measure λ is said to be a left (or right) transverse to B. A subset of G is called (left or right) shy if it is subset of a Borel (left or right) shy set. Also, we preserve notions of transverse measures for subsets of Borel (left or right) shy sets.

It can be mentioned especially the deep result of Topsøe and Hoffmann-Jørgens [135] and Mycielski [83] asserted *that the class of (left or right) shy sets in a Polish group is an σ-ideal.* Starting from this point, there are many observations devoted to construct of criterions of shyness for subsets in various Polish groups. Among of them the result of Dougherty [31] can be mentioned especially which asserts that *all compact subsets of a non-locally compact Polish group that has an invariant metric are shy.* For example, this result follows that if G is a Polish group with a two sided invariant metric that is σ-compact but not compact (e.g., $\mathbb{R}$ or $\mathbb{Z}$) and X is the set of eventually constant sequences in $G^{\mathbb{N}}$, then X is shy. Indeed it is easy to check that X is σ-compact and hence Dougherty's theorem implies that it is shy. A question whether compact sets are (left or right) shy sets in a non-locally compact Polish group, still remains open (see, [129], Problem 3, p. 43).

The purpose of the present chapter is to extend Dougherty criterion of shyness for subsets(see [31], Proposition 12) in the product of a countable family of unimodular Polish groups that are not compact.

The chapter is organized as follows.

In Section 5.2 we give extension of the notion of generators of shy sets introduced for Polish topological vector spaces [96] to all Polish groups. We prove that the class of generators of shy sets in an abelian Polish group G is non-empty if G contains any uncountable locally compact Hausdorff topological subgroup. This result (see, Theorem 5.1) extends one result obtained for Polish topological vector spaces in [96](see, Theorem 2.1, p. 238). In addition, we show that if G has an invariant metric then every generator of shy sets on G is non-σ-finite.

In Section 5.3, we construct left (or right)-invariant generators of shy sets on the product of locally compact Hausdorff topological groups. Analogously, we give an example of a two sided invariant generator of shy sets in the product of unimodular Polish groups. For every compact set K in the product of a countable family of non-compact unimodular Polish groups, we construct a two sided invariant generator of shy sets μ such that $\mu(K) = 0$

In Section 5.4 we present the main result of our chapter and construct such an example of a shy set in the product of a countable family of unimodular Polish groups (that are not compact) which is not described by Dougherty above mentioned criterion of shyness for subsets.

5.2 On Generators of Shy Sets in Polish Groups

Let G be a complete metric group, by which we mean a group with a complete metric for which the transformation (from $G \times G$ onto G) which sends (x,y) into $x^{-1}y$ is continuous. Let $B(G)$ denotes the σ-algebra of Borel subsets of G.

A set $X \subseteq G$ will be called shy if there exists a probability measure m over G such that $m(gXh) = 0$ for all $g,h \in G$. Such an m will be called left and right transversal to X.

A set $X \subseteq G$ will be called *prevalent* iff the complement of X is shy.

We now present some important facts about transversality and shyness established in [83] for complete metric groups. They are generalizations of Facts 3.1–3.3 established in [50].

Fact 5.2.1. *Every shy Borel set X in a separable complete metric group (equivalently, Polish group) has a left and right transversal measure which is finite with compact support. Furthermore, the support of this measure can be taken to have arbitrarily small diameter.*

Proof. Let m be a left and right transversal measure to a Borel set $X \subseteq G$. Then it can be restricted to a compact set U of finite and positive measure, and the restriction is certainly also a left and right transversal measure to X. Also, since U is compact it can be covered for each "$\varepsilon > 0$ by a finite number of balls of radius ε", and at least one of these balls must intersect U in a set of positive measure. The intersection of U with the closure of this ball is compact, and the restriction of m to this set is also a left and right transversal measure to X. □

Fact 5.2.2 ([83], Theorem 1, p. 31). *If G is a metric locally compact group, which has a basis of cardinality less than the first real-valued measurable cardinal and $Y \subseteq G$, then the following three conditions are equivalent:*

(i) *Y is shy;*

(ii) *some (or all) Haar measure of Y vanishes;*

(iii) *there exists a Borel probability measure on G which is left (or right) transversal to Y.*

Proof. Condition (ii) makes sense because all Haar measures are absolutely continuous relative to each other. We must be careful because the Haar measures are not Borel measures in the usual sense(they are defined on the σ-ring generated by compact sets and completed). Nevertheless it is easy to check that (ii)→(i)→(iii). It remains to show that (iii)→(ii). Let μ be any left Haar measure on G and m be left transversal to Y as in (iii). By Fact 5.2.1, we can assume that m has a compact support, and we define

$$\nu(X) = \int_G m(g^{-1}X)\mu(dg).$$

So $\nu(Y) = 0$ and, in view of the above, it suffices to prove that ν is a Haar measure. In fact ν is left invariant since

$$\nu(hX) = \int_G m(g^{-1}hX)\mu(dg) = \int_G m((h^{-1}g)^{-1}X)\mu(dg)$$

$$= \int_G m(k^{-1}X)\mu(hdk) = \nu(X).$$

□

Fact 5.2.3 ([83], Theorem 2, p. 32). *If Y_1 and Y_2 are shy, then $Y_1 \cup Y_2$ is also shy, i.e., shy sets constitute an ideal.*

Proof. Given Borel sets $Y_1, Y_2 \in \mathcal{B}(G)$ containing the original sets and measures μ and ν left and right transversal to Y_1 and Y_2 respectively, we must then find a measure which is a left and right transversal to both Y_1 and Y_2. We define

$$\mu(X) = \mu \times \nu\{(x,y) \in G^2 : xy \in X\}.$$

We will show that m is left and right transversal to Y_1 and Y_2. We have

$$m(gY_1h) = \int_G \mu(gY_1hy^{-1})\nu(dy) = 0;$$

and

$$m(gY_2h) = \int_G \nu(x^{-1}gY_2h)\mu(dx) = 0;$$

Hence $m(g(Y_1 \cup Y_2)h) = 0$. □

Fact 5.2.4 ([83], Theorem 3, p. 32). *The union of a countable collection of shy sets is shy in Polish group.*

Proof. Given a countable collection of shy subsets of G, let $(Y_k)_{k\in\mathbb{N}}$ be shy Borel sets containing the original sets. For $n = 1,2,\ldots$, let m_n be a left and right transversal to Y_n. By Fact 5.2.1, we can assume without loss of generality that each m_n is finite and supported on a compact set U_n with diameter at most 2^{-n}. By normalizing and translating the measures, we can also assume that $m_n(U_n) = 1$ for all n and that the unity of G belongs to U_n. Notice that if $y_n \in U_n$ for $n = 1,2,\ldots$ then the infinite product $g_1g_2\cdots$ in the sense of group multiplication converges (by the assumptions about diameters of the U_n's).

The infinite Cartesian product $U^{\Pi} = U_1 \times U_2 \times \cdots$ is compact by the well known Tychonoff theorem and has a product measure $m^{\Pi} = m_1 \times m_2 \times \cdots$ defined on its Borel subsets, with $m^{\Pi}(U^{\Pi}) = 1$. Since G is complete and each vector in U_n lies at most 2^{-n} away from zero, there is a continuous mapping from U^{Π} into G defined by

$$(g_1, g_2, \ldots) \longmapsto g_1g_2\cdots.$$

The image U of U^{Π} under this mapping is compact, and m^{Π} induces a measure m supported on U, given by

$$m(X) = m^{\Pi}(\{(g_1, g_2, \ldots) \in U^{\Pi} : g_1g_2\cdots \in X\}).$$

We will be done if we show that m is a left and right transversal to every $Y_1 \cup Y_2 \cup \cdots$. We have

$$g_1g_2\cdots \in hY_nk \Longleftrightarrow g_n \in (g_1\cdots g_{n-1})^{-1}hY_nk(g_{n+1}g_{n+2}\cdots)^{-1}.$$

Since the Cartesian product of measures is associative (and commutative), we can write

$$m^{\Pi} = m_n \times \nu_n^{\Pi}$$

with

$$\nu_n^{\Pi} = m_1 \times m_{n-1} \times m_{n+1} \times \cdots.$$

Let $\prod^n U_n = U_1 \times \cdots \times U_{n-1} \times U_{n+1} \times U_{n+2} \times \cdots$. Then, since m_n is left and right transversal to Y_n,

$$m(hY_nk) = \int_{\prod^n U_n} m_n((g_1\cdots g_{n-1})^{-1}hY_nk(g_{n+1}g_{n+2}\cdots)^{-1})\nu_n^{\Pi}(dg) = 0.$$

So m is a left and right transversal to each Y_n and hence also to $Y_1 \cup Y_2 \cup \cdots$. □

The next assertion gives a positive answer to the Problem 4.2.1 for abelian Polish groups which contain any uncountable locally-compact Hausdorff topological subgroup(compare that approach with the approach introduced in Section 4).

Theorem 5.2.1. *Let $(G,+)$ be an abelian Polish group which admits the following representation $G = G_0 + G_1$, where G_0 is an uncountable locally compact Hausdorff topological group and G_1 is such a group that $G_0 \cap G_1 = \{\mathbf{0}\}$, where $\{\mathbf{0}\}$ denotes the zero of the group $(G,+)$. Then there exists a semi-finite inner regular invariant generator λ of shy sets in G. The generator λ is non-σ-finite iff the G_1 is uncountable.*

Proof. Let μ be a Haar measure on G_0. For every $X \in B(G)$ we set

$$\lambda(X) = \sum_{g_1 \in G_1} \mu(X + g_1 \cap G_0).$$

Let us show that λ is a measure. Let $(X_k)_{k\in\mathbb{N}}$ be a family of pairwise disjoint Borel sets in G. Then we get

$$\lambda(\cup_{k\in\mathbb{N}} X_k) = \sum_{g_1\in G_1} \mu((\cup_{k\in\mathbb{N}} X_k) + g_1 \cap G_0) =$$

$$= \sum_{g_1\in G_1} \mu(\cup_{k\in\mathbb{N}} (X_k + g_1) \cap G_0) = \sum_{g_1\in G_1} \mu(\cup_{k\in\mathbb{N}} (X_k + g_1 \cap G_0)) =$$

$$= \sum_{g_1\in G_1} \sum_{k\in\mathbb{N}} \mu((X_k + g_1 \cap G_0)) = \sum_{k\in\mathbb{N}} \sum_{g_1\in G_1} \mu((X_k + g_1 \cap G_0)) =$$

$$= \sum_{k\in\mathbb{N}} \lambda(X_k).$$

Let us show that the measure λ is quasi-finite.

Since μ is a regular Borel measure on G_0 there exists a compact set $F \subset G_0$ such that $0 < \mu(F) < \infty$. We have

$$\lambda(F) = \sum_{g_1\in G_1} \mu(F + g_1 \cap G_0) = \mu(F \cap G_0) = \mu(F).$$

The measure λ is invariant. Indeed, for $f \in G$ there exists $f_0 \in G_0$ and $f_1 \in G_1$ such that $f = f_0 + f_1$. For $X \in B(G)$, we have

$$\lambda(X + f) = \sum_{g_1\in G_1} \mu((X + f) + g_1 \cap G_0) = \sum_{g_1\in G_1} \mu(X + f_1 + g_1 + f_0 \cap G_0) =$$

$$= \sum_{g_1\in G_1} \mu(X + f_1 + g_1 \cap G_0) = \sum_{f_1+g_1\in G_1+f_1} \mu(X + f_1 + g_1 \cap G_0) =$$

$$= \sum_{f_1+g_1\in G_1} \mu(X + f_1 + g_1 \cap G_0) = \sum_{h_1\in G_1} \mu(X + h_1 \cap G_0) = \lambda(X).$$

Now let us show that λ is a generator.

Let S be a subset of G with $\overline{\lambda}(S) = 0$. Since $\overline{\lambda}$ is a completion of λ, there exists a Borel set S' for which $S \subseteq S'$ and $\lambda(S') = 0$. Taking in to account that F is a compact set in G with $0 < \lambda(F) < \infty$ and applying a simple consequence of the invariance of the measure λ stating that $\lambda(S' + v) = 0$ for $v \in G$ we deduce that λ is transverse to a Borel set S'. This means that S' is a Borel shy set. Finally, since S is a subset of a Borel shy set S', we conclude that S is a shy set. One can observe that the generator λ is non-σ-finite iff $\mathrm{card}(G_1) \geq \omega$.

Let us show that the generator λ is inner regular. For this, we are to show that

$$(\forall X)(\forall \varepsilon)(0 < \lambda(X) < \infty \;\&\; \varepsilon > 0 \rightarrow (\exists F_\varepsilon)(F_\varepsilon \subseteq X \;\&\; F_\varepsilon \text{ is compact } \&\; \lambda(X \setminus F_\varepsilon) < \varepsilon)).$$

Since $0 < \lambda(X) < \infty$, there exists $I_0 \subset G_1$ such that $\mathrm{card}(I_0) \leq \aleph_0$ and

$$0 < \lambda(X) = \sum_{g\in I_0} \mu(X + g \cap G_0) < \infty.$$

We set $I_0 = (g_m)_{m \in N}$. Let n_0 be a natural number such that

$$\sum_{1 \le m \le n_0} \mu(X + g_m \cap G_0) > \lambda(X) - \frac{\varepsilon}{2}.$$

For $\varepsilon > 0$ there exists a compact set F_m such that $F_m \subseteq X + g_m$ and

$$\mu((X + g_m) \setminus F_m) < \frac{\varepsilon}{2^{m+1}}$$

for $1 \le m \le n_0$.

We set $F_\varepsilon = \cup_{1 \le k \le n_0}(F_k - g_k)$. It is obvious that F_ε is compact in V.

Finally, we get

$$\lambda(X \setminus \cup_{1 \le k \le n_0}(F_k - g_k)) = \sum_{g \in I_0} \mu((X \setminus \cup_{1 \le k \le n_0}(F_k - g_k)) + g \cap G_0) =$$

$$= \sum_{m \in N} \mu(((X \setminus \cup_{1 \le k \le n_0}(F_k - g_k)) + g_m) \cap G_0) =$$

$$= \sum_{1 \le m \le n_0} \mu(((X \setminus \cup_{1 \le k \le n_0}(F_k - g_k)) + g_m) \cap G_0) +$$

$$+ \sum_{m > n_0} \mu(((X \setminus \cup_{1 \le k \le n_0}(F_k - g_k)) + g_m) \cap G_0) \le$$

$$\le \sum_{1 \le m \le n_0} \mu(((X \setminus (F_m - g_m)) + g_m) \cap G_0) + \sum_{m > n_0} \mu(X + g_m \cap G_0) \le$$

$$\le \sum_{1 \le m \le n_0} \mu(((X + g_m) \setminus F_m) \cap G_0) + \sum_{m > n_0} \mu(X + g_m \cap G_0) \le$$

$$\le \sum_{1 \le m \le n_0} \frac{\varepsilon}{2^{m+1}} + \frac{\varepsilon}{2} \le \sum_{m \in N} \frac{\varepsilon}{2^{m+1}} + \frac{\varepsilon}{2} = \frac{\varepsilon}{2} + \frac{\varepsilon}{2} = \varepsilon.$$

Let us show that λ is semi-finite. Indeed, if $\lambda(X) > 0$, then there exists $g_0 \in G_1$ such that $0 < \mu(X + g_0 \cap G_0) < \infty$. Using the property of inner regularity of μ we deduce that there exists a compact set $F \subseteq X$ with $0 < \lambda(F) < \infty$. The latter relation means that λ is semi-finite. □

Theorem 5.2.2. *Let G be a non-locally compact abelian Polish group with an invariant metric which contains any uncountable locally compact Hausdorff topological group. Then the class of generators of shy sets in G is non-empty and each its element is non-σ-finite.*

Proof. By Theorem 5.2.1 we claim that the class of generators of shy sets in G is non-empty. Now assume that λ be any σ-finite element of this class. Then there exists a countable family of compact sets $\{K_n : n \in N\}$ such that

$$\lambda(G \setminus \cup_{n \in N} K_n) = 0.$$

By Dougherty's result (see, [31], Proposition 12, p. 87) we claim that every compact set in G is shy. By using Fact 5.2.4 we deduce the set $\cup_{n \in \mathbb{N}} K_n$ is shy in G. By definition of a generator of shy sets in G, we claim that the set $G \setminus \cup_{n \in N} K_n$ being a set of λ-measure zero, is shy in G. Using again Fact 5.2.4 we establish that G is shy, which is a required contradiction because there does not exist a Borel measure on G which is transverse to the set G. □

The following corollary is immediate consequence of Theorems 5.2.1–5.2.2.

Corollary 5.2.1. *There always exists a semi-finite inner regular generator of shy sets in the Polish topological vector space G and no any generator has a property of uniqueness if $dim(G) > 1$. In addition, if G is infinite-dimensional then every generator of shy sets in G is non-σ-finite.*

Remark 5.2.1. The result of Corollary 5.2.1 has been obtained in [96] (see, Theorem 2.1, p. 238; Theorem 2.2, p. 241).

By the scheme considered in the proof of Theorem 5.2.2, one can prove the following

Proposition 5.2.1. *Every generator of shy sets in a non-locally compact Polish group G with an invariant metric is non-σ-finite.*

Theorem 5.2.3. *Every quasi-finite two sided quasi-invariant Borel measure μ defined on a Polish group G is a generator of shy sets in G.*

Proof. Let $\overline{\mu}(S) = 0$ for $S \subset G$. Since $\overline{\mu}(S) = 0$, there exists a Borel set S' for which $S \subseteq S'$ and $\mu(S') = 0$. By using the property of two sided quasi-invariance of the Borel measure μ, we have

$$(\forall f, g)(f, g \in G \rightarrow \mu(fS'g) = 0).$$

This relation together with the property of quasi-finiteness of μ implies that the Borel measure μ is transverse to the Borel set S' and therefore S' is a Borel shy set. S is a shy set as a subset of the Borel shy set S'. □

Corollary 5.2.2. *Since every two sided invariant Borel measure is at the same time two sided quasi-invariant, as a simple consequence of Theorem 5.2.3 we get the validity of Lemma 4.2.2 asserted that every quasi-finite two-sided invariant Borel measure in a Polish group G is a generator of shy sets in G.*

5.3 On Generators of Shy Sets in the Product of Locally Compact Hausdorff Topological Groups

We begin this section by a certain observation belonging to Baker [3].

Let $(X_i, M_i, \mu_i)_{i \in \mathbb{N}}$ be a sequence of measure spaces and for each i ($i \in \mathbb{N}$), let ρ_i be a metric on the set X_i. Assume that the following conditions are satisfied.

(i) Each (X_i, ρ_i) is a locally compact metric space.

(ii) Each M_i contains the family, $\mathcal{B}(X_i)$, of Borel subsets of X_i and μ_i is a regular Borel measure on X_i (cf. [125], Def. 2.15).

(iii) For all i, and for every $\delta > 0$, there exists a sequence (A_j) of Borel subsets of X_i such that $d_i(A_j) < \delta$ and $X_i = \cup_{j=1}^{\infty} A_j$, where $d_i(A_j)$ is the diameter of A_j in X_i.

Dcfine $X = \prod_{i=1}^{\infty} X_i$ and let equip X with the product topology. Let denote by $\mathcal{R}$ the family of all rectangles $R \subseteq X$ of the form $R = \prod_{i=1}^{\infty} R_i$, $R_i \in \mathcal{B}(X_i)$, and

$$0 \leq \prod_{i=1}^{\infty} \mu_i(R_i) := \lim_{n \to \infty} \prod_{i=1}^{n} \mu_i(R_i) < +\infty.$$

For $R \in \mathcal{R}$, we define $\tau(R) = \prod_{i=1}^{\infty} \mu_i(R_i)$.
Let τ^* be the set function on the powerset $\mathcal{P}(X)$ defined by

$$\tau^*(E) = \inf\left\{\sum \tau(R_j) : R_j \in \mathcal{R} \,\&\, E \subseteq \cup R_j\right\}$$

for $E \subseteq X$.

Also, let use the convention that $0 \cdot +\infty = +\infty \cdot 0 = 0$, $+\infty \cdot +\infty = +\infty$, and that the infimum taken over the empty set has the value $+\infty$.

Lemma 5.3.1 ([3], Theorem I, p. 2579). *The set function τ^* is an outer measure on X. Let M be the σ-algebra of subsets of X which are measurable with respect to τ^*, and let μ be the measure on X obtained by restricting τ^* to $\mathcal{B}(X)$. Then, for all $R = \prod_{i=1}^{\infty} R_i \in \mathcal{R}$, we have $\mu(R) = \prod_{i=1}^{\infty} \mu_i(R_i)$. If each space X_i contains disjoint subsets A_i, B_i such that $\mu_i(A_i) = \mu_i(B_i) = 1$, then the measure μ is not σ-finite. Finally, assume that each (X_i, ρ_i) is an M_i-measurable group. If each μ_i is left-invariant measure on M_i, then μ is left-invariant Borel measure on X. Similarly, if each μ_i is right-invariant measure on M_i, then μ is right-invariant Borel measure on X.*

The following lemma is an immediate consequence of Lemma 5.3.1.

Lemma 5.3.2. *Under conditions of Lemma 5.3.1, if each X_i is a Polish group and μ_i is a two sided invariant Borel measure on X_i $(i \in \mathbb{N})$, then μ is a two sided invariant Borel measure in X.*

Proof. By the definition of the measure μ, for $g = (g_i)_{i\in\mathbb{N}}, f = (f_i)_{i\in\mathbb{N}} \in X$ and for $E \in \mathcal{B}(X)$, we get

$$\mu(gEf) = \inf\left\{\sum \tau(R_j) : R_j \in \mathcal{R} \,\&\, gEf \subseteq \cup R_j\right\} =$$

$$= \inf\left\{\sum \tau(R_j) : R_j \in \mathcal{R} \,\&\, E \subseteq \cup g^{-1}R_j f^{-1}\right\} =$$

$$= \inf\left\{\sum \tau(g^{-1}R_j f^{-1}) : g^{-1}R_j f^{-1} \in \mathcal{R} \,\&\, E \subseteq \cup g^{-1}R_j f^{-1}\right\} =$$

$$= \inf\left\{\sum \tau(R_j) : R_j \in \mathcal{R} \,\&\, E \subseteq \cup R_j\right\} = \mu(E).$$

□

Recall that a locally compact Hausdorff topological group G whose left Haar measure coincides with its right Haar measure is called a unimodular group. Clearly, every abelian locally compact σ-compact Hausdorff topological groups are unimodular. There exist an examples of unimodular groups that are not abelian (see, for example [71]). The definition of a unimodular group follows that its every Haar measure is two sided invariant. Indeed, if λ is a left Haar measure, then it must be its right Haar measure. Correspondingly, for every two elements $f, g \in G$ and for every Borel subset E in G we get $\mu(fEg) = \mu(fE) = \mu(E)$.

Theorem 5.3.1. *For $i \in \mathbb{N}$, let G_i be a unimodular Polish group that is not compact and let μ_i be a Haar measure on G_i. Then the Borel measure μ defined by Lemma 5.3.1 is a two sided invariant generator of shy sets on the product-group $G = \prod_{i\in\mathbb{N}} G_i$.*

Proof. By Lemma 5.3.2, we claim that the Borel measure μ is a two-sided invariant Borel measure in G. An application of Corollary 5.2.2 ends the proof of Theorem 5.3.1. □

By Lemmas 4.2.2 and 5.3.1 we can prove the following assertion.

Theorem 5.3.2. *For $i \in \mathbb{N}$, let G_i be a locally compact Hausdorff topological group that is not compact and let μ_i be a left (or right) Haar measure on G_i. Then the Borel measure μ defined by Lemma 5.3.1 is a left (or right) invariant generator of left (or right) shy sets on the product-group $G = \prod_{i \in \mathbb{N}} G_i$.*

Lemma 5.3.3. *Let G_1 and G_2 be two Polish groups. Then X is (left or right) shy in G_1 iff $X \times G_2$ is (left or right) shy in $G_1 \times G_2$.*

Lemma 5.3.4. *For $k \in \mathbb{N}$, let G_k be a unimodular Polish group that is not compact and let μ_k be a Haar measure on G_k. Let $(Y_k)_{k \in N}$ be such a family that $0 \le \mu_k(Y_k) < +\infty$. Then $\prod_{k \in \mathbb{N}} Y_k$ is shy generated by any two sided invariant generator of shy sets on the product-group $G = \prod_{k \in \mathbb{N}} G_k$.*

Proof. For $k \in \mathbb{N}$, the μ_k is not finite and we can choose such a Borel set $Z_k \subset G_k$ that $2^k \mu_k(Y_k) < \mu_k(Z_k)$ and $\mu_k(Z_k) > 0$. For $k \in \mathbb{N}$, we set $\lambda_k = \frac{\mu_k}{\mu_k(Z_k)}$. Let μ be a two sided $\prod_{k \in \mathbb{N}} G_k$-invariant Borel measure on G defined by the family $(\lambda_k)_{k \in \mathbb{N}}$(see Lemma 13.3.2). Then we get

$$\mu\Big(\prod_{k \in \mathbb{N}} Y_k\Big) = \prod_{k \in \mathbb{N}} \lambda_k(Y_k) = \prod_{k \in \mathbb{N}} \frac{\mu_k(Y_k)}{\mu_k(Z_k)} \le \prod_{k \in \mathbb{N}} \frac{1}{2^k} = 0.$$

By Lemma 4.2.2, μ is a generator of shy sets on the product-group G. □

Now it is not hard to prove the following lemma.

Lemma 5.3.5. *For $k \in \mathbb{N}$, let G_k be a locally compact Hausdorff topological group that is not compact and let μ_k be a left (or right) Haar measure on G_k. Let $(Y_k)_{k \in N}$ be such a family that $0 \le \mu_k(Y_k) < +\infty$ for $k \in \mathbb{N}$. Then $\prod_{k \in \mathbb{N}} Y_k$ is left(or right)-shy set generated by any left (or right) invariant generator of left (or right) shy sets on the product-group $G = \prod_{k \in \mathbb{N}} G_k$.*

Lemma 5.3.6. *For $k \in \mathbb{N}$, let G_k be a locally compact Hausdorff topological group. Let K be a compact subset in $\prod_{k \in \mathbb{N}} G_k$. Then there exists a family $(Y_k)_{k \in \mathbb{N}}$ such that:*

1) *Y_k is compact in G_k;*
2) *$K \subset \prod_{k \in \mathbb{N}} Y_k$;*

We have the following corollaries of Lemmas 5.3.4–5.3.6.

Corollary 5.3.1. *For $k \in \mathbb{N}$, let G_k be a unimodular Polish group that is not compact. Then any compact subset (and hence any K_σ subset) of the product-group $\prod_{k \in \mathbb{N}} G_k$ is shy generated by any two sided invariant generator of shy sets on the entire group.*

Proof. Let K be any compact subset in $\prod_{k \in \mathbb{N}} G_k$. Let $(Y_k)_{k \in \mathbb{N}}$ be a family of subsets mentioned in Lemma 5.3.6. For $k \in \mathbb{N}$, let λ_k be a Haar measure defined on G_k. Since λ_k is not finite, we claim that $0 \le \lambda_k(Y_k) < +\infty$ for $k \in \mathbb{N}$. By Lemma 5.3.4 we deduce that the set $\prod_{k \in \mathbb{N}} Y_k$ is shy in $\prod_{k \in \mathbb{N}} G_k$ generated by any two sided invariant generator μ of shy sets on $\prod_{k \in \mathbb{N}} G_k$. The latter relation implies that the set K, being a subset of the Borel shy set $\prod_{k \in \mathbb{N}} Y_k$, is shy generated by the two sided invariant generator of shy sets μ. □

Corollary 5.3.2. *For $k \in \mathbb{N}$, let G_k be an uncountable locally compact Hausdorff topological group that is not compact. Then any compact subset (and hence any K_σ subset) of $\prod_{k \in \mathbb{N}} G_k$ is left shy and right shy generated by any left invariant and right invariant generators of shy sets on the entire group, respectively.*

5.4 On a Certain Example of a Shy Set in the Product of Unimodular Polish Groups that are not Compact

The main result of the present section is formulated as follows.

Theorem 5.4.1. *For $k \in \mathbb{N}$, let $(G_k)_{k\in\mathbb{N}}$ be a family of unimodular Polish groups that are not compact. Let λ_k be a Haar measure on G_k. Let $\mathcal{H}$ be a class of Borel measurable rectangles of the form $\prod_{k\in\mathbb{N}} Y_k$ such that*

$$card(\{k : 0 \leq \lambda_k(Y_k) < +\infty\}) = \omega.$$

If a subset $X \subseteq \prod_{i\in\mathbb{N}} G_i$ is covered by the union of countable family of elements of $\mathcal{H}$ then X is shy in the product-group $G = \prod_{i\in\mathbb{N}} G_i$.

Proof. Firstly, let us show that every element of the class $\mathcal{H}$ is shy in G. Indeed, let $\prod_{k\in\mathbb{N}} Y_k \in \mathcal{H}$. We set $A = \{k : 0 \leq \lambda_k(Y_k) < +\infty\}$. Then by Lemma 5.3.4, we claim that the set $\prod_{k\in A} Y_k$ is shy in $\prod_{k\in A} G_k$. By Lemma 5.3.3, the set $\prod_{k\in A} Y_k \times \prod_{i\in\mathbb{N}\setminus A} G_i$ is shy in $\prod_{k\in\mathbb{N}} G_k$. Also, $\prod_{k\in\mathbb{N}} Y_k$ is shy as a subset of the shy set $\prod_{k\in A} Y_k \times \prod_{i\in\mathbb{N}\setminus A} G_i$. Following [83](see, Theorem 3, p. 32) the union of the countable family of elements of $\mathcal{H}$ is shy in G and we claim that X is shy in G. □

In order to show that Theorem 5.4.1 extends Proposition 12 (cf. [31], p. 87) in the product of unimodular Polish groups that are not compact we present the following example.

Example 5.4.1. Let $\mathbb{R}$ be an abelian Polish group of real numbers over the usual operation of addition. Then a space of real-valued sequences $\mathbb{R}^\infty$ equipped with usual product topology stands the product of a countable family of unimodular Polish groups that are not compact.

We set

$$(\forall k)\Big(k \in \mathbb{N} \rightarrow \Delta_k = \sum_{i=1}^{\infty} \Delta_k^{(i)}\Big),$$

where $\Delta_k^{(i)} = [3i; 3i + \frac{1}{2^i}]$ for $i \in \mathbb{N}$.

Let consider a Borel set $E = \prod_{k=1}^{\infty} \Delta_k$.

Let $(\mu_k)_{k\in\mathbb{N}}$ be a family of linear Lebesgue measures on $\mathbb{R}_k := \mathbb{R}$.

Clearly, $0 < \mu_k(\Delta_k) < +\infty$ for $k \in \mathbb{N}$. Hence, by Theorem 13.4.1 we claim that $\prod_{k\in\mathbb{N}} \Delta_k$ is shy in $\prod_{k\in\mathbb{N}} G_k$.

Now let us show that E is not σ-compact.

Let $\mathcal{F}$ be a family of compact subsets in $\mathbb{R}^{\mathbb{N}}$ of the form $\prod_{k\in\mathbb{N}} F_k$, where F_k is compact in $\mathbb{R}_k$.

If we assume that E is covered by the union of countable family of compact sets in $\mathbb{R}^\infty$, then by Lemma 5.3.6 we can deduce that E is covered also by the union of any countable family $(R_j)_{j\in\mathbb{N}}$ of elements of $\mathcal{F}$, where $R_j = \prod_{k=1}^{\infty} F_k^{(j)}$ for $j \in \mathbb{N}$ and $F_k^{(j)}$ is compact in $\mathbb{R}_k$.

We have

$$E = \prod_{k\in N} \Delta_k = \prod_{k\in\mathbb{N}} \Big(\sum_{i=1}^{\infty} \Delta_k^{(i)}\Big).$$

For $j \in \mathbb{N}$ we choose $\Delta_j^{(i_j)}$ such that $F_j^{(j)} \cap \Delta_j^{(i_j)} = \emptyset$. Then $\prod_{j \in \mathbb{N}} \Delta_j^{(i_j)} \subset E$ and $(\forall j)(j \in \mathbb{N} \rightarrow R_j \cap \prod_{k=1}^{\infty} \Delta_k^{(i_k)} = \emptyset)$. We have obtained a required contradiction and thus, E is not σ-compact in $\mathbb{R}^{\infty}$.

Chapter 6

On Ordinary and Standard "Lebesgue Measures" in Separable Banach Spaces

6.1 Introduction

By using an additional set-theoretical axiom asserted that *each subset of* $\mathbb{R}$ *is Lebesgue measurable* an example of an invariant measure firstly has been constructed on the powerset of a Banach space with absolutely convergent Schauder basis such that the constructed measure takes the value 1 on the standard rectangle(see [92, Th. 7.3]). A version of Lebesgue measure on every separable Banach space that has a Schauder basis firstly has been constructed in [42, Th. 12] without any additional set-theoretical assumption. More lately has been demonstrated that the completion $\overline{\mu}_Q$ of the Yamasaki-Kharazishvili measure μ_Q has the uniqueness property in the class of all L_1-invariant σ-finite measures in X with domain $\mathrm{dom}(\overline{\mu}_Q)$ (see comment in [43, p. 121]). The latter result has been extended by [44, Th. 4.2] to all infinite-dimensional separable Banach spaces. In [44], the problem of invariant measures for infinite-dimensional separable Banach spaces has been considered. Here has been demonstrated that their approach is distinct from that of Oxtoby [88]. The methods developed by [2] and [42], have been used for the construction of a translation invariant Borel measure in an infinite-dimensional separable Banach space with absolutely convergent Markushevich basis which gets the numerical value 1 on the standard parallelepiped defined by that basis.

In [98], new concepts of the Lebesgue measure on $\mathbb{R}^\infty$ have been proposed. Here has been demonstrated that Baker's both measures [2], [3], Mankiewicz and Preiss - Tišer generators [96] and the measure [91] are not an α-standard Lebesgue measure on $\mathbb{R}^\infty$ for $\alpha = (1, 1, \cdots)$.

Let α be an infinite parameter set and let $(\alpha_i)_{i \in I}$ be its any partition such that α_i is a non-empty finite subset for every $i \in I$. For $j \in \alpha$, let μ_j be a σ-finite Borel measure defined on a Polish metric space (E_j, ρ_j). In [106], a new concept of a standard $(\alpha_i)_{i \in I}$-product of measures $(\mu_j)_{j \in \alpha}$ has been introduced and its existence has been proved. As consequence, "a standard $(\alpha_i)_{i \in I}$-Lebesgue measure" on the Borel σ-algebra of subsets of

$\mathbb{R}^\alpha$ for every infinite parameter set α has been constructed such that it is invariant under a group generated by shifts and canonical permutations. A certain interesting application of the standard Lebesgue measure m^α in $\mathbb{R}^\alpha$ for a construction of uniform measures in the Banach space ℓ^α for an arbitrary parameter set α has been considered.

The purpose of the present chapter is to introduce a new concept of α-standard and α-ordinary "Lebesgue measures" in infinite-dimensional separable Banach spaces.

The chapter is organized as follows.

In Section 6.2 we give some basic definitions and results from measure theory and functional analysis. In Section 6.3 we present our main results.

6.2 Basic Definitions and Results from Measure Theory and Functional Analysis

a) *Continuous mappings.*

Lemma 6.2.1 [35, Theorem 2.3.6]. *Let $(Y_i)_{i\in I}$ be a family of topological spaces and $Y = \prod_{i\in I} Y_i$ be their cartesian product endowed with product topology. Let $p_j : Y \to Y_j$ be the j-projection for $j \in I$, defined by*

$$p_j((y_i)_{i\in I}) = y_j \quad (y_i)_{i\in I} \in Y.$$

If X is a topological space, then a mapping $f : X \to Y$ is continuous if and only if each composition $p_j \circ f$, $j \in I$, is continuous.

Lemma 6.2.2 [55, Theorem 15.1(Lusin-Souslin)]. *Let X, Y be Polish spaces and $f : X \to Y$ be continuous. If $A \subseteq X$ is Borel and $f|_A$ is injective then $f(A)$ is Borel.*

b) *Markushevich basis.*

Let X be an infinite-dimensional separable Banach space. A sequence $(x_k)_{k=1}^\infty \subset X$ is called *minimal* if each vector x_k is not contained in the closed linear span of $(x_l)_{l\neq k}$. A sequence in X is called *fundamental* if its closed linear span coincides with X. It is easy to verify that for a fundamental minimal sequence $(x_k)_{k=1}^\infty$ there exists a unique sequence $(x_k^*)_{k=1}^\infty$ of continuous linear functionals satisfying the condition $x_k^*(x_l) = \delta_{kl}$ $(k, l \in \mathbb{N})$. This sequence is called *biorthogonal* to $(x_k)_{k=1}^\infty$. Thus, if (x_k) is minimal and fundamental, then to each $x \in X$ there corresponds a generalized Fourier series

$$\sum_{k=1}^\infty x_k^*(x) x_k.$$

The vector x is uniquely determined by this series if and only if the biorthogonal sequence $(x_k^*)_{k=1}^\infty$ is *total* (that is for each $x \neq 0$ there exists $k \in \mathbb{N}$ such that $x_k^*(x) \neq 0$). A fundamental minimal sequence with a total biorthogonal sequence is called the *Markushevich basis* (M-basis in short). By the Markushevich theorem, for every countably-dimensional dense subspace L of a separable Banach space X there is an M-basis $(x_k, x_k^*)_{k=1}^\infty$ of X such that the linear span $\text{lin}(x_i)_1^\infty = L$. [130, p. 226]. Conversely, each Banach space with M-basis is separable. We call an M-basis *absolutely convergent* if $\sum_{k=1}^\infty \|x_k\| < \infty$. The following statement follows immediately from the Markushevich theorem mentioned above.

Lemma 6.2.3 [44, Lemma 2.3]. *Every infinite-dimensional separable Banach space has absolutely convergent M-basis.*

Lemma 6.2.4 [44, Lemma 2.4]. *Let (x_k, x_k^*) be an absolutely convergent M-basis in a Banach space X. Then for every bounded scalar sequence (a_k) the series $\sum_{k=1}^{\infty} a_k x_k$ is absolutely convergent to some element $x \in X$ and moreover $x_k^*(x) = a_k$ for all k.*

Let $(x_k, x_k^*)_{k=1}^{\infty}$ be an absolutely convergent M-basis of a Banach space X and define a rectangle P by:

$$P = \{x \in X : |x_k^*(x)| \leq 1/2 \ \text{ for all } \ k \in \mathbb{N}\}. \tag{6.2.1}$$

Obviously, P is a compact subset in X.

In the sequel, unlike N. Bourbaki well known notion, under N we understand a set $\{1, 2, \cdots\}$. We denote by $\mathbb{R}^{\mathbb{N}}$ the vector space of all real-valued sequences equipped with the product topology. We denote by $e_k = (0, \ldots, 0, 1, 0, \ldots)$, where 1 is in the k-th position, $k = 1, 2, \ldots$, the unit vectors of $\mathbb{R}^{\mathbb{N}}$.

If we define the operator $T : X \to \mathbb{R}^{\mathbb{N}}$ by

$$Tx = (x_k^*(x))_{k=1}^{\infty}, \tag{6.2.2}$$

then T is clearly linear, injective (because the M-basis is total), continuous (by Lemma 6.2.1) and $Tx_k = e_k$ for all k.

Lemma 6.2.5 [44, Lemma 2.5]. *The subspace $S := T(X) \in \mathcal{B}(\mathbb{R}^{\mathbb{N}})$ and the operator $T : X \to S$ is a Borel isomorphism.*

c) *Ordinary and Standard "Lebesgue measures" in $\mathbb{R}^{\mathbb{N}}$.*

Let $(\beta_j)_{j \in \mathbb{N}} \in [0, +\infty]^N$.

We say that a number $\beta \in [0, +\infty]$ is an ordinary product of numbers $(\beta_j)_{j \in \mathbb{N}}$ if

$$\beta = \lim_{n \to \infty} \prod_{i=1}^{n} \beta_i.$$

An ordinary product of numbers $(\beta_j)_{j \in \mathbb{N}}$ is denoted by $(\mathbf{O}) \prod_{i \in \mathbb{N}} \beta_i$.

A standard product of the family of numbers $(\beta_i)_{i \in \mathbb{N}}$ is denoted by $(\mathbf{S}) \prod_{i \in \mathbb{N}} \beta_i$ and defined as follows:

$(\mathbf{S}) \prod_{i \in \mathbb{N}} \beta_i = 0$ if $\sum_{i \in \mathbb{N}^-} \ln(\beta_i) = -\infty$, where $N^- = \{i : ln(\beta_i) < 0\}$ [1], and $(\mathbf{S}) \prod_{i \in \mathbb{N}} \beta_i = e^{\sum_{i \in \mathbb{N}} \ln(\beta_i)}$ if $\sum_{i \in \mathbb{N}^-} \ln(\beta_i) \neq -\infty$.

Let $\alpha = (n_k)_{k \in \mathbb{N}} \in \mathbb{N}^{\mathbb{N}}$. We set

$$F_0 = [0, n_0] \cap \mathbb{N},\ F_1 = [n_0{+}1, n_0{+}n_1] \cap \mathbb{N},\ \ldots, F_k = [n_0{+}\cdots{+}n_{k-1}{+}1, n_0{+}\cdots{+}n_k] \cap \mathbb{N}, \ldots.$$

We say that a number $\beta \in [0, +\infty]$ is an ordinary α-product of numbers $(\beta_i)_{i \in \mathbb{N}}$ if β is an ordinary product of numbers $(\prod_{i \in F_k} \beta_i)_{k \in \mathbb{N}}$. An ordinary α-product of numbers $(\beta_i)_{i \in \mathbb{N}}$ is denoted by $(\mathbf{O}, \alpha) \prod_{i \in \mathbb{N}} \beta_i$.

We say that a number $\beta \in [0, +\infty]$ is a standard α-product of numbers $(\beta_i)_{i \in \mathbb{N}}$ if β is a standard product of numbers $(\prod_{i \in F_k} \beta_i)_{k \in \mathbb{N}}$. A standard α-product of numbers $(\beta_i)_{i \in \mathbb{N}}$ is denoted $(\mathbf{S}, \alpha) \prod_{i \in \mathbb{N}} \beta_i$.

[1] We set $ln(0) = -\infty$

Let $\alpha=(n_k)_{k\in\mathbb{N}}\in\mathbb{N}^{\mathbb{N}}$. Let $(\alpha)O\mathcal{R}$ be the class of all infinite-dimensional measurable α-rectangles $R=\prod_{i\in\mathbb{N}}R_i(R_i\in\mathcal{B}(\mathbb{R}^{n_i}))$ for which an ordinary product of numbers $(m^{n_i}(R_i))_{i\in\mathbb{N}}$ exists and is finite.

We say that a measure λ being the completion of a translation-invariant Borel measure is an ordinary α-Lebesgue measure on $\mathbb{R}^\infty$(or, shortly, O(α)LM) if for every $R\in(\alpha)O\mathcal{R}$ we have

$$\lambda(R)=(\mathbf{O})\prod_{k\in\mathbb{N}}m^{n_k}(R_k).$$

Let $\alpha=(n_k)_{k\in\mathbb{N}}\in\mathbb{N}^{\mathbb{N}}$. Let $(\alpha)\mathcal{SR}$ be the class of all infinite-dimensional measurable α-rectangles $R=\prod_{i\in\mathbb{N}}R_i(R_i\in\mathcal{B}(\mathbb{R}^{n_i}))$ for which a standard product of numbers $(m^{n_i}(R_i))_{i\in\mathbb{N}}$ exists and is finite.

We say that a measure λ being the completion of a translation-invariant Borel measure is a standard α-Lebesgue measure on $\mathbb{R}^\infty$(or, shortly, S(α)LM) if for every $R\in(\alpha)\mathcal{SR}$ we have

$$\lambda(R)=(\mathbf{S})\prod_{k\in\mathbb{N}}m^{n_k}(R_k).$$

Lemma 6.2.6 [98, Proposition 1]. *Note that for every $\alpha=(n_k)_{k\in\mathbb{N}}\in\mathbb{N}^{\mathbb{N}}$ the following strict inclusion*

$$(\alpha)O\mathcal{R}\subset(\alpha)\mathcal{SR}$$

holds.

Lemma 6.2.7 [92, Theorem 1]. *For every $\alpha=(n_i)_{i\in\mathbb{N}}\in\mathbb{N}^{\mathbb{N}}$, there exists a Borel measure μ_α on $\mathbb{R}^\infty$ which is $O(\alpha)LM$.*

Lemma 6.2.8 [92, Theorem 2]. *For every $\alpha=(n_i)_{i\in\mathbb{N}}\in\mathbb{N}^{\mathbb{N}}$, there exists a Borel measure ν_α on $\mathbb{R}^\infty$ which is $(\alpha)LM$.*

Let μ_1 and μ_2 be two measures defined on the measurable space $(\mathbb{E},\mathbb{S})$.

Following [46, p. 124], we say that the μ_1 is absolutely continuous with respect to the μ_2, in symbols $\mu_1\ll\mu_2$, if

$$(\forall X)(X\in\mathbb{S}\ \&\ \mu_2(X)=0\rightarrow\mu_1(X)=0).$$

Following [46, p. 126], two measures μ_1 and μ_2 for which both $\mu_1\ll\mu_2$ and $\mu_2\ll\mu_1$ are called equivalent, in symbols $\mu_1\equiv\mu_2$.

Lemma 6.2.9 [92, Theorem 3]. *For every $\alpha=(n_i)_{i\in\mathbb{N}}\in\mathbb{N}^{\mathbb{N}}$, we have $\nu_\alpha\ll\mu_\alpha$ and the measures ν_α and μ_α are not equivalent.*

Let $\alpha=(n_i)_{i\in\mathbb{N}}\in\mathbb{N}^{\mathbb{N}}$ such that $n_i=n_j$ for every $i,j\in\mathbb{N}$. We set $F_i=(a_1^{(i)},\dots,a_{n_0}^{(i)})$ for every $i\in\mathbb{N}$. Let f be any permutation of $\mathbb{N}$ such that for every $i\in\mathbb{N}$ there exists $j\in\mathbb{N}$ such that $f(a_k^{(i)})=a_k^{(j)}$ for $1\le k\le n_0$. Then a map $A_f:\mathbb{R}^\infty\rightarrow\mathbb{R}^\infty$ defined by $A_f((z_k)_{k\in\mathbb{N}})=(z_{f(k)})_{k\in\mathbb{N}}$ for $(z_k)_{k\in\mathbb{N}}\in\mathbb{R}^{\mathbb{N}}$, is called a canonical α-permutations of $\mathbb{R}^{\mathbb{N}}$.

A group of transformations generated by all α-permutations and shifts of $\mathbb{R}^\infty$, is denoted by $\mathcal{G}_\alpha$.

Lemma 6.2.10 [92, Corollary 1]. *For every $\alpha=(n_i)_{i\in\mathbb{N}}\in\mathbb{N}^{\mathbb{N}}$ for which $n_i=n_j(i,j\in\mathbb{N})$, the measure ν_α is $\mathcal{G}_\alpha$-invariant.*

Lemma 6.2.11 [99, Theorem 3]. *Let* $\alpha = (n_i)_{i\in\mathbb{N}} \in \mathbb{N}^{\mathbb{N}}$, *and let* $T^{n_i} : \mathbb{R}^{n_i} \to \mathbb{R}^{n_i}$ $(i > 1)$ *be a family of linear transformations with Jacobians* $\Delta_i \neq 0$ *and* $0 < \prod_{i=1}^{\infty}\Delta_i < \infty$. *Let* $T^N : \mathbb{R}^{\mathbb{N}} \to \mathbb{R}^{\mathbb{N}}$ *be a map defined by*

$$T^{\mathbb{N}}(x) = (T^{n_1}(x_1,\ldots,x_{n_1}), T^{n_2}(x_{n_1+1},\ldots,x_{n_1+n_2}),\ldots),$$

where $x = (x_i)_{i\in N} \in \mathbb{R}^{\mathbb{N}}$. *Then for each* $E \in \mathcal{B}(\mathbb{R}^{\mathbb{N}})$, *we have*

$$\mu_\alpha(T^N(E)) = \Big(\prod_{i=1}^{\infty}\Delta_i\Big)\mu_\alpha(E).$$

d) *Ordinary and Standard "Lebesgue measures" in an infinite dimensional separable Banach space.*

Let X be an infinite-dimensional separable Banach space and (x_k, x_k^*) be an absolutely convergent M-basis. Let $T : X \to \mathbb{R}^{\mathbb{N}}$ be a linear operator defined by (6.2.2).

Let $\alpha = (n_k)_{k\in\mathbb{N}} \in \mathbb{N}^{\mathbb{N}}$. We say that a translation-invariant Borel measure μ on X is an ordinary α-Lebesgue measure on X(or, shortly, O(α)LM(X)) if for every $R \in (\alpha)\mathcal{OR}$ for which $R \cap T(X) \in (\alpha)\mathcal{OR}$ we have

$$\mu(T^{-1}(R)) = \mu_\alpha(R \cap T(X)),$$

where μ_α becomes from Lemma 6.2.7.

We say that a translation-invariant Borel measure μ on X is a standard α-Lebesgue measure on X(or, shortly, S(α)LM(X)) if for every $R \in (\alpha)\mathcal{SR}$ for which $R \cap T(X) \in (\alpha)\mathcal{SR}$ we have

$$\mu(T^{-1}(R)) = \nu_\alpha(R \cap T(X)).$$

6.3 Main Results

Let X be an infinite-dimensional separable Banach space and (x_k, x_k^*) be an absolutely convergent M-basis.

Theorem 6.3.1. *For every* $\alpha = (n_i)_{i\in\mathbb{N}} \in \mathbb{N}^{\mathbb{N}}$, *there exists a non-zero Borel measure* ψ_α *in* X *which is O(α)LM(X).*

Proof. For $Y \in \mathcal{B}(X)$ we set $\psi_\alpha(Y) = \mu_\alpha(T(Y))$, where μ_α becomes from Lemma 6.2.7 and T is defined by (6.2.2).

Obviously, ψ_α is a non-zero Borel measure because for parallelepiped P defined by (6.2.1), we have

$$\psi_\alpha(P) = \nu_\alpha(T(P)) = \nu_\alpha([-\frac{1}{2},\frac{1}{2}]^{\mathbb{N}}) = 1.$$

The latter relation follows from the evident fact that $[-\frac{1}{2},\frac{1}{2}]^{\mathbb{N}}$ belongs to the class O(α)LM for every $\alpha = (n_i)_{i\in\mathbb{N}} \in \mathbb{N}^{\mathbb{N}}$.

Let us show that ψ_α is translation invariant measure. Indeed, for $Y \in \mathcal{B}(X)$ and $h \in X$, we have

$$\psi_\alpha(Y+h) = \mu_\alpha(T(Y+h)) = \mu_\alpha(T(Y)+T(h)) = \mu_\alpha(T(Y)) = \psi_\alpha(Y).$$

Let us show that ψ_α is O(α)LM(X). Let $R \in (\alpha)\mathcal{OR}$ for which $R \cap T(S) \in (\alpha)\mathcal{OR}$. Then we have

$$\psi_\alpha(T^{-1}(R)) = \psi_\alpha(T^{-1}(R \cap T(X))) = \mu_\alpha(T(T^{-1}(R \cap T(X))) = \mu_\alpha(R \cap T(X)).$$

□

Theorem 6.3.2. *For every $\alpha = (n_i)_{i\in\mathbb{N}} \in \mathbb{N}^{\mathbb{N}}$, there exists a non-zero Borel measure η_α in X which is S(α)LM(X).*

Proof. For $Y \in \mathcal{B}(X)$ we set $\eta_\alpha(Y) = \nu_\alpha(T(Y))$, where ν_α becomes from Lemma 6.2.8 and T is defined by (6.2.2). Obviously, η_α is non-zero Borel measure because for parallelepiped P defined by (6.2.1), we have

$$\eta_\alpha(P) = \nu_\alpha(T(P)) = \nu_\alpha([-\frac{1}{2}, \frac{1}{2}]^{\mathbb{N}}) = 1.$$

The latter relation follows from the evident fact that $[-\frac{1}{2}, \frac{1}{2}]^{\mathbb{N}}$ belongs to the class S(α)LM for every $\alpha = (n_i)_{i\in\mathbb{N}} \in \mathbb{N}^{\mathbb{N}}$.

Let us show that η_α is translation invariant measure. Indeed, for $Y \in \mathcal{B}(X)$ and $h \in X$, we have

$$\eta_\alpha(Y+h) = \nu_\alpha(T(Y+h)) = \nu_\alpha(T(Y)+T(h)) = \nu_\alpha(T(Y)) = \eta_\alpha(Y).$$

Let us show that η_α is S(α)LM(X). Let $R \in (\alpha)\mathcal{SR}$ for which $R \cap T(S) \in (\alpha)\mathcal{SR}$. Then we have

$$\eta_\alpha(T^{-1}(R)) = \eta_\alpha(T^{-1}(R \cap T(X))) = \nu_\alpha(T(T^{-1}(R \cap T(X))) = \nu_\alpha(R \cap T(X)).$$

□

Theorem 6.3.3. *For every $\alpha = (n_i)_{i\in\mathbb{N}} \in \mathbb{N}^{\mathbb{N}}$ we have $\eta_\alpha \ll \psi_\alpha$ and η_α and ψ_α are not equivalent.*

Proof. Let us show that $\eta_\alpha \ll \psi_\alpha$. Indeed, let $\psi_\alpha(Y) = 0$ for $Y \in \mathcal{B}(X)$. Then we have

$$\psi_\alpha(Y) = \mu_\alpha(T(Y)) = 0.$$

By Lemma 6.2.9 we know that $\nu_\alpha \ll \mu_\alpha$ for every $\alpha = (n_i)_{i\in\mathbb{N}} \in \mathbb{N}^{\mathbb{N}}$ which implies that $\nu_\alpha(T(Y)) = 0$. By the definition of η_α we deduce that the value $\nu_\alpha(T(Y))$ coincides with the value $\eta_\alpha(Y)$ which implies that $\eta_\alpha(Y) = 0$.

Now we have to show that η_α and ψ_α are not equivalent. For $i \in \mathbb{N}$, we set

$$D_i = \prod_{j\in F_i}[0, e^{\frac{(-1)^i}{i \times n^i}}].$$

It is obvious that $m^{n_i}(D_i) = e^{\frac{(-1)^i}{i}}$ for $i \in \mathbb{N}$.

Note that the sequence of positive real numbers $\{e^{\frac{(-1)^i}{i\times n^i}} : i \in \mathbb{N}\}$ is bounded from above by the number $e^{\frac{1}{2}}$.

Since $(x_k, x_k^*)_{k=1}^{\infty}$ is an absolutely convergent M-basis of a Banach space X, by Lemma 6.2.4 we deduce that the series $\sum_i \sum_{j \in F_i} a_j x_j$ is absolutely convergent to some element $x \in X$ when $|a_j| \leq e^{\frac{(-1)^i}{i\times n^i}}$ for $j \in F_i$. By that reason we have that $T(D) = \prod_{i\in\mathbb{N}} D_i$, where D is defined by

$$D = \{\sum_i \sum_{j\in F_i} a_j x_j : 0 \leq a_j \leq e^{\frac{(-1)^i}{i\times n^i}} \text{ for } j \in F_i\}.$$

On the one hand we have

$$\eta_\alpha(D) = \nu_\alpha(T(D)) = \nu_\alpha(\prod_{i\in\mathbb{N}} D_i) = 0,$$

because

$$(\mathbf{S})\prod_{i\in\mathbb{N}} m^{n_i}(D_i) = (\mathbf{S})\prod_{i\in\mathbb{N}} e^{\frac{(-1)^i}{i}} = 0.$$

On the other hand we have

$$\psi_\alpha(D) = \mu_\alpha(T(D)) = \mu_\alpha(\prod_{i\in\mathbb{N}} D_i) = 2,$$

because

$$(\mathbf{O})\prod_{i\in\mathbb{N}} m^{n_i}(D_i) = (\mathbf{O})\prod_{i\in\mathbb{N}} e^{\frac{(-1)^i}{i}} = 2.$$

The latter relations imply that η_α and ψ_α are not equivalent. □

Remark 6.3.1. Let X be a Banach space with an absolutely convergent M-basis (x_k, x_k^*). By using [44, Proposition 3.1]), one can prove that η_α and ψ_α are invariant inner regular semi-finite non-σ-finite Borel measures in X for every $\alpha = (n_i)_{i\in\mathbb{N}} \in \mathbb{N}^{\mathbb{N}}$ such that $\eta_\alpha(P) = \psi_\alpha(P) = 1$. By using Theorem 6.3.3, we give a new proof of [44, Theorem 4.4]) asserted that *the class of invariant inner regular semi-finite non-σ-finite Borel measures in X which takes the value 1 on the set P fails the uniqueness property.*

Let $\alpha = (n_i)_{i\in\mathbb{N}} \in \mathbb{N}^{\mathbb{N}}$ such that $n_i = n_j$ for every $i, j \in \mathbb{N}$. We set $F_i = (a_1^{(i)}, \ldots, a_{n_0}^{(i)})$ for every $i \in \mathbb{N}$. Let f be any permutation of $\mathbb{N}$ such that for every $i \in \mathbb{N}$ there exists $j \in \mathbb{N}$ such that $f(a_k^{(i)}) = a_k^{(j)}$ for $1 \leq k \leq n_0$. If a map $\tilde{A}_f : X \to X$ defined by $\tilde{A}_f(\sum_k x_k x_k^*) = \sum_k x_{f(k)} x_k^*$ is convergent for all $x = \sum_k x_k x_k^* \in X$, then $\tilde{A}_f$ is called a canonical α-permutations of X.

We denote by $\tilde{\mathcal{G}}_\alpha$ a group generated by shifts of X and by such canonical α-permutations $\tilde{A}_f$ of X for which the following condition

$$(\forall x)(x \in X \to \tilde{A}_f(x) = T^{-1}(A_f(T(x))))$$

holds.

Theorem 6.3.4. *For every $\alpha = (n_i)_{i\in\mathbb{N}} \in \mathbb{N}^{\mathbb{N}}$ for which that $n_i = n_j (i, j \in \mathbb{N})$, the measure η_α is $\tilde{\mathcal{G}}_\alpha$-invariant.*

Proof. By Theorem 6.3.2 we know that the measure η_α is translation invariant. We have to show that η_α is invariant under action of each canonical α-permutation (of X) $\tilde{A}_f$. Indeed, by definition of the canonical α-permutation of X and by Lemma 6.2.10, we have

$$(\forall Y)(Y \in \mathcal{B}(X) \to \eta_\alpha(\tilde{A}_f(Y)) = \nu_\alpha(T(\tilde{A}_f(Y))) =$$

$$\nu_\alpha(A_f(T(Y))) = \nu_\alpha(T(Y)) = \eta_\alpha(Y)).$$

This ends the proof of Theorem 6.3.4. □

Let $\alpha = (n_i)_{i\in\mathbb{N}} \in \mathbb{N}^{\mathbb{N}}$, and $L^{n_i} := \{\sum_{j\in F_i} a_j x_j : (a_j)_{j\in F_i} \in \mathbb{R}^{n_i}\}$. Let $T^{n_i} : \mathbb{R}^{n_i} \to \mathbb{R}^{n_i}$ $(i > 1)$ be a family of linear transformations.
Let $T^N : \mathbb{R}^{\mathbb{N}} \to \mathbb{R}^{\mathbb{N}}$ be a map defined by

$$T^{\mathbb{N}}(x) = (T^{n_1}(x_1,\dots,x_{n_1}), T^{n_2}(x_{n_1+1},\dots,x_{n_1+n_2}),\dots),$$

where $x = (x_i)_{i\in N} \in \mathbb{R}^{\mathbb{N}}$.

Let $\tilde{T}^{n_i} : L^{n_i} \to L^{n_i}$ be defined by

$$\tilde{T}^{n_i}(\sum_{j\in F_i} a_j x_j) = \sum_{j\in F_i} Pr_j(T^{n_i}((a_j)_{j\in F_i}))$$

for $i \in \mathbb{N}$, where Pr_j denotes j-th projection in X defined by $Pr_j(x) = x_j^*(x)x_j$.

Let $\tilde{T}^N : X \to X$ be a map defined by

$$\tilde{T}^N(x) = \sum_i Pr_{L^{n_i}}(x),$$

where $Pr_{L^{n_i}}$ denotes a usual projection on the vector subspace L^{n_i} defined by $Pr_{L^{n_i}}(x) = \sum_{j\in F_i} x_j^*(x)x_j$.

Theorem 6.3.5 *Let $\alpha = (n_i)_{i\in\mathbb{N}} \in \mathbb{N}^{\mathbb{N}}$, and let $T^{n_i} : \mathbb{R}^{n_i} \to \mathbb{R}^{n_i}$ $(i > 1)$ be a family of linear transformations with Jacobians $\Delta_i \neq 0$ and $0 < \prod_{i=1}^{\infty} \Delta_i < \infty$. If $\tilde{T}^N$ is defined for every $x \in X$ such that the condition*

$$\Big(\forall x\Big)\Big(x \in X \to \tilde{T}^N(x) = T^{-1}(T^N(T(x)))\Big)$$

holds, then for each $Y \in \mathcal{B}(X)$ the following change of variable formula for the measure ψ_α

$$\psi_\alpha(\tilde{T}^N(Y)) = \Big(\prod_{i=1}^{\infty}\Delta_i\Big)\psi_\alpha(Y)$$

is valid.

Proof. By the definition of the measure ψ_α and by the assumption of the theorem we get

$$\psi_\alpha(\tilde{T}^N(Y)) = \mu_\alpha(T(\tilde{T}^N(Y))) = \mu_\alpha(T^N(T(Y))).$$

By using Lemma 6.2.11, we get

$$\psi_\alpha(\tilde{T}^N(Y)) = \mu_\alpha(T^N(T(Y))) = \Big(\prod_{i=1}^{\infty}\Delta_i\Big)\mu_\alpha(T(Y)) = \Big(\prod_{i=1}^{\infty}\Delta_i\Big)\psi_\alpha(Y).$$

This ends the proof of theorem. □

6.4 On a Certain Carrier of Ordinary and Standard Lebesgue Measures in Separable Banach Spaces

Remark 6.4.1. Let X be a separable Banach space. Following [50], a set Y is called shy if it is a subset of a Borel set Y' for which $\mu(Y'+x)=0$ for every $x\in X$ and some Borel probability measure μ such that $\mu(K)=\mu(X)$ for some compact K. A measure μ is said to be a transverse (or a transverse measure). Following [96], a Borel measure μ in X is called a generator of shy sets in X if the validity of the condition $\overline{\mu}(Y)=0$ for $Y\subset X$ implies that Y is shy in X, where $\overline{\mu}$ denotes the usual completion of the Borel measure μ. Following [96, Corollary 2.1], every quasifinite translation invariant Borel measure in X is a generator of shy sets in X. By using the latter result, we deduce that ν_α and ψ_α are generators of shy sets in X for each $\alpha=(n_i)_{i\in\mathbb{N}}\in\mathbb{N}^{\mathbb{N}}$.

The purpose of the present section is to study the topological structure of the carrier [2] of the ordinary and standard Lebesgue measures in separable Banach spaces.

The main result is contained in the following assertion.

Theorem 6.4.1. *For each $\alpha=(n_i)_{i\in\mathbb{N}}\in\mathbb{N}^{\mathbb{N}}$, η_α(or ψ_α) is generator of shy sets in separable Banach space X and a carrier of that generator may be choice a meagre set which is not covered by countable family of compact sets in X.*

Proof. Since η_α(or ψ_α) is quasifinite translation invariant Borel measure in $\mathbb{R}^{\mathbb{N}}$, by [96](see Corollary 2.1, p. 241) we deduce that ν_α(or ψ_α) is generator of shy sets.

Following [89](see Theorem 16.5) the real axis $\mathbb{R}$ may be written as the disjoint union of a shy set X_1 and a meagre set X_2. We set $Z_1:=X_1\times\mathbb{R}^{\mathbb{N}\setminus\{1\}}$ and $Z_2:=X_2\times\mathbb{R}^{\mathbb{N}\setminus\{1\}}$.

It is obvious that $\nu_\alpha(Z_1)=0$ (or $\mu_\alpha(Z_1)=0$) which implies that Z_2 is carrier of ν_α(or μ_α). On the one hand, Z_2 is a meagre set in $\mathbb{R}^{\mathbb{N}}$ which is not covered by a countable union of compact sets in $\mathbb{R}^{\mathbb{N}}$. Indeed, assume the contrary and $Z_2=\cup_{k\in\mathbb{N}}F_k$, where F_k is compact in $\mathbb{R}^{\mathbb{N}}$, then by [50](see Fact 8, p. 226) we will get that F_k is a shy set in $\mathbb{R}^{\mathbb{N}}$ for each $k\in\mathbb{N}$. Since ν_α (or μ_α) is generator of shy sets we claim that Z_1 is a shy set in $\mathbb{R}^{\mathbb{N}}$. By [50](see Fact $3''$, p. 224), we deduce that $Z_1\cup\cup_{k\in\mathbb{N}}F_k=\mathbb{R}^{\mathbb{N}}$ is a shy set in $\mathbb{R}^{\mathbb{N}}$, which is contradiction because for each quasifinite Borel measure μ in $\mathbb{R}^{\mathbb{N}}$ we have that $\mu(\mathbb{R}^{\mathbb{N}}+t)=\mu(\mathbb{R}^{\mathbb{N}})>0$ for each $t\in\mathbb{R}^{\mathbb{N}}$.

Now it is obvious that $T^{-1}(Z_2)$ is carrier of η_α(or ψ_α) which satisfies all conditions in Theorem 6.4.1.

□

Corollary 6.4.1. *Every infinite dimensional separable Banach space may be written as the disjoint union of a shy set and a meagre set.*

Corollary 6.4.2. *In every infinite dimensional separable Banach space X there is a meagre set D which is a prevalent and is not covered by a countable union of compact sets.*

Corollary 6.4.3. *In every infinite dimensional separable Banach space X there is a meagre set D which is not covered by a countable union of compact sets and for which the following*

[2] Let μ be a Borel measure defined in G. A Borel subset $A\subseteq G$ is called carrier of μ if $\mu(G\setminus A)=0$.

condition holds

$$(\forall h)(h \in X \to D \cap (D+h) \text{ is prevalent}).$$

Corollary 6.4.4. *In every infinite dimensional separable Banach space X there is a meagre set D such that for arbitrary $h \in X$ there is a point $y \in D$ such that an infinite arithmetic progression $y, y+h, \cdots$ belongs to D.*

Proof. Let $D := Z_2$, where Z_2 comes from Theorem 4.1. Let consider sets $\{D - ih : i \in \mathbb{N}\}$. Since $\cap_{i \in \mathbb{N}}(D - ih)$ again is prevalent, there is $y \in \cap_{i \in \mathbb{N}}(D - ih)$. Since $y \in D - ih$ for $i \in \mathbb{N}$, there are $y_i \in D(i \in \mathbb{N})$ such that $y = y_i - ih$ for $i \in \mathbb{N}$ which means that $y_i = y + ih$. Hence an infinite arithmetic progression $y, y+h, \cdots$ belongs to D. □

Corollary 6.4.5. *In every infinite dimensional separable Banach space X there is a meagre set D such that for arbitrary $h \in X$ a set D_h defined by*

$$D_h = \{y : y \in D \ (\forall i)(i \in \mathbb{N} \to y + ih \in D\}$$

is prevalent in X.

Proof. Let $D := Z_2$, where Z_2 comes from Theorem 6.4.1. Let consider sets $\{D - ih : i \in \mathbb{N}\}$. Note that $\cap_{i \in \mathbb{N}}(D - ih)$ is prevalent. We set $B_h = \cap_{i \in \mathbb{N}}(D - ih)$. Let $y \in B_h$. Since $y \in D - ih$ for $i \in \mathbb{N}$, there is $y_i \in D(i \in \mathbb{N})$ such that $y = y_i - ih$ for $i \in \mathbb{N}$ which means that $y_i = y + ih$. Hence an infinite arithmetic progression $y, y+h, \cdots$ belongs to D. Since $B_h \subseteq D_h$ we deduce that D_h is prevalent in X. □

Corollary 6.4.6. *In every infinite dimensional separable Banach space X there is a meagre set D such that for an arbitrary sequence of elements $h = (h_k) \in X$ a set D_h defined by*

$$D_h = \{y : y \in D \ (\forall i)(i \in \mathbb{N} \to y + \sum_{k=1}^{i} h_k \in D\}$$

is prevalent in X.

Proof. Let D comes from Theorem 6.4.1. Let consider sets $\{D - \sum_{k=1}^{i} h_k : i \in \mathbb{N}\}$. Note that $\cap_{i \in \mathbb{N}}(D - \sum_{k=1}^{i} h_k)$ is prevalent. We set $B_h = \cap_{k \in \mathbb{N}}(D - \sum_{k=1}^{i} h_k)$. Let $y \in B_h$. Since $y \in D - \sum_{k=1}^{i} h_k$ for $i \in \mathbb{N}$, there is $y_i \in D(i \in \mathbb{N})$ such that $y = y_i - \sum_{k=1}^{i} h_k$ for $k \in \mathbb{N}$ which means that $y_i = y + \sum_{k=1}^{i} h_k$. Hence an infinite sequence $(y + \sum_{k=1}^{i} h_k)_{i \in \mathbb{N}}$ belongs to D. Since $B_h \subseteq D_h$ we deduce that D_h is prevalent in X. □

Corollary 6.4.6 admits the following reformulation.

Corollary 6.4.7 *In every infinite dimensional separable Banach space X there is a meagre set D such that for an arbitrary $h \in X$, a set $X \setminus D_h$ is of η_α(or ψ_α)-measure zero(and, hence shy in X), where*

$$D_h = \{y : y \in D \ (\forall i)(i \in \mathbb{N} \to y + ih \in D\}$$

.

Chapter 7

On a Standard Product of an Arbitrary Family of σ-Finite Borel Measures with Domain in Polish Spaces

Let $(X_i, \mathbf{B}_i, \mu_i)$ $(i \in \mathbb{N})$ be a family of regular Borel measure spaces, where X_i is a Hausdorff topological space. In [74] has been proved an existence of a Borel measure μ on $\prod_{i \in \mathbb{N}} X_i$ (with respect to the product topology) such that if $K_i \subseteq X_i$ is compact for all $i \in \mathbb{N}$ and $\prod_{i \in \mathbb{N}} \mu_i(K_i)$ converges, then $\mu(\prod_{i \in \mathbb{N}} K_i) = \prod_{i \in \mathbb{N}} \mu_i(K_i)$. Note that a special case of this result (for the case where $X_i = R$ and m_i is Lebesgue measure) was proved only fairly recently in [2]. Slightly late, in [3] has been improved result [74] as follows: there exists of a Borel measure λ on $\prod_{i \in \mathbb{N}} X_i$ such that if $R_i \subseteq X_i$ is measurable for $i \in \mathbb{N}$ and $\prod_{i \in \mathbb{N}} \mu_i(R_i)$ converges, then $\lambda(\prod_{i \in \mathbb{N}} R_i) = \prod_{i \in \mathbb{N}} \mu_i(R_i)$.

Note that both above mentioned constructions in the case when multiplied measures coincide with a concrete σ-finite Borel measure μ in a Hausdorff topological space X give a measure $\mu^{\mathbb{N}}$ which is not invariant under permutations of the $X^{\mathbb{N}}$; To left this defect we introduce a notion of a *standard product of measures* and prove their existence under some assumptions. Our approach, unlike [74], [2], [3], is based on the notion of a standard product of a family of real numbers. Main results of the chapter are the theorem about existence of a *standard product of measures* and its invariance under action of some group of transformations. In the case when multiplied measures coincide with Lebesgue measure on $\mathbb{R}$ our product occurs to be invariant under permutations of the $\mathbb{R}^{\mathbb{N}}$(see, [99]) unlike Baker's measures [2], [3]. In addition, our construction is essentially different from the points of view of [74] and [3], because it allows us to construct a *standard product of measures* for an arbitrary (no only for countable) family of σ-finite Borel measures with domains in Polish spaces.

Suppose that X is a topological space. The Borel sets $\mathcal{B}(X)$ are the σ-algebra generated by the open sets of a topological space X, and the Baire sets $\mathcal{B}_0(X)$ are the smallest σ-algebra making all real-valued continuous functions measurable. In 1957 (see [77]), Mařík proved that all normal, countably paracompact spaces have the following property: Every Baire measure extends to a regular Borel measure. Spaces which have this property have come to be known as Mařík spaces.

The present chapter devoted to applications of properties of some Mařík spaces for definition of a product of an infinite family of σ-finite Borel measures with domain in Polish spaces.

In order to do it we recall some important notions and well-known results from general topology and probability theory.

X is a Hausdorff space iff distinct points in X have disjoint neighborhoods. X is a regular space if and only if, given any closed set F and any point x that does not belong to F, there exists a neighbourhood U of x and a neighbourhood V of F that are disjoint. X is a normal space if and only if, given any disjoint closed sets E and F, there are neighbourhoods U of E and V of F that are also disjoint. X is a regular Hausdorff space if and only if it is both regular and Hausdorff. X is a completely regular space if and only if, given any closed set F and any point x that does not belong to F, there is a continuous function f from X to the real line $\mathbb{R}$ such that $f(x)$ is 0 and $f(y)$ is 1 for every y in F. X is a Tychonoff space, if and only if it is both completely regular and Hausdorff.

Lemma 7.1 ([134]). *The following statements about a product of nonempty metric spaces are equivalent:*

(i) *The product is normal.*

(ii) *At most $\aleph_0$ of the factor spaces are noncompact.*

Lemma 7.2 ([141]). *Every normal regular space is completely regular, and every normal Hausdorff space is Tychonoff.*

Recall that a Borel measure μ, defined on a Hausdorff topological space (X,τ), is called Radon if

$$(\forall Y)(Y \in \mathcal{B}(X)\ \&\ 0 \le \mu(Y) < +\infty \rightarrow \mu(Y) = \sup_{\substack{K \subseteq Y \\ \text{K is compact in X}}} \mu(K))\ \diamondsuit$$

and called dense, if the condition $\diamondsuit$ holds for $Y = X$.

A family $(U_i)_{i\in I}$ of open subsets in (X,τ) is called a generalized sequence if

$$(\forall i_1)(\forall i_2)(i_1 \in I\ \&\ i_2 \in I \rightarrow (\exists i_3)(i_3 \in I \rightarrow (U_{i_1} \subset U_{i_3}\ \&\ U_{i_2} \subset U_{i_3}))).$$

A Borel probability measure μ defined on X is called τ-smooth if, for an arbitrary generalized sequence $(U_i)_{i\in I}$, the condition

$$\mu\Big(\bigcup_{i\in I} U_i\Big) = \sup_{i\in I} \mu(U_i)$$

is valid.

A Baire probability measure μ on X is called τ_0-smooth if, for an arbitrary generalized sequence $(U_i)_{i\in I}$ of open Baire subsets in X, for which $\bigcup_{i\in I} U_i$ is also a Baire subset, the condition

$$\mu\Big(\bigcup_{i\in I} U_i\Big) = \sup_{i\in I} \mu(U_i)$$

is valid.

The following lemma plays a key role in our future investigations.

Lemma 7.3 ([138], Theorem 3.3, p. 42). *Let X be completely regular topological space, μ be a Baire probability measure defined on the σ-algebra $\mathcal{B}_0(X)$. Then*

(a) *if μ is τ_0-smooth, there exists a unique τ-smooth Borel extension on X.*

(b) *if the space X is Hausdorff and μ is dense on $\mathcal{B}_0(X)$, then μ admits a unique Radon extension on $\mathcal{B}(X)$.*

Let $(E_j,\tau_j)_{j\in\alpha}$ be a family of Hausdorff topological spaces. We denote by $(\prod_{j\in\alpha}E_j,\tau)$ the Tychonoff product of the family of topological spaces $(E_j,\tau_j)_{j\in\alpha}$.

Lemma 7.4. *Let $(E_j,\rho_j)_{j\in\alpha}$ be a family of non-empty Polish metric spaces that at most $\aleph_0$ of them are noncompact and let μ_j be a Borel probability measure on E_j for $j\in\alpha$. Then the product measure $\prod_{j\in\alpha}\mu_j$ is τ_0-smooth and dense on $\prod_{j\in\alpha}E_j$.*

Proof. It is obvious that $\mathrm{dom}(\prod_{j\in\alpha}\mu_j)=\mathcal{B}_0(\prod_{j\in\alpha}E_j)$.

Let $(U_i)_{i\in I}$ be an arbitrary generalized sequence of open Baire subsets in $\prod_{j\in\alpha}E_j$, for which $\bigcup_{i\in I}U_i$ is also a Baire subset.

The latter relation getting together with an assumption of Lemma 17.4 stated that at most $\aleph_0$ of the family $(E_j,\rho_j)_{j\in\alpha}$ are noncompact follows that there exists a countable subset $\alpha_0\subseteq\alpha$ and $U_{\alpha_0}\in\mathcal{B}(\prod_{j\in\alpha_0}E_j)$ such that

$$\bigcup_{i\in I}U_i=U_{\alpha_0}\times\Big(\prod_{j\in\alpha\setminus\alpha_0}E_j\Big)$$

and

$$(\forall j)(j\in\alpha\setminus\alpha_0\rightarrow(E_j,\rho_j)\text{ is compact}).$$

By inner regularity of the Borel probability measure $\prod_{j\in\alpha_0}\mu_j$ with domain in Polish space there exists an increasing family of compact sets $(F_k)_{k\in\mathbb{N}}$ such that $F_k\subseteq U_{\alpha_0}$ and

$$\lim_{n\rightarrow\infty}\Big(\prod_{j\in\alpha_0}\mu_j\Big)(F_n)=\Big(\prod_{j\in\alpha_0}\mu_j\Big)(U_{\alpha_0}).$$

We set $D_n=F_n\times\prod_{j\in\alpha\setminus\alpha_0}E_j$ for $n\in\mathbb{N}$. It is obvious that $(D_n)_{n\in\mathbb{N}}$ is an increasing family of compact subsets in $\prod_{j\in\alpha}E_j$ such that

$$\lim_{n\rightarrow\infty}\Big(\prod_{j\in\alpha}\mu_j\Big)(D_n)=\Big(\prod_{j\in\alpha}\mu_j\Big)(\cup_{i\in I}U_i).$$

It is obvious that $(U_i)_{i\in I}$ is covering of D_n for every $n\in\mathbb{N}$. Hence, using the definition of the generalized sequence of open sets in topological space, we can construct such a sequence $(i_n)_{n\in\mathbb{N}}$ of indices of I that the sequence $(U_{i_n})_{n\in\mathbb{N}}$ will be increasing and $D_n\subseteq U_{i_n}$ for $n\in\mathbb{N}$. We have

$$\Big(\prod_{j\in\alpha}\mu_j\Big)(D_n)\leq\Big(\prod_{j\in\alpha}\mu_j\Big)(U_{i_n})$$

for every $n\in\mathbb{N}$. Hence

$$\Big(\prod_{j\in\alpha}\mu_j\Big)(\cup_{i\in I}U_i)=\lim_{n\rightarrow\infty}\Big(\prod_{j\in\alpha}\mu_j\Big)(D_n)\leq$$

$$\leq \lim_{n\to\infty}\Big(\prod_{j\in\alpha}\mu_j\Big)(U_{i_n}) \leq \Big(\prod_{j\in\alpha}\mu_j\Big)(\cup_{i\in I}U_i).$$

The latter relation means that the condition

$$\Big(\prod_{j\in\alpha}\mu_j\Big)\Big(\bigcup_{i\in I}U_i\Big) = \sup_{i\in I}\Big(\prod_{j\in\alpha}\mu_j\Big)(U_i)$$

holds. Thus, the measure $\prod_{j\in\alpha}\mu_j$ is τ_0-smooth on $\prod_{j\in\alpha}E_j$.

Let us show that the measure $\prod_{j\in\alpha}\mu_j$ is dense.

We set

$$\alpha_1 = \{j : \ E_j \text{ is not compact}\}.$$

It is clear that for the Borel measure $\prod_{j\in\alpha_1}\mu_j$ there exists an increasing sequence of compact subsets $(F_k)_{k\in\mathbb{N}}$ in $\prod_{j\in\alpha_1}E_j$ that

$$\lim_{k\to+\infty}\Big(\prod_{j\in\alpha_1}\mu_j\Big)(F_k) = 1.$$

Now it is easy that $(F_k \times \prod_{j\in\alpha\setminus\alpha_1}E_j)_{k\in\mathbb{N}}$ is an increasing sequence of compact subsets in $\prod_{j\in\alpha}E_j$ such that

$$\lim_{k\to+\infty}\Big(\prod_{j\in\alpha}\mu_j\Big)\Big(F_k \times \prod_{j\in\alpha\setminus\alpha_1}E_j\Big) = 1. \qquad \square$$

Lemma 7.5. *Let $(E_j,\rho_j)_{j\in\alpha}$ be a family of non-empty Polish metric spaces that at most $\aleph_0$ of them are noncompact and let μ_j be a Borel probability measure on E_j for $j \in \alpha$. Then there exists a unique τ-smooth and Radon extension of the Baire measure $\prod_{j\in\alpha}\mu_j$ from the σ-algebra $\mathcal{B}_0(\prod_{j\in\alpha}E_j)$ to the σ-algebra $\mathcal{B}(\prod_{j\in\alpha}E_j)$.*

Proof. By Lemma 7.1, $\prod_{j\in\alpha}E_j$ is normal. Hence, in order to show that $\prod_{j\in\alpha}E_j$ is regular it is sufficient to show that every point $(x_j)_{j\in\alpha}$ is closed in $\prod_{j\in\alpha}E_j$. But the latter relation follows from the Tychonoff well-known theorem because the point $(x_j)_{j\in\alpha}$ can be considered as a product of compact sets $(\{x_j\})_{j\in\alpha}$. Thus, is normal and regular. By Lemma 7.2 we claim that the $(\prod_{j\in\alpha}E_j,\tau)$ is a completely regular topological space. Applications of Lemma 7.3 and Lemma 7.4 end the proof of Lemma 7.5. $\square$

We have the following lemma.

Lemma 7.6 ([92], Lemma 4.4, p. 67)**.** *Let (E_1,τ_1) and (E_2,τ_2) be two topological spaces. Denote by $B(E_1)$ and $B(E_2)$ (correspondingly, $B(E_1 \times E_2)$) the class of all Borel subsets generated by the topologies τ_1 and τ_2 (correspondingly, $\tau_1 \times \tau_2$). If at least one of these topological spaces has a countable base, then the equality*

$$B(E_1) \times B(E_2) = B(E_1 \times E_2)$$

holds.

The following definition extends Definition 1.2 for an uncountable family of non-negative real numbers $(\beta_j)_{j\in\alpha} \in [0,+\infty]^{\alpha}$.

Definition 7.1. A standard product of the family of numbers $(\beta_j)_{j\in\alpha}$ is denoted by $(\mathbf{S})\prod_{j\in\alpha}\beta_j$ and defined as follows:

$(\mathbf{S})\prod_{j\in\alpha}\beta_j = 0$ if $\sum_{i\in\alpha^-}\ln(\beta_j) = -\infty$, where $\alpha^- = \{j : ln(\beta_j) < 0\}$ [1], and

$(\mathbf{S})\prod_{j\in\alpha}\beta_j = e^{\sum_{j\in\alpha}\ln(\beta_j)}$ if $\sum_{j\in\alpha^-}\ln(\beta_j) \neq -\infty$.

Remark 7.1. Let $(E_j,\rho_j)_{j\in\alpha}$ be again a sequence of non-empty Polish metric spaces that at most $\aleph_0$ of them are noncompact. Let $(\mu_j)_{j\in\alpha}$ be a sequence of Borel non-zero diffused finite measures with $\mathrm{dom}(\mu_j) = \mathcal{B}(E_j)$ for $j \in \alpha$ and

$$0 < (\mathbf{S})\prod_{j\in\alpha}\mu_j(E_j) < +\infty.$$

By Lemmas 7.5, we claim that there exists a unique τ-smooth and Radon Borel extension λ of the Baire probability measure $\prod_{j\in\alpha}\frac{\mu_j}{\mu_j(E_j)}$. A Borel measure

$$(\mathbf{S})\prod_{j\in\alpha}\mu_j(E_j)\times\lambda$$

is called a standard product of the family of finite Borel measures $(\mu_j)_{j\in\alpha}$ and is denoted by $(\mathbf{S})\prod_{j\in\alpha}\mu_j$.

We put

$$\tau_i = \prod_{j\in\alpha_i}\mu_j.$$

Lemma 7.7. *Let α be again an arbitrary infinite parameter set and let $(\alpha_i)_{i\in I}$ be its any partition such that α_i is a non-empty finite subset of the α for every $i \in I$. Let μ_j be a σ-finite diffused Borel measure defined on a Polish space (E_j,ρ_j) for $j \in \alpha$.*

We denote by $\mathcal{R}_{(\alpha_i)_{i\in I}}$ the family of all measurable rectangles $R \subseteq \prod_{j\in\alpha}E_j$ of the form $\prod_{i\in I}R_i$ with the property $0 \leq (\mathbf{S})\prod_{i\in I}\tau_i(R_i) < \infty$ such that at most $\aleph_0$ of them are noncompact(i.e., card$\{i : i\in I$ & R_i is not compact in $\prod_{j\in\alpha_i}E_j\} \leq \aleph_0$.)

We suppose that there exists $R_0 = \prod_{i\in I}R_i^{(0)} \in \mathcal{R}_{(\alpha_i)_{i\in I}}$ such that

$$0 < (\mathbf{S})\prod_{i\in I}\tau_i(R_i^{(0)}) < \infty.$$

For $X \in \mathcal{B}(R)$, we set $\mu_R(X) = 0$ if

$$(\mathbf{S})\prod_{i\in I}\tau_i(R_i) = 0,$$

and

$$\mu_R(X) = (\mathbf{S})\prod_{i\in I}\tau(R_i)\times\Big(\prod_{i\in I}\frac{\tau_{iR_i}}{\tau_i(R_i)}\Big)(X)$$

otherwise, where $\frac{\tau_{iR_i}}{\tau_i(R_i)}$ is a Borel probability measure defined on R_i as follows

$$(\forall X)\Big(X\in\mathcal{B}(R_i)\rightarrow\frac{\tau_{iR_i}}{\tau_i(R_i)}(X) = \frac{\tau_i(Y\cap R_i)}{\tau_i(R_i)}\Big).$$

Then the family of measures $(\mu_R)_{R\in\mathcal{R}}$ is consistent.

[1]We set $ln(0) = -\infty$

Proof. Let $R_1 = \prod_{i\in I} R_i^{(1)}$ and $R_2 = \prod_{i\in I} R_i^{(2)}$ be two elements of the class $\mathcal{R} = \mathcal{R}_{(\alpha_i)_{i\in I}}$.

Without loss of generality it can be assumed that $0 < (\mathbf{S})\prod_{i\in I}\tau_i(R_i^{(1)}) < \infty$ and $0 < (\mathbf{S})\prod_{i\in I}\tau_i(R_i^{(2)}) < \infty$.

We will show that $\mu_{R_1}(X) = \mu_{R_2}(X)$ for every $X \in \mathcal{B}(R_1 \cap R_2)$. In this case it is sufficient to show that $\mu_{R_1}(Y) = \mu_{R_2}(Y)$ for every elementary measurable rectangle $Y = \prod_{i\in I} Y_i$ in $R_1 \cap R_2$. Note here that under an elementary measurable rectangle $Y = \prod_{i\in I} Y_i$ in $R_1 \cap R_2$ we assume a subset of $R_1 \cap R_2$ such that $Y_i \in \mathcal{B}(R_i^{(1)} \cap R_i^{(2)})$ for every $i \in \mathbb{N}$ and, in addition, there exists a finite subset I_0 of I such that $Y_i = R_i^{(1)} \cap R_i^{(2)}$ for $i \in I \setminus I_0$.

For every $i \in I$ and for every $Y_i \in \mathcal{B}(R_i^{(1)} \cap R_i^{(2)})$ we have

$$\tau_i(Y_i \cap R_i^{(1)} \cap R_i^{(2)}) = \tau_i(Y_i \cap R_i^{(1)}) = \tau_i(Y_i \cap R_i^{(2)}).$$

The latter relation implies that

$$(\mathbf{S})\prod_{i\in I}\tau_i(Y_i \cap R_i^{(1)} \cap R_i^{(1)}) = (\mathbf{S})\prod_{i\in I}\tau_i(Y_i \cap R_i^{(1)}) = (\mathbf{S})\prod_{i\in I}\tau_i(Y_i \cap R_i^{(1)}).$$

Hence we get

$$\mu_{R_1}\Big(\prod_{i\in I} Y_i\Big) = (\mathbf{S})\prod_{i\in I}\tau_i(Y_i \cap R_i^{(1)}) = (\mathbf{S})\prod_{i\in I}\tau_i(Y_i \cap R_i^{(1)} \cap R_i^{(2)})) =$$

$$= (\mathbf{S})\prod_{i\in I}\tau_i(Y_i \cap R_i^{(2)}) = \mu_{R_2}\Big(\prod_{i\in I} Y_i\Big).$$

Since a class $\mathcal{A}(R_1 \cap R_2)$ of all finite disjoint unions of elementary measurable rectangles in $R_1 \cap R_2$ is a ring, and since, by definition, the class $\mathcal{B}_0(R_1 \cap R_2)$ of Baire subsets of $R_1 \cap R_2$ is a minimal σ-ring generated by the ring $\mathcal{A}(R_1 \cap R_2)$, we claim (cf. [46] Theorem B, p. 27) that the class of all those sets of $R_1 \cap R_2$ for which this equality holds coincides with the class $\mathcal{B}_0(R_1 \cap R_2)$.

Since restrictions of μ_{R_1} and μ_{R_2} to the class $\mathcal{B}_0(R_1 \cap R_2)$ coincide, and $R_1 \cap R_2$ is a product of non-empty Polish metric spaces that at most $\aleph_0$ of them are noncompact, by Lemma 7.5 we claim that their Borel extensions coincide such that extended Borel measure is unique, τ-smooth and Radon. The latter relation means that the family of measures $(\mu_R)_{R\in\mathcal{R}}$ is consistent and Lemma 7.7 is proved. □

Let α be again an arbitrary infinite parameter set and let $(\alpha_i)_{i\in I}$ be its any partition such that α_i is a non-empty finite subset of the α for every $i \in I$. Let μ_j be a σ-finite continuous Borel measure defined on a Polish space (E_j, ρ_j) for $j \in \alpha$.

We denote by $\mathcal{R}_{(\alpha_i)_{i\in I}}$ the family of all measurable rectangles $R \subseteq \prod_{j\in\alpha} E_j$ of the form $\prod_{i\in I} R_i$ with the property $0 \le (\mathbf{S})\prod_{i\in I}\tau_i(R_i) < \infty$ such that at most $\aleph_0$ of them are noncompact.

We suppose that there exists $R_0 = \prod_{i\in I} R_i^{(0)} \in \mathcal{R}_{(\alpha_i)_{i\in I}}$ such that

$$0 < (\mathbf{S})\prod_{i\in I}\tau_i(R_i^{(0)}) < \infty.$$

We say that a Borel measure $\nu_{(\alpha_i)_{i\in I}}$ defined on $\mathcal{B}(\prod_{j\in\alpha} E_j)$ is called a standard

$(\alpha_i)_{i\in I}$-product of the family of σ-finite continuous Borel measures $(\mu_j)_{j\in\alpha}$ if for every

$$R = \prod_{i\in I} R_i \in \mathcal{R}_{(\alpha_i)_{i\in I}}$$

we have

$$\nu_{(\alpha_i)_{i\in I}}(R) = (\mathbf{S})\prod_{i\in I} \tau_i(R_i),$$

where $\tau_i = \prod_{j\in\alpha_i} \mu_j$ for $i \in I$.

Theorem 7.1. *Let μ_j be a σ-finite continuous Borel measure defined on a Polish space (E_j, ρ_j) for $j \in \alpha$. Let α be again an arbitrary infinite parameter set and let $(\alpha_i)_{i\in I}$ be its any partition such that α_i is a non-empty finite subset of the α for every $i \in I$, and let us suppose that there exists $R_0 = \prod_{i\in I} R_i^{(0)} \in \mathcal{R}_{(\alpha_i)_{i\in I}}$ such that*

$$0 < (\mathbf{S})\prod_{i\in I} \tau_i(R_i^{(0)}) < \infty.$$

Then there exists a standard $(\alpha_i)_{i\in I}$-product of the family $(\mu_j)_{j\in\alpha}$.

Proof. We denote by $\mathcal{R}_{(\alpha_i)_{i\in I}}$ the family of all measurable rectangles $R \subseteq \prod_{j\in\alpha} E_j$ of the form $\prod_{i\in I} R_i$ with the property $0 \le (\mathbf{S})\prod_{i\in I} \tau_i(R_i) < \infty$ such that at most $\aleph_0$ of them are noncompact.

For $X \in \mathcal{B}(R)$, we set $\mu_R(X) = 0$ if

$$(\mathbf{S})\prod_{i\in I} \tau_i(R_i) = 0,$$

and

$$\mu_R(X) = (\mathbf{S})\prod_{i\in I} \tau(R_i) \times \Big(\prod_{i\in I} \frac{\tau_{iR_i}}{\tau_i(R_i)}\Big)(X)$$

otherwise, where $\frac{\tau_{iR_i}}{\tau_i(R_i)}$ is a Borel probability measure defined on R_i as follows

$$(\forall X)\Big(X \in \mathcal{B}(R_i) \to \frac{\tau_{iR_i}}{\tau_i(R_i)}(X) = \frac{\tau_i(Y \cap R_i)}{\tau_i(R_i)}\Big).$$

By Lemma 7.7, the family of measures $(\mu_R)_{R\in\mathcal{R}}$ is consistent. We set

$$\nu_{(\alpha_i)_{i\in I}} = \mu_{\mathcal{R}_{(\alpha_i)_{i\in I}}},$$

where the measure $\mu_{\mathcal{R}_{(\alpha_i)_{i\in I}}}$ is defined by Lemma 1.1.

This ends the proof of Theorem 7.1. □

In the sequel we denote a standard $(\alpha_i)_{i\in I}$-product of the family $(\mu_j)_{j\in\alpha}$ by

$$(\mathbf{S}, (\alpha_i)_{i\in I})\prod_{j\in\alpha} \mu_j.$$

Here we present a certain example of the family of σ-finite continuous Borel measures $(\mu_j)_{j\in\mathbb{N}}$ defined on the real axis $\mathbb{R}$ and of two different partitions $(\alpha_i)_{i\in\mathbb{N}}$ and $(\beta_i)_{i\in\mathbb{N}}$ of the $\mathbb{N} \setminus \{0\}$ for which

$$(\mathbf{S}, (\alpha_i)_{i\in\mathbb{N}})\prod_{j\in\mathbb{N}} \mu_j \neq (\mathbf{S}, (\beta_i)_{i\in\mathbb{N}})\prod_{j\in\mathbb{N}} \mu_j.$$

Example 7.1. We set $\alpha = \mathbb{N}$. For $j \in \mathbb{N}$, let l_j be a linear Lebesgue measure on $\mathbb{R}$. Let $\alpha_i = \{i+1\}$ and $\beta_i = \{2i+1, 2(i+1)\}$ for $i \in \mathbb{N}$.

We set

$$Y_i = \left[0, \frac{1}{2}\right] \times [0,2].$$

It is obvious that

$$\left((\mathbf{S}, (\beta_i)_{i\in\mathbb{N}}) \prod_{j\in\mathbb{N}} l_j\right)\left(\prod_{i\in\mathbb{N}} Y_i\right) = 1$$

and

$$\left((\mathbf{S}, (\alpha_i)_{i\in\mathbb{N}}) \prod_{j\in\mathbb{N}} l_j\right)\left(\prod_{i\in\mathbb{N}} Y_i\right) = 0.$$

In connection with Theorems 7.1 and Example 7.1 we state the following

Problem 7.1. Under assumptions of Theorem 7.1, describe all pairs of partitions $(\alpha_i)_{i\in I}$ and $(\beta_i)_{i\in I}$ of the α for which $(\mathbf{S}, (\alpha_i)_{i\in I}) \prod_{j\in\alpha} \mu_j = (\mathbf{S}, (\beta_i)_{i\in I}) \prod_{j\in\alpha} \mu_j$.

The next statement is an immediate consequence of Theorem 7.1.

Theorem 7.2. *Under assumptions of Theorem* 7.1, *if each measure μ_j is G_j-left-and-right-invariant, where G_j denotes a group of Borel transformations of the E_j for $j \in \alpha$, then the measure $(\mathbf{S}, (\alpha_i)_{i\in I}) \prod_{j\in\alpha} \mu_j$ is $\prod_{j\in\alpha} G_j$-left-and-right-invariant.*

Proof. We set $G = \prod_{j\in\alpha} G_j$. Let us show that the measure $(\mathbf{S}, (\alpha_i)_{i\in I}) \prod_{j\in\alpha} \mu_j$ is G-left-and-right-invariant. Indeed, let $g, f \in G$ and $X \in \mathcal{B}(\prod_{j\in\alpha} E_j)$.

If X is not covered by countable family of elements of $\mathcal{R}_{(\alpha_i)_{i\in I}}$ then such will be gXf because the class $\mathcal{R}_{(\alpha_i)_{i\in I}}$ is left-and-right-invariant, i.e., $g\mathcal{R}_{(\alpha_i)_{i\in I}} f = \mathcal{R}_{(\alpha_i)_{i\in I}}$ for every $g, f \in G$. Hence, by the definition of the measure $((\alpha_i)_{i\in I}) \prod_{j\in\alpha} \mu_j$ we have

$$\left((\mathbf{S}, (\alpha_i)_{i\in I}) \prod_{j\in\alpha} \mu_j\right)(gXf) = +\infty.$$

Now let X be covered by the family $(A_k)_{k\in\mathbb{N}}$ of elements of $\mathcal{R}_{(\alpha_i)_{i\in I}}$. Then gXf will be covered by the family $(gA_k f)_{k\in\mathbb{N}}$ of elements of $\mathcal{R}_{(\alpha_i)_{i\in I}}$. Under Convention 1.1, we get

$$\left((\mathbf{S}, (\alpha_i)_{i\in I}) \prod_{j\in\alpha} \mu_j\right)(gXf) = \sum_{n=1}^{\infty} \lambda_{gA_n f}((gA_n f \setminus \cup_{k=1}^{n-1} gA_k f) \cap gXf) =$$

$$= \sum_{n=1}^{\infty} \lambda_{gA_n f}(g((A_n f \setminus \cup_{k=1}^{n-1} A_k f) \cap Xf)) =$$

$$= \sum_{n=1}^{\infty} \lambda_{A_n f}((A_n f \setminus \cup_{k=1}^{n-1} A_k f) \cap Xf) =$$

$$= \sum_{n=1}^{\infty} \lambda_{A_n f}((A_n \setminus \cup_{k=1}^{n-1} A_k) \cap X) f) =$$

$$= \sum_{n=1}^{\infty} \lambda_{A_n}((A_n \setminus \cup_{k=1}^{n-1} A_k) \cap X) = \left((\mathbf{S}, (\alpha_i)_{i\in I}) \prod_{j\in\alpha} \mu_j\right)(X).$$

□

By the scheme used in the proof of Theorem 7.2, one can prove the following assertion.

Theorem 7.3. *Under assumption of Theorem 7.1, if each measure μ_j is G_j-left-invariant, where G_j denotes a group of Borel transformations of the E_j for $j \in \alpha$, then the measure $(\mathbf{S},(\alpha_i)_{i\in I})\prod_{j\in\alpha}\mu_j$ is a $\prod_{j\in\alpha} G_j$-left-invariant.*

Observation 7.1. *Under conditions of Theorem 7.1, the measure $(\mathbf{S},(\alpha_i)_{i\in I})\prod_{j\in\alpha}\mu_j$ is Radon.*

Proof. Let $0 < \big((\mathbf{S},(\alpha_i)_{i\in I})\prod_{j\in\alpha}\mu_j\big)(X) < \infty$. Under Convention 1.1, it means that $X \in \mathcal{B}(\prod_{j\in\alpha} E_j)$ is covered by any countable family $(A_n)_{n\in\mathbb{N}}$ of elements of $\mathcal{R}_{(\alpha_i)_{i\in I}}$ such that

$$\Big((\mathbf{S},(\alpha_i)_{i\in I})\prod_{j\in\alpha}\mu_j\Big)(X) = \sum_{n=1}^{\infty}\lambda_{A_n}((A_n\setminus\cup_{k=1}^{n-1}A_k)\cap X).$$

Since the measure λ_{A_n} is Radon, we can choose a compact set

$$F_n \subseteq (A_n\setminus\cup_{k=1}^{n-1}A_k)\cap X$$

such that

$$\lambda_{A_n}\big(((A_n\setminus\cup_{k=1}^{n-1}A_k)\cap X)\setminus F_n\big) < \frac{\varepsilon}{2^{n+1}}$$

for $n \in \mathbb{N}$.

Also, we can choose a natural number n_ε such that

$$\sum_{n=n_\varepsilon+1}^{\infty}\lambda_{A_n}((A_n\setminus\cup_{k=1}^{n-1}A_k)\cap X) < \frac{\varepsilon}{2}.$$

Finally, we get

$$\Big((\mathbf{S},(\alpha_i)_{i\in I})\prod_{j\in\alpha}\mu_j\Big)(X\setminus\cup_{s=1}^{n_\varepsilon}F_s) =$$

$$= \sum_{n=1}^{\infty}\lambda_{A_n}((A_n\setminus\cup_{k=1}^{n-1}A_k)\cap(X\setminus\cup_{s=1}^{n_\varepsilon}F_s)) =$$

$$= \sum_{n=1}^{n_\varepsilon}\lambda_{A_n}((A_n\setminus\cup_{k=1}^{n-1}A_k)\cap(X\setminus\cup_{s=1}^{n_\varepsilon}F_s)) +$$

$$= \sum_{n=n_\varepsilon+1}^{\infty}\lambda_{A_n}((A_n\setminus\cup_{k=1}^{n-1}A_k)\cap(X\setminus\cup_{s=1}^{n_\varepsilon}F_s)) \leq$$

$$= \sum_{n=1}^{n_\varepsilon}\lambda_{A_n}(((A_n\setminus\cup_{k=1}^{n-1}A_k)\cap X)\setminus F_n) +$$

$$= \sum_{n=n_\varepsilon+1}^{\infty}\lambda_{A_n}((A_n\setminus\cup_{k=1}^{n-1}A_k)\cap X) \leq \frac{\varepsilon}{2}+\frac{\varepsilon}{2} = \varepsilon.$$ □

Remark 7.2. For $j \in \alpha$, we set $E_j = \mathbb{R}$ and $\mu_j = m$, where m denotes a linear Lebesgue measure on $\mathbb{R}$.

Let $(\alpha_i)_{i\in I}$ be any partition of α such that α_i is a non-empty finite for every $i \in I$.

It is clear that $\prod_{j\in\alpha}[a_j,b_j] \in \mathcal{R}_{(\alpha_i)_{i\in I}}$ if $0 \le (\mathbf{S})\prod_{i\in I} m^{\alpha_i}(\prod_{j\in\alpha_i}[a_j,b_j]) < \infty$, where m^{α_i} is a Lebesgue measure on $\mathbb{R}^{\alpha_i}$.

Then the measure $(\mathbf{S},(\alpha_i)_{i\in I})\prod_{j\in\alpha}\mu_j$ has the following property

$$\Big((\mathbf{S},(\alpha_i)_{i\in I})\prod_{j\in\alpha}\mu_j\Big)\Big(\prod_{j\in\alpha}[a_j,b_j]\Big) = (\mathbf{S})\prod_{j\in\alpha}(b_i - a_i).$$

The measure $(\mathbf{S},(\alpha_i)_{i\in I})\prod_{j\in\alpha}\mu_i$ is called a standard "$(\alpha_i)_{i\in I}$-Lebesgue measure" on $\mathbb{R}^{\alpha}$.

When $\mathrm{card}(\alpha_i) = 1$ for every $i \in I$, then $(\mathbf{S},(\alpha_i)_{i\in I})\prod_{j\in\alpha}\mu_i$ is called a standard Lebesgue measure on $\mathbb{R}^{\alpha}$ and is denoted by m^{α}.

Let f be any permutation of α. A mapping $A_f : \mathbb{R}^{\alpha} \to \mathbb{R}^{\alpha}$, defined by $A_f((x_i)_{i\in\alpha}) = (x_{f(i)})_{i\in\alpha}$ for $(x_i)_{i\in\alpha} \in \mathbb{R}^{\alpha}$, is called a canonical permutation of $\mathbb{R}^{\alpha}$.

Note that in our situation $\mathcal{R}_{(\alpha_i)_{i\in I}}$ is the family of all measurable rectangles $R \subseteq \mathcal{B}(\mathbb{R}^{\alpha})$ of the form $\prod_{i\in\alpha} Y_i$ with the property $0 \le (\mathbf{S})\prod_{i\in\alpha} m(Y_i) < \infty$ such that at most $\aleph_0$ of them are noncompact (i.e., card $\{i : i \in I \ \& \ Y_i$ is not compact in $\prod_{i\in\alpha_i} E_i\} \le \aleph_0$.). It is obvious to see that a measure m^{α} is invariant under a group $\mathcal{P}(\mathbb{R}^{\alpha})$ generated by shifts and canonical permutations of $\mathbb{R}^{\alpha}$ and

$$m^{\alpha}\Big(\prod_{i\in\alpha} Y_i\Big) = (\mathbf{S})\prod_{i\in\alpha} m(Y_i).$$

Chapter 8

On Strict Standard and Strict Ordinary Products of Measures and Some of their Applications

8.1 Introduction

The purpose of the chapter is to discuss the concept of strict ordinary and strict standard products of quasi-finite measures which is rather different from concepts [98], [106], which do not yield constructions of Mankiewicz and Preiss-Tišer generators on $\mathbf{R}^\infty$ [96](see, [98], Examples 1-2, pp. 220-221).

The present chapter is organized as follows.

In Section 8.2, we give an auxiliary lemma concerned with an existence of a measure defined by the consistent family of finite measures. This lemma differs from the Lemma 1 established in [98].

In Section 8.3, we consider a concept of strict ordinary and strict standard products of a countable family of (no only σ-finite) measures, prove their existence and study the connection between such measures.

In Section 8.4, we consider a concept of strict standard product of an arbitrary family of (no only σ-finite) Borel measures with domain in Polish spaces.

In Section 8.5, we consider some auxiliary propositions in Solovay model.

In Section 8.6, we give infinite version of Rademacher theorem in R^∞. In particular, we show that if $f : \mathbf{R}^\infty \to R$ is a Lipschitz function and $R^{(N)}$ is a group of all eventually zero sequences, then, in Solovay model, f is λ-almost everywhere $R^{(N)}$-Gâteaux differentiable on $\mathbf{R}^\infty$.

In Section 8.7, for an arbitrary infinite parameter set α, we consider the problem of the existence of uniform measures on a Banach space ℓ^α of all real-valued bounded functions. We show that such measures always exist on ℓ^α and they are not defined uniquely by their values on balls for an infinite parameter set α.

In Section 8.8 we give an affirmative solution of Question 6.4 [106] in Solovay model and demonstrate that this question is independent from the theory $ZF + DC$.

8.2 An Auxiliary Lemma

Let (E,S) be a measurable space and let $\mathcal{R}$ be any subclass of the σ-algebra S. Let $(\mu_B)_{B\in\mathcal{R}}$ be such a family of σ-finite measures that for $B\in\mathcal{R}$ we have $dom(\mu_B)=S\cap\mathcal{P}(B)$, where $\mathcal{P}(B)$ denotes the power set of the set B.

Definition 8.2.1. A family $(\mu_B)_{B\in\mathcal{R}}$ is called to be consistent if

$$(\forall X)(\forall B_1,B_2)(X\in S\ \&\ B_1,B_2\in\mathcal{R}\rightarrow\mu_{B_1}(X\cap B_1\cap B_2)=\mu_{B_2}(X\cap B_1\cap B_2)).$$

For $X\in S$ and for a countable family $(A_n)_{n\geq -1}$ of elements of $\mathcal{R}$ with $A_{-1}=A_0=\emptyset$, we put

$$\mu_{\cup_{n\in N}A_n}(X\cap\cup_{n\in N}A_n)=\sum_{n\in N}\mu_{A_n}((A_n\setminus\cup_{k=0}^{n-1}A_k)\cap X).$$

If $(A_n)_{n\geq 1}$ and $(B_n)_{n\geq 1}$ are two countable families of elements of $\mathcal{R}$ with $A_{-1}=A_0=B_{-1}=B_0=\emptyset$, then for $X\in S\cap\cup_{n\in N}A_n\cap\cup_{n\in N}B_n$ we have

$$\mu_{\cup_{n\in N}A_n}(X)=\mu_{\cup_{n\in N}B_n}(X).$$

Indeed, we have

$$\mu_{\cup_{n\in N}A_n}(X)=\sum_{n\in N}\mu_{A_n}\big((A_n\setminus\cup_{k=0}^{n-1}A_k)\cap X\big)$$

$$=\sum_{n\in N}\mu_{A_n}\big((A_n\setminus\cup_{k=0}^{n-1}A_k)\cap\big(\cup_{m\in N}\big(B_m\setminus\cup_{l=0}^{m-1}B_l\big)\big)\cap X\big)$$

$$=\sum_{n\in N}\mu_{A_n}\big(\cup_{m\in N}\big((A_n\setminus\cup_{k=0}^{n-1}A_k)\cap\big(B_m\setminus\cup_{l=0}^{m-1}B_l\big)\big)\cap X\big)$$

$$=\sum_{n\in N}\sum_{m\in N}\mu_{A_n}\big((A_n\setminus\cup_{k=0}^{n-1}A_k)\cap\big(B_m\setminus\cup_{l=0}^{m-1}B_l\big)\cap X\big)$$

$$=\sum_{m\in N}\sum_{n\in N}\mu_{A_n}\big((A_n\setminus\cup_{k=0}^{n-1}A_k)\cap\big(B_m\setminus\cup_{l=0}^{m-1}B_l\big)\cap X\big)$$

$$=\sum_{m\in N}\sum_{n\in N}\mu_{B_m}\big((A_n\setminus\cup_{k=0}^{n-1}A_k)\cap\big(B_m\setminus\cup_{l=0}^{m-1}B_l\big)\cap X\big)$$

$$=\sum_{m\in N}\mu_{B_m}\big(\cup_{n\in N}\big(A_n\setminus\cup_{k=0}^{n-1}A_k\big)\cap\big(B_m\setminus\cup_{l=0}^{m-1}B_l\big)\cap X\big)$$

$$=\sum_{m\in N}\mu_{B_m}\big(\big(B_m\setminus\cup_{l=0}^{m-1}B_l\big)\cap X\big)=\mu_{\cup_{n\in N}B_n}(X).$$

Lemma 8.2.1 ([98], Lemma 1). *Let $(\mu_B)_{B\in\mathcal{R}}$ be a consistent family of σ-finite measures on (E,S). If $X\in S$ is covered by a countable family $(A_n)_{n\in N}$ of elements of the $\mathcal{R}$ such that $A_{-1}=A_0=\emptyset$, then we put*

$$\lambda_{\mathcal{R}}^{(1)}(X)=\mu_{\cup_{n\in N}A_n}(X),$$

and $\lambda_{\mathcal{R}}^{(1)}(X)=+\infty$, otherwise.

Then $\lambda_{\mathcal{R}}^{(1)}$ is such a measure on (E,S) that

(i) $\lambda_{\mathcal{R}}^{(1)}(B)=\mu_B(B)$ for every $B\in\mathcal{R}$;

(ii) if there exists a non-countable family of pairwise disjoint sets $\{B_i : i \in I\} \subseteq \mathcal{R}$ *such that* $0 < \mu_{B_i}(B_i) < \infty$, *then the measure* $\lambda^{(1)}_{\mathcal{R}}$ *is non-σ-finite;*

(iii) if G is a group of measurable transformations of E such that $G(\mathcal{R}) = \mathcal{R}$ *and*

$$(\forall B)(\forall X)(\forall g)\big((B \in \mathcal{R} \,\&X \in S \cap \mathcal{P}(B) \,\&\, g \in G\big) \to \mu_{g(B)}(g(X)) = \mu_B(X)),$$

where $\mathcal{P}(B)$ *denotes a power set of the set B, then the measure* $\lambda^{(1)}_{\mathcal{R}}$ *is G-invariant.*

Lemma 8.2.2. *Let* $(\mu_B)_{B \in \mathcal{R}}$ *be a consistent family of σ-finite measures on a measurable space* (E,S). *We set*

$$(\forall X)(X \in S \to \lambda^{(2)}_{\mathcal{R}}(X)) = \sup\{\mu_{\cup_{n\in N}A_n}(X \cap \cup_{n\in N}A_n) :$$

$$(A_n)_{n\ge -1} \in \mathcal{R}^{\{-1\}\cup N} \,\&\, A_{-1} = A_0 = \emptyset\}.$$

Then the $\lambda^{(2)}_{\mathcal{R}}$ *is the measure on* (E,S) *such that*

(i) $\lambda^{(2)}_{\mathcal{R}}(B) = \mu_B(B)$ *for every* $B \in \mathcal{R}$;

(ii) if there exists a non-countable family of pairwise disjoint sets $\{B_i : i \in I\} \subseteq \mathcal{R}$ *such that* $0 < \mu_{B_i}(B_i) < \infty$, *then the measure* $\lambda^{(2)}_{\mathcal{R}}$ *is non-σ-finite;*

(iii) if G is a group of measurable transformations of E such that $G(\mathcal{R}) = \mathcal{R}$ *and*

$$(\forall B)(\forall X)(\forall g)\big((B \in \mathcal{R} \,\&X \in S \cap \mathcal{P}(B) \,\&\, g \in G\big) \to \mu_{g(B)}(g(X)) = \mu_B(X)),$$

where $\mathcal{P}(B)$ *denotes a power set of the set B, then the measure* $\lambda^{(2)}_{\mathcal{R}}$ *is G-invariant.*

(iv) for $X \in S$, $\lambda^{(2)}_{\mathcal{R}}(X) = 0$ *iff* $(\forall B)(B \in \mathcal{R} \to \mu_B(X \cap B) = 0)$.

Proof. Let us prove that the functional $\lambda^{(2)}_{\mathcal{R}}$ is σ-additive.

Let $(X_k)_{k\in N}$ be a countable family of pairwise disjoint elements of S.

Case I. $\lambda^{(2)}_{\mathcal{R}}(\cup_{k\in N}X_k) = A < +\infty$.

Then, for $i \in N$, there exists a family of elements $(A^{(i)}_n)_{n\ge -1} \in \mathcal{R}^{\{-1\}\cup N}$ with $A^{(i)}_{-1} = A^{(i)}_1 = \emptyset$ such that

$$A - \frac{1}{1+i} < \mu_{\cup_{n\in N}A^{(i)}_n}(\cup_{k\in N}X_k \cap \cup_{n\in N}A^{(i)}_n) < A.$$

Then we have that

$$\mu_{\cup_{i,n\in N}A^{(i)}_n}(\cup_{k\in N}X_k \cap \cup_{i,n\in N}A^{(i)}_n) = A.$$

It is obvious that

$$(\forall B)(B \in \mathcal{R} \to \mu_{\cup_{i,n\in N}A^{(i)}_n}((B \setminus \cup_{i,n\in N}A^{(i)}_n) \cap \cup_{k\in N}X_k) = 0).$$

The later relation follows from the definition of the $\lambda^{(2)}_{\mathcal{R}}(\cup_{k\in N}X_k)$. Indeed, if for any $B_0 \in \mathcal{R}$, we have

$$\mu_{\cup_{i,n\in N}A^{(i)}_n}((B_0 \setminus \cup_{n\in N}A^{(i)}_n) \cap \cup_{k\in N}X_k) = \delta > 0,$$

then

$$\mu_{\cup_{i,n\in N}A_n^{(i)}\cup B_0}(\cup_{k\in N}X_k \cap (B_0 \cup \cup_{i,n\in N}A_n^{(i)})) = \mu_{\cup_{i,n\in N}A_n^{(i)}}(\cup_{k\in N}X_k \cap \cup_{i,n\in N}A_n^{(i)})$$

$$+\mu_{B_0\setminus\cup_{i,n\in N}A_n^{(i)}}((B_0 \setminus \cup_{i,n\in N}A_n^{(i)}) \cap \cup_{k\in N}X_k) = A + \delta > A.$$

As a simple consequence, we get

$$\mu_{\cup_{i,n\in N}A_n^{(i)}}(X_k \cap \cup_{i,n\in N}A_n^{(i)})$$

$$= \sup\{\mu_{\cup_{n\in N}A_n}(X_k \cap \cup_{n\in N}A_n) : (A_n)_{n\geq -1} \in \mathcal{R}^{\{-1\}\cup N} \,\&\, A_{-1} = A_0 = \emptyset\}.$$

Hence, we have

$$A = \lambda_{\mathcal{R}}^{(2)}(\cup_{k\in N}X_k) = \mu_{\cup_{i,n\in N}A_n^{(i)}}(\cup_{k\in N}X_k \cap \cup_{n\in N}A_n^{(i)}) = \sum_{k\in N}\mu_{\cup_{i,n\in N}A_n^{(i)}}(X_k \cap \cup_{n\in N}A_n^{(i)})$$

$$= \sum_{k\in N}\sup\{\mu_{\cup_{n\in N}A_n}(X_k \cap \cup_{n\in N}A_n) : (A_n)_{n\geq -1} \in \mathcal{R}^{\{-1\}\cup N} \,\&\, A_{-1} = A_0 = \emptyset\} = \sum_{k\in N}\lambda_{\mathcal{R}}^{(2)}(X_k).$$

Case II. $\lambda_{\mathcal{R}}^{(2)}(\cup_{k\in N}X_k) = +\infty$.

Then, for $i \in N$, there exists a family of elements $(A_n^{(i)})_{n\geq -1} \in \mathcal{R}^{\{-1\}\cup N}$ with $A_{-1}^{(i)} = A_0^{(i)} = \emptyset$ such that

$$\mu_{\cup_{n\in N}^{(i)}A_n}(\cup_{k\in N}X_k \cap \cup_{n\in N}A_n^{(i)}) > i.$$

Then we have

$$\mu_{\cup_{i,n\in N}A_n^{(i)}}(\cup_{k\in N}X_k \cap \cup_{i,n\in N}A_n^{(i)}) = +\infty.$$

We also have

$$+\infty = \lambda_{\mathcal{R}}^{(2)}(\cup_{k\in N}X_k) = \mu_{\cup_{i,n\in N}A_n^{(i)}}(\cup_{k\in N}X_k \cap \cup_{n\in N}A_n^{(i)})$$

$$= \sum_{k\in N}\mu_{\cup_{i,n\in N}A_n^{(i)}}(X_k \cap \cup_{n\in N}A_n^{(i)}) \leq \sum_{k\in N}\sup\{\mu_{\cup_{n\in N}A_n}(X_k \cap \cup_{n\in N}A_n) \,\&$$

$$(A_n)_{n\geq -1} \in \mathcal{R}^{\{-1\}\cup N} \,\&\, A_{-1} = A_0 = \emptyset\} = \sum_{k\in N}\lambda_{\mathcal{R}}^{(2)}(X_k).$$

Proof of the item (i)**.** We set $C_k = B$ for $k \geq 1$ and $C_{-1} = C_0 = \emptyset$. Then the family $(C_k)_{k\geq 1}$ covers B and by the definition of $\lambda_{\mathcal{R}}^{(2)}$ we have

$$\lambda_{\mathcal{R}}^{(2)}(B) = \sup\{\mu_{\cup_{n\in N}A_n}(X \cap \cup_{n\in N}A_n) \,\&$$

$$(A_n)_{n\geq -1} \in \mathcal{R}^{\{-1\}\cup N} \,\&\, A_{-1} = A_0 = \emptyset\}$$

$$\geq \mu_{\cup_{n\in N}C_n}(B \cap \cup_{n\in N}C_n) = \mu_B(B).$$

On the other hand, for $(A_n)_{n\geq -1} \in \mathcal{R}^{\{-1\}\cup N}$ with $A_{-1} = A_0 = \emptyset$, we have

$$\mu_{\cup_{n\in N}A_n}(B \cap \cup_{n\in N}A_n) = \sum_{n\in N}\mu_{A_n}((A_n \setminus \cup_{k=0}^{n-1}A_k) \cap B)$$

$$= \sum_{n \in N} \mu_B((A_n \setminus \cup_{k=0}^{n-1} A_k) \cap B) = \mu_B(\cup_{n \in N} A_n \cap B) \leq \mu_B(B),$$

which implies that $\lambda_{\mathcal{R}}^{(2)}(B) \leq \mu_B(B)$.

The proof of the item (ii) is obvious and we omit it.

Proof of the item (iii)**.** Let G be a group of measurable transformations of E such that $G(\mathcal{R}) = \mathcal{R}$ and

$$(\forall B)(\forall X)(\forall g)\big((B \in \mathcal{R} \,\&\, X \in B \cap S \,\&\, g \in G\big) \to \mu_{g(B)}(g(X)) = \mu_B(X)).$$

We are to show that the measure $\lambda_{\mathcal{R}}^{(2)}$ is G-invariant.

For $X \in S$, we have

$$\lambda_{\mathcal{R}}^{(2)}(g(X)) = \sup\{\mu_{\cup_{n \in N} A_n}(g(X) \cap \cup_{n \in N} A_n) \,\&$$

$$(A_n)_{n \geq -1} \in \mathcal{R}^{\{-1\} \cup N} \,\&\, A_{-1} = A_0 = \emptyset\}$$

$$= \sup\{\sum_{n \in N} \mu_{A_n}((A_n \setminus \cup_{k=0}^{n-1} A_k) \cap g(X)) \,\&$$

$$(A_n)_{n \geq -1} \in \mathcal{R}^{\{-1\} \cup N} \,\&\, A_{-1} = A_0 = \emptyset\}$$

$$= \sup\{\sum_{n \in N} \mu_{g^{-1}(A_n)}((g^{-1}(A_n) \setminus \cup_{k=0}^{n-1} g^{-1}(A_k)) \cap X) :$$

$$(g^{-1}(A_n))_{n \geq -1} \in \mathcal{R}^{\{-1\} \cup N} \,\&\, g^{-1}(A_{-1}) = g^{-1}(A_0) = \emptyset\}$$

$$= \sup\{\sum_{n \in N} \mu_{B_n}((B_n \setminus \cup_{k=0}^{n-1} B_k) \cap X) : (B_n)_{n \geq -1} \in \mathcal{R}^{\{-1\} \cup N} \,\&\, B_{-1} = B_0 = \emptyset\}$$

$$= \sup\{\mu_{\cup_{n \in N} B_n}(X \cap \cup_{n \in N} B_n) : (B_n)_{n \geq -1} \in \mathcal{R}^{\{-1\} \cup N} \,\&\, B_{-1} = B_0 = \emptyset\} = \lambda_{\mathcal{R}}^{(2)}(X).$$

Proof of the item (iv)**.** $(\Rightarrow)$. If we assume the contrary then for any $B_0 \in \mathcal{R}$ we get $\mu_{B_0}(X \cap B_0) = \delta > 0$.

Let $C_{-1} = C_0 = \emptyset$ and $C_n = B_0$ for $n > 0$. Then we get

$$\lambda_{\mathcal{R}}^{(2)}(X) = \sup\{\mu_{\cup_{n \in N} A_n}(X \cap \cup_{n \in N} A_n) :$$

$$(A_n)_{n \geq -1} \in \mathcal{R}^{\{-1\} \cup N} \,\&\, A_{-1} = A_0 = \emptyset\}$$

$$\geq \mu_{\cup_{n \in N} C_n}(X \cap \cup_{n \in N} C_n) = \delta > 0,$$

which contradict to the condition $\lambda_{\mathcal{R}}^{(2)}(X) = 0$.

$(\Leftarrow)$. The validity of the condition

$$(\forall B)(B \in \mathcal{R} \to \mu_B(X \cap B) = 0)$$

implies that, for every $(A_n)_{n \geq -1} \in \mathcal{R}^{\{-1\} \cup N}$ with $A_{-1} = A_0 = \emptyset$, we have

$$\mu_{\cup_{n \in N} A_n}(X \cap \cup_{n \in N} A_n) = 0.$$

Hence,

$$\lambda_{\mathcal{R}}^{(2)}(X) = \sup\{\mu_{\cup_{n \in N} A_n}(X \cap \cup_{n \in N} A_n) :$$

$$(A_n)_{n \geq -1} \in \mathcal{R}^{\{-1\} \cup N} \,\&\, A_{-1} = A_0 = \emptyset\} = 0.$$

□

Remark 8.2.1. Let $E = R^2$ and $S = \mathcal{B}(R^2)$. We set $B_y = R \times \{y\}$ for $y \in R$ and

$$\mathcal{R} = \{B_y : y \in R\}.$$

Let μ_{B_y} be a linear Borel measure defined on B_y. Then the measure $\lambda_{\mathcal{R}}^{(1)}$ defined by Lemma 8.2.1 does not satisfy the condition of the Lemma 8.2.3. Indeed, for a set $X = \{0\} \times R$ we have $\mu_{B_y}(X \cap B_y) = 0$ for $y \in R$, but $\lambda_{\mathcal{R}}^{(1)}(X) = +\infty$, because the X is not covered by countable family of elements of $\mathcal{R}$. On the other hand, for the measure $\lambda_{\mathcal{R}}^{(2)}$ defined by Lemma 8.2.2 we have $\lambda_{\mathcal{R}}^{(2)}(X) = 0$. This example shows us that measures $\lambda_{\mathcal{R}}^{(1)}$ and $\lambda_{\mathcal{R}}^{(2)}$ are different.

Remark 8.2.2. Let $(\mu_B)_{B \in \mathcal{R}}$ be a consistent family of σ-finite measures on a measurable space (E, S). Let $\overline{\mu}_B$ be a completion of the μ_B for $B \in \mathcal{R}$. We set $S_B = dom(\overline{\mu}_B) \cup \mathcal{P}(E \setminus B)$. Further, we set

$$\tilde{S} = \cap_{B \in \mathcal{R}} S_B$$

and $\tilde{\mu}_B$ be a measure defined by

$$\tilde{\mu}_B(X) = \overline{\mu}_B(X)$$

for $X \in \tilde{S}$.

Then it is not hard to show that a family of measures $(\tilde{\mu}_B)_{B \in \mathcal{R}}$ is a consistent family of σ-finite measures on a measurable space $(E, \tilde{S})$. We denote by $\tilde{\lambda}_{\mathcal{R}}^{(1)}$ and $\tilde{\lambda}_{\mathcal{R}}^{(2)}$ the measures defined Lemmas 8.2.1 and 8.2.2, respectively. Note that both measures $\tilde{\lambda}_{\mathcal{R}}^{(1)}$ and $\tilde{\lambda}_{\mathcal{R}}^{(2)}$ are complete, but we do not know whether the following equalities

$$\tilde{\lambda}_{\mathcal{R}}^{(1)} = \overline{\lambda_{\mathcal{R}}^{(1)}}$$

and

$$\tilde{\lambda}_{\mathcal{R}}^{(2)} = \overline{\lambda_{\mathcal{R}}^{(2)}}$$

hold.

In this context we consider the following example: *Let $(\mu_{B_y})_{y \in R}$ be a family of measures on R^2 defined in Remark* 8.2.1. *Then, under Continuum hypothesis (CH), the $\tilde{S}$ strict contains the class of all Lebesgue measurable subsets of the R^2. Indeed, if (A, B) is Sierpinski partition of the R^2 such that $card(A \cap B_y) \leq \omega$ for every $y \in R$, then $A \in \tilde{S}$ and $\tilde{\lambda}_{\mathcal{R}}(A) = 0$.*

On the other hand, in Solovay model, the $\tilde{S}$ coincides with the class of Lebesgue measurable subsets of the R^2.

Hence, a question asking whether the $\tilde{S}$ coincides with the class of Lebesgue measurable subsets of the R^2 is not solvable within the theory $ZF + DC$.

The later fact is equivalent to the assertion that a question asking whether the following equality

$$\sum_{y \in R} \overline{b_y} = \overline{\sum_{y \in R} b_y}$$

holds, where, for $y \in R$, the b_y denotes a Borel measure on R^2 defined by

$$b_y(X) = \mu_{B_y}(X \cap B_y),$$

is not solvable within the theory $ZF + DC$.

Example 8.2.1. Let $\mathcal{R}$ be the class of all infinite dimensional rectangles $R \in \mathcal{B}(\mathbf{R}^\infty)$ of the form

$$R = \prod_{i=1}^{\infty}(a_i, b_i), \quad -\infty < a_i < b_i < +\infty$$

with $0 \leq \prod_{i=1}^{\infty}(b_i - a_i) < +\infty$, where

$$\prod_{i=1}^{\infty}(b_i - a_i) := \lim_{n\to\infty}\prod_{i=1}^{n}(b_i - a_i).$$

Let λ_R be defined by $\lambda_R(X) = \prod_{i=1}^{\infty} m(X_i)$ for every elementary rectangular set $X = \prod_{i\in N} X_i \in \mathcal{B}(R)$. By Charatheodory theorem, there exists a unique extension μ_R of the λ_R to whole $\mathcal{B}(R)$.

Since the family of measures $(\mu_R)_{R\in\mathcal{R}}$ is consistent and $\mathcal{R}$ is translation-invariant, by Lemma 8.2.1, we claim that the $\lambda_{\mathcal{R}}^{(1)}$ is "Lebesgue measure" [2]. Analogously, by Lemma 8.2.2, we claim that the $\lambda_{\mathcal{R}}^{(2)}$ coincides with Baker generator λ(cf. [92], Theorem 5.1, p. 249).

Example 8.2.2. Let $\mathcal{R}$ be the class of all infinite dimensional rectangles $R \in \mathcal{B}(\mathbf{R}^\infty)$ of the form

$$R = \prod_{i=1}^{\infty} R_i, \quad R_i \in \mathcal{B}(\mathbf{R})$$

such that

$$0 \leq \prod_{i=1}^{\infty} m(R_i) := \lim_{n\to\infty}\prod_{i=1}^{n} m(R_i) < \infty,$$

where m denotes a one-dimensional classical Borel measure on $\mathbf{R}$.

Let λ_R be defined by $\lambda_R(X) = \prod_{i=1}^{\infty} m(X_i)$ for every elementary rectangular set $X = \prod_{i\in N} X_i \in \mathcal{B}(R)$. By Charatheodory theorem, there exists a unique extension μ_R of the λ_R to whole $\mathcal{B}(R)$.

Using same argument as in Example 8.2.1, by Lemma 8.2.1 we claim that $\lambda_{\mathcal{R}}^{(1)}$ is "Lebesgue measure" II [3]. Note here that $\lambda_{\mathcal{R}}^{(2)}$ also is "Lebesgue measure" on $\mathbf{R}^\infty$ which is different from the "Lebesgue measure" II.

Example 8.2.3. The Mankiewicz generator G_M [96] is a usual completion of the functional μ defined by

$$\mu(X) = \sum_{a\in\ell_1^{\perp}} \mu_{[0,1]^N}(X - a \cap B_{[0,1]^N})$$

for every $X \in \mathcal{B}(\mathbf{R}^\infty)$, where

(i) $\mu_{[0,1]^N}$ denotes Kharazishvili quasi-generator of shy sets on $\mathbf{R}^\infty$ (see, [96]),

(ii) $B_{[0,1]^N} = \cup_{n\in N}\left(\mathbf{R}^n \times [0,1]^{N\setminus\{1,\dots,n\}}\right)$,

(iii) $\ell_1^{\perp}$ denotes a linear compliment of the vector subspace ℓ_1 in $\mathbf{R}^\infty$.

This measure G_M is an $\mathcal{G}$-invariant and has a property that X is a standard cube null set iff X is of G_M-measure zero for every $X \subset \mathbf{R}^\infty$, where $\mathcal{G}$ denotes a group generated by shipts and canonical permutations of $\mathbf{R}^\infty$.

We set

$$\mathcal{R} = \{R : R = [0,1]^N + a \text{ for any } a \in \mathbf{R}^\infty\}$$

and

$$\mu_R(X) = \lambda(X - a \cap [0,1]^N)$$

for every $X \in \mathcal{B}(\mathbf{R}^\infty)$, where $\lambda = \mu^N$ and μ is a linear probability Lebesgue measure on $[0,1]$.

Then the class $\mathcal{R}$ and the family $(\mu_R)_{R\in\mathbf{R}}$ are invariant under the group $\mathcal{G}$, where $\mathcal{G}$ denotes a group generated by all shifts and canonical permutations of the $\mathbf{R}^\infty$.

Note that the family $(\mu_R)_{R\in\mathbf{R}}$ satisfies all conditions of Lemmas 8.2.1-8.2.2 which implies that the measures $\lambda^{(1)}_{\mathcal{R}}$ and $\lambda^{(2)}_{\mathcal{R}}$, are $\mathcal{G}$-invariant. Note here that the measure $\lambda^{(2)}_{\mathcal{R}}$, unlike the measure $\lambda^{(1)}_{\mathcal{R}}$, coincides with Mankiewicz's generator G_M.

Example 8.2.4. Let $(L_i)_{i\in I}$ be a family of all different n-dimensional vector subspaces of $\mathbf{R}^\infty$ and let $\ell^{(i)}_n$ be an n-dimensional Lebesgue measure on L_i. For $i \in I$, let denote by $L_i^\perp$ a linear compliment of L_i. Then a functional $G_{P\&T}$, defined by

$$G^{(n)}_{P\&T}(X) = \sum_{i\in I}\sum_{a\in L_i^\perp} \ell^{(i)}_n(X - a \cap L_i)$$

for $X \in \mathcal{B}(\mathbf{R}^\infty)$, is an $\mathcal{G}$-invariant (see, Example 8.2.3) Borel measure and $G_{P\&T}(Y) = 0$ iff Y is n-dimensional null in the sense of Preiss and Tišer for every $Y \subset \mathbf{R}^\infty$.

We set

$$\mathcal{R} = \{L_i + a : a \in \mathbf{R}^\infty, i \in I\}$$

and

$$\mu_{L_i+a}(X) = \ell^{(i)}_n(X - a \cap L_i)$$

for every $X \in \mathcal{B}(\mathbf{R}^\infty)$.

Since the class $\mathcal{R}$ and the family $(\mu_R)_{R\in\mathbf{R}}$, being $\mathcal{G}$-invariant, satisfy all conditions of Lemmas 8.2.1-8.2.2, we deduce that the measures $\lambda^{(1)}_{\mathcal{R}}$ and $\lambda^{(2)}_{\mathcal{R}}$ are $\mathcal{G}$-invariant.

Note here that the measure $\lambda^{(2)}_{\mathcal{R}}$, unlike the measure $\lambda^{(1)}_{\mathcal{R}}$, coincides with Preiss and Tišer generator $G_{P\&T}$.

8.3 On Strict Ordinary and Standard Products of Infinite Family of σ-Finite Measures

Suppose that $(\beta_i)_{i\in\mathbf{N}} \in [0,+\infty]^{\mathbf{N}}$.

Let us recall definitions of ordinary and standard products of non-negative real numbers $(\beta_i)_{i\in\mathbf{N}}$.

Definition 8.3.1. We say that a number $\beta \in [0,+\infty]$ is an ordinary product of numbers $(\beta_j)_{j\in\mathbf{N}}$ if

$$\beta = \lim_{n\to\infty}\prod_{i=1}^{n}\beta_i.$$

An ordinary product of numbers $(\beta_j)_{j\in\mathbf{N}}$ is denoted by $(\mathbf{O})\prod_{i\in\mathbf{N}}\beta_i$.

Definition 8.3.2. A standard product of the family of numbers $(\beta_i)_{i\in\mathbf{N}}$ is denoted by $(\mathbf{S})\prod_{i\in\mathbf{N}}\beta_i$ and defined as follows: $(\mathbf{S})\prod_{i\in\mathbf{N}}\beta_i = 0$ if $\sum_{i\in\mathbf{N}^-}\ln(\beta_i) = -\infty$, where $\mathbf{N}^- = \{i : ln(\beta_i) < 0\}$ [1], and $(\mathbf{S})\prod_{i\in\mathbf{N}}\beta_i = e^{\sum_{i\in\mathbf{N}}\ln(\beta_i)}$ if $\sum_{i\in\mathbf{N}^-}\ln(\beta_i) \neq -\infty$.

Let $(E_i, S_i)_{i\in\mathbf{N}}$ be a sequence of measure spaces. Let $(\nu_i)_{i\in\mathbf{N}}$ be a sequence of finite measures with $dom(\nu_i) = S_i$ for $i \in \mathbf{N}$. We put $\mathcal{C}(\prod_{i\in\mathbf{N}} E_i) = \prod_{i\in\mathbf{N}} S_i$.

Definition 8.3.3. Suppose that

$$0 \leq (\mathbf{O})\prod_{i\in\mathbf{N}}\nu_i(E_i) < +\infty.$$

For $X \in \mathcal{C}(\prod_{i\in\mathbf{N}} E_i)$, we set

$$\Big((\mathbf{O})\prod_{i\in\mathbf{N}}\nu_i\Big)(X) = \Big((\mathbf{O})\prod_{i\in\mathbf{N}}\nu_i(E_i)\Big)\Big[\prod_{i\in\mathbf{N}}\frac{\nu_i}{\nu_i(E_i)}\Big](X),$$

if

$$0 < (\mathbf{O})\prod_{i\in\mathbf{N}}\nu_i(E_i) < +\infty,$$

and

$$\Big((\mathbf{O})\prod_{i\in\mathbf{N}}\nu_i\Big)(X) = 0$$

if

$$(\mathbf{O})\prod_{i\in\mathbf{N}}\nu_i(E_i) = 0.$$

A measure $(\mathbf{O})\prod_{i\in\mathbf{N}}\nu_i$ is called an ordinary product of the family of finite measures $(\nu_i)_{i\in\mathbf{N}}$.

Definition 8.3.4. Suppose that

$$0 \leq (\mathbf{S})\prod_{i\in\mathbf{N}}\nu_i(E_i) < +\infty.$$

For $X \in \mathcal{C}(\prod_{i\in\mathbf{N}} E_i)$, we set

$$\Big((\mathbf{S})\prod_{i\in\mathbf{N}}\nu_i\Big)(X) = \Big((\mathbf{S})\prod_{i\in\mathbf{N}}\nu_i(E_i)\Big)\Big[\prod_{i\in\mathbf{N}}\frac{\nu_i}{\nu_i(E_i)}\Big](X),$$

if

$$0 < (\mathbf{S})\prod_{i\in\mathbf{N}}\nu_i(E_i) < +\infty,$$

and

$$\Big((\mathbf{S})\prod_{i\in\mathbf{N}}\nu_i\Big)(X) = 0$$

if

$$(\mathbf{S})\prod_{i\in\mathbf{N}}\nu_i(E_i) = 0.$$

A measure $(\mathbf{S})\prod_{i\in\mathbf{N}}\nu_i$ is called a standard product of the family of finite measures $(\nu_i)_{i\in\mathbf{N}}$.

---

[1] We set $ln(0) = -\infty$

Remark 8.3.1. If $\nu_i(E_i) > 0$ for $i \in \mathbf{N}$, then by Anderson well known results there exists a product of probability measures $\prod_{i\in\mathbf{N}} \frac{\nu_i}{\nu_i(E_i)}$. Coefficient $(\mathbf{O})\prod_{i\in\mathbf{N}}\nu_i(E_i)$ and $(\mathbf{S})\prod_{i\in\mathbf{N}}\nu_i(E_i)$ make only some correction. If $\nu_i(E_i) = e^{\frac{(-1)^i}{i+1}}$ for $i \in \mathbf{N}$, then $(\mathbf{S})\prod_{i\in\mathbf{N}}\nu_i$ is identically a zero measure, whenever $(\mathbf{O})\prod_{i\in\mathbf{N}}\nu_i$ is a non-zero measure. This example shows us that notions of ordinary and standard products are different.

Now let $(\mu_i)_{i\in\mathbf{N}}$ be a sequence of non-zero σ-finite measures with $dom(\mu_i) = \mathcal{S}_i$ for $i \in \mathbf{N}$.

We denote by $\mathcal{R}(O)$ the family of all measurable rectangles $R \subseteq \prod_{i\in\mathbf{N}} E_i$ of the form $\prod_{i\in\mathbf{N}} Y_i$ with the property

$$0 \le (\mathbf{O})\prod_{i\in\mathbf{N}}\mu_i(Y_i) < \infty.$$

Analogously, we denote by $\mathcal{R}(\mathcal{S})$ the family of all measurable rectangles $R \subseteq \prod_{i\in\mathbf{N}} E_i$ of the form $\prod_{i\in\mathbf{N}} Y_i$ with the property

$$0 \le (\mathbf{S})\prod_{i\in\mathbf{N}}\mu_i(Y_i) < \infty.$$

We set

$$(\mu_i|_{Y_i})(Y) = \mu_i(Y_i \cap Y)$$

for every $Y \in Y_i \cap \mathcal{S}_i$.

Let $C(\prod_{i\in\mathbf{N}} E_i)$ denotes the product of σ-algebras $(S_i)_{i\in\mathbf{N}}$.

Case 1. Let $R = \prod_{i\in\mathbf{N}} Y_i \in \mathcal{R}(O)$.

For $X \in C(\prod_{i\in\mathbf{N}} E_j) \cap R$ we set

$$\lambda_R^{ordinary}(X) = (\mathbf{O})\prod_{i\in\mathbf{N}}(\mu_i|_{Y_i}))(X \cap R),$$

if $0 < (\mathbf{O})\prod_{i\in\mathbf{N}}\mu_i(Y_i) < +\infty$, and

$$\lambda_R^{ordinary}(X) = 0,$$

otherwise.

Case 2. Let $R = \prod_{i\in\mathbf{N}} Y_i \in \mathcal{R}(\mathcal{S})$.

For $X \in C(\prod_{i\in\mathbf{N}} E_j) \cap R$ we set

$$\lambda_R^{standard}(X) = (\mathbf{S})\prod_{i\in\mathbf{N}}(\mu_i|_{Y_i}))(X \cap R),$$

if $0 < (\mathbf{S})\prod_{i\in\mathbf{N}}\mu_i(Y_i) < +\infty$, and

$$\lambda_R^{standard}(X) = 0,$$

otherwise.

Let $\mathcal{R}(O)^+$ denotes the family of all measurable rectangles $R \subseteq \prod_{i\in\mathbf{N}} E_i$ of the form $\prod_{i\in\mathbf{N}} Y_i$ with the property

$$0 < (\mathbf{O})\prod_{i\in\mathbf{N}}\mu_i(Y_i) < \infty.$$

Let $\mathcal{R}(O)^0$ denotes the family of all measurable rectangles $R \subseteq \prod_{i\in\mathbf{N}} E_i$ of the form $\prod_{i\in\mathbf{N}} Y_i$ with the property

$$0 = (\mathbf{O})\prod_{i\in\mathbf{N}}\mu_i(Y_i).$$

Let $\mathcal{R}(S)^+$ denotes the family of all measurable rectangles $R \subseteq \prod_{i\in\mathbf{N}} E_i$ of the form $\prod_{i\in\mathbf{N}} Y_i$ with the property

$$0 < (\mathbf{S})\prod_{i\in\mathbf{N}}\mu_i(Y_i) < \infty.$$

Let $\mathcal{R}(S)^0$ denotes the family of all measurable rectangles $R \subseteq \prod_{i\in\mathbf{N}} E_i$ of the form $\prod_{i\in\mathbf{N}} Y_i$ with the property

$$0 = (\mathbf{S})\prod_{i\in\mathbf{N}}\mu_i(Y_i).$$

For $R \in \mathcal{R}(O)$, let us denote by $\overline{\lambda}_R^{ordinary}$ the completion of the $\lambda_R^{ordinary}$. We set

$$S^{ordinary} = \cap_{R\in\mathcal{R}(O)}\Big(dom\overline{\lambda}_R^{ordinary} \cup \mathcal{P}\Big(\prod_{i\in\mathbf{N}} E_i \setminus R\Big)\Big).$$

Analogously, for $R \in \mathcal{R}(S)$, let us denote by $\overline{\lambda}_R^{standard}$ the completion of the $\lambda_R^{standard}$. We set

$$S^{standard} = \cap_{R\in\mathcal{R}(O)}\Big(dom\overline{\lambda}_R^{standard} \cup \mathcal{P}\Big(\prod_{i\in\mathbf{N}} E_i \setminus R\Big)\Big).$$

Definition 8.3.5. The measure μ is said to be a strict ordinary product of the family of measures $(\mu_i)_{i\in\mathbf{N}}$ if the following three conditions hold:

(i) for every $R = \prod_{i\in\mathbf{N}} Y_i \in \mathcal{R}(O)$ we have

$$\mu\Big(\prod_{i\in\mathbf{N}} Y_i\Big) = (\mathbf{O})\prod_{i\in\mathbf{N}}\mu_i(Y_i);$$

(ii)

$$(\forall X)(X \in dom(\mu) \to ((\mu(X) = 0) \Leftrightarrow (\forall R)(R \in \mathcal{R}(O)^+ \to \mu_R(R\cap X) = 0)));$$

(iii) $S^{ordinary} \subseteq dom(\mu)$.

Definition 8.3.6. The measure μ is said to be a strict standard product of the family of measures $(\mu_i)_{i\in\mathbf{N}}$ if the following three conditions hold:

(i) for every $R = \prod_{i\in\mathbf{N}} Y_i \in \mathcal{R}(S)$ we have

$$\mu\Big(\prod_{i\in\mathbf{N}} Y_i\Big) = (\mathbf{S})\prod_{i\in\mathbf{N}}\mu_i(Y_i);$$

(ii)

$$(\forall X)(X \in dom(\mu) \to ((\mu(X) = 0) \Leftrightarrow (\forall R)(R \in \mathcal{R}(S)^+ \to \mu_R(R\cap X) = 0)));$$

(iii) $S^{standard} \subseteq dom(\mu)$.

Lemma 8.3.1. *The family of measures* $(\overline{\lambda}_R^{ordinary})_{R\in\mathcal{R}(O)}$, *defined on the* σ*-algebra* $S^{ordinary}$ *is consistent, i.e., for every* $R_1, R_2 \in \mathcal{R}(O)$ *the following equality*

$$(\forall X)(X \in S^{ordinary} \rightarrow \overline{\lambda}_{R_1}^{ordinary}(X\cap R_1\cap R_2) = \overline{\lambda}_{R_2}^{ordinary}(X\cap R_1\cap R_2))$$

holds.

Proof. Let $R_1 = \prod_{i\in\mathbf{N}} R_i^{(1)}$ and $R_2 = \prod_{i\in\mathbf{N}} R_i^{(2)}$ be two elements of the class $\mathcal{R}(O)$.

Without loss of generality it can be assumed that $0 < (\mathbf{O})\prod_{i\in\mathbf{N}} m_i(R_i^{(1)}) < \infty$ and $0 < (\mathbf{O})\prod_{i\in\mathbf{N}} m_i(R_i^{(2)}) < \infty$.

Firstly, let us show that $\overline{\lambda}_{R_1}^{ordinary}(X) = \overline{\lambda}_{R_2}^{ordinary}(X)$ for every $X \in \mathcal{C}(R_1\cap R_2)$. In this case it is sufficient to show that $\lambda_{R_1}^{ordinary}(Y) = \lambda_{R_2}^{ordinary}(Y)$ for every elementary measurable rectangle $Y = \prod_{i\in\mathbf{N}} Y_i$ in $R_1\cap R_2$. Note here that under an elementary measurable rectangle $Y = \prod_{i\in\mathbf{N}} Y_i$ in $R_1\cap R_2$ we assume a subset of $R_1\cap R_2$ such that $Y_i \in \mathcal{C}(R_i^{(1)}\cap R_i^{(2)})$ for every $i\in\mathbf{N}$ and, in addition, there exists a finite subset $I_0 \subset \mathbf{N}$ such that $Y_i = R_i^{(1)}\cap R_i^{(2)}$ for $i \in I\setminus I_0$.

For $i\in\mathbf{N}$ and $Y_i \in \mathcal{C}(R_i^{(1)}\cap R_i^{(2)})$, we have

$$m_i(Y_i\cap R_i^{(1)}\cap R_i^{(2)}) = m_i(Y_i\cap R_i^{(1)}) = m_i(Y_i\cap R_i^{(2)}).$$

The latter relation implies that

$$(\mathbf{O})\prod_{i\in\mathbf{N}} m_i(Y_i\cap R_i^{(1)}\cap R_i^{(2)}) = (\mathbf{O})\prod_{i\in\mathbf{N}} m_i(Y_i\cap R_i^{(1)})$$

$$= (\mathbf{O})\prod_{i\in\mathbf{N}} m_i(Y_i\cap R_i^{(1)}).$$

Analogously, we have

$$(\mathbf{O})\prod_{i\in\mathbf{N}} m_i(Y_i\cap R_i^{(1)}\cap R_i^{(2)}) = (\mathbf{O})\prod_{i\in\mathbf{N}} m_i(Y_i\cap R_i^{(2)})$$

$$= (\mathbf{O})\prod_{i\in\mathbf{N}} m_i(Y_i\cap R_i^{(2)}).$$

Thus, we get

$$\lambda_{R_1}^{ordinary}\Big(\prod_{i\in\mathbf{N}} Y_i\Big) = (\mathbf{O})\prod_{i\in\mathbf{N}} m_i(Y_i\cap R_i^{(1)}) = (\mathbf{O})\prod_{i\in\mathbf{N}} m_i(Y_i\cap R_i^{(1)}\cap R_i^{(1)})$$

$$= (\mathbf{O})\prod_{i\in\mathbf{N}} m_i(Y_i\cap R_i^{(2)}) = \lambda_{R_2}^{ordinary}\Big(\prod_{i\in I} Y_i\Big).$$

Since a class $\mathcal{A}(R_1\cap R_2)$ of all finite disjoint unions of elementary measurable rectangles in $R_1\cap R_2$ is a ring, and since, by definition, the class $\mathcal{C}(R_1\cap R_2)$ of cylindrical sets of $R_1\cap R_2$ is a minimal σ-ring generated by the ring $\mathcal{A}(R_1\cap R_2)$, we claim (cf. [46], Theorem B, p. 27) that the class of all those sets of $R_1\cap R_2$ for which this equality holds coincides with the class $\mathcal{C}(R_1\cap R_2)$.

Hence, for $X \in \mathcal{C}(R_1 \cap R_2)$ we have

$$\lambda_{R_1}^{ordinary}(X) = \lambda_{R_2}^{ordinary}(X) = \lambda_{R_1 \cap R_2}^{ordinary}(X).$$

Now let $X \in S^{ordinary}$. Then $R_1 \cap R_2 \cap X$ also belongs to the $S^{ordinary}$ because $S^{ordinary}$ is a σ-algebra and it contains R_1, R_2, X , respectively.

Also,

$$R_1 \cap R_2 \cap X \in dom(\overline{\lambda}_{R_1}^{ordinary}) \cap dom(\overline{\lambda}_{R_2}^{ordinary}) \cap dom(\overline{\lambda}_{R_1 \cap R_2}^{ordinary}).$$

By using properties of completions of $\lambda_{R_1}^{ordinary}, \lambda_{R_2}^{ordinary}$ and $\lambda_{R_1 \cap R_2}^{ordinary}$, respectively, we claim that there exists cylindrical subsets B_1 and B_2 of $R_1 \cap R_2$ such that

(i) $\overline{\lambda}_{R_1}^{ordinary}(B_1 \Delta (R_1 \cap R_2 \cap X)) = 0$,

(ii) $\overline{\lambda}_{R_2}^{ordinary}(B_2 \Delta (R_1 \cap R_2 \cap X)) = 0$,

It is clear

(iii) $\overline{\lambda}_{R_1}^{ordinary}(B_1 \setminus (R_1 \cap R_2)) = 0$,

(iv) $\overline{\lambda}_{R_2}^{ordinary}(B_2 \setminus (R_1 \cap R_2)) = 0$.

We have

$$R_1 \cap R_2 \cap X = (B_1 \setminus X_1) \cup X_2$$

and

$$R_1 \cap R_2 \cap X = (B_2 \setminus Y_1) \cup Y_2,$$

where

$$X_1 \subseteq Z_1 \subseteq R_1 \cap R_2, X_2 \subseteq Z_2 \subseteq R_1 \cap R_2,$$
$$\lambda_{R_1}^{ordinary}(Z_1) = \lambda_{R_1}^{ordinary}(Z_2) = 0$$

and

$$Y_1 \subseteq W_1 \subseteq R_1 \cap R_2, Y_2 \subseteq W_2 \subseteq R_1 \cap R_2, \lambda_{R_2}^{ordinary}(W_1) = \lambda_{R_1}^{ordinary}(W_2) = 0.$$

Assume the contrary and let

$$\overline{\lambda}_{R_1}^{ordinary}(B_1 \setminus B_2) > 0.$$

This means that

$$\emptyset = (R_1 \cap R_2 \cap X) \setminus (R_1 \cap R_2 \cap X) = \big((B_1 \setminus X_1) \cup X_2\big) \setminus \big((B_2 \setminus Y_1) \cup Y_2\big) \neq \emptyset$$

because

$$\overline{\lambda}_{R_1}^{ordinary}\big((B_1 \setminus X_1) \cup X_2\big) \setminus \big((B_2 \setminus Y_1) \cup Y_2\big) = \overline{\lambda}_{R_1}^{ordinary}(B_1 \setminus B_2) > 0.$$

Since $\lambda_{R_1}^{ordinary}$ and $\lambda_{R_2}^{ordinary}$ are consistent, their values on cylindrical subsets Z_1, Z_2, W_1, W_2 are equal to zero. The validity of the inclusion $B_1 \Delta B_2 \subseteq Z_1 \cup Z_2 \cup W_1 \cup W_2$ implies that

$$\lambda_{R_1}^{ordinary}(B_1 \Delta B_2) = \lambda_{R_2}^{ordinary}(B_1 \Delta B_2) = 0.$$

The later relation means that

$$\lambda_{R_1}^{ordinary}(B_1) = \lambda_{R_1}^{ordinary}(B_2)$$

and

$$\lambda_{R_2}^{ordinary}(B_1) = \lambda_{R_2}^{ordinary}(B_2).$$

Finally we get

$$\overline{\lambda}_{R_1}^{ordinary}(R_1 \cap R_2 \cap X) = \lambda_{R_1}^{ordinary}(B_1)$$

$$= \lambda_{R_2}^{ordinary}(B_1) = \lambda_{R_2}^{ordinary}(B_2) = \overline{\lambda}_{R_2}^{ordinary}(R_1 \cap R_2 \cap X).$$

Thus the consistency of the family of measures $(\overline{\lambda}_R^{ordinary})_{R \in \mathcal{R}(O)}$ is proved. □

By scheme used in the proof of Lemma 8.3.1, we can prove

Lemma 8.3.2. *The family of measures* $(\overline{\lambda}_R^{standard})_{R \in \mathcal{R}(S)}$, *defined on the* σ*-algebra* $S^{standard}$ *is consistent, i.e., for every* $R_1, R_2 \in \mathcal{R}(S)$ *the following equality*

$$(\forall X)(X \in S^{standard} \to \overline{\lambda}_{R_1}^{standard}(X \cap R_1 \cap R_2) = \overline{\lambda}_{R_2}^{standard}(X \cap R_1 \cap R_2))$$

holds.

Theorem 8.3.1. *For* $i \in \mathbf{N}$, *let* μ_i *be a measure defined on a measure space* (E_i, S_i). *Assume, that there exists* $R_0 = \prod_{i \in \mathbf{N}} Z_i \in \mathcal{R}(O)$ *with*

$$0 < (\mathbf{O}) \prod_{i \in \mathbf{N}} \mu_i(Z_i) < +\infty.$$

Then there exists a complete measure $\overline{\nu}^{ordinary}$ *on* $S^{ordinary}$ *such that:*

(a) for every $R = \prod_{i \in \mathbf{N}} Y_i \in \mathcal{R}(O)$ *we have*

$$\overline{\nu}^{ordinary}\Big(\prod_{i \in \mathbf{N}} Y_i\Big) = (\mathbf{O}) \prod_{i \in \mathbf{N}} \mu_i(Y_i);$$

(b) if each μ_i *is a probability measure on* E_i, *then* $\overline{\nu}^{ordinary}$ *coincides with the completion of the* $\prod_{i \in \mathbf{N}} \mu_i$;

(c) if cardinality of indices i, for which E_i *contains disjoint measurable subsets* A_i, B_i *such that* $\mu_i(A_i) = \mu_i(B_i) = 1$, *is not finite then the measure* $\overline{\nu}^{ordinary}$ *is not* σ*-finite;*

(d) the domain of the $\overline{\nu}^{ordinary}$ *contains the class of all universally measurable subsets of the* $\prod_{i \in \mathbf{N}} E_i$.

Proof. By Lemma 8.3.1 the family of measures $(\overline{\lambda}_R^{ordinary})_{R \in \mathcal{R}(O)}$ is consistent. Application of Lemma 8.2.1 yields that the measure $\overline{\nu}^{ordinary} = \mu_{\mathcal{R}(O)}$ satisfies conditions (a) and (c).

Now, let (E_i, S_i, μ_i) be a probability space for $i \in \mathbf{N}$. Then $\prod_{i \in \mathbf{N}} E_i$ belongs to the class $\mathcal{R}(O)$ and we have $\overline{\nu}^{ordinary}(A) = \mu_{\mathcal{R}(O)}(A) = \prod_{i \in \mathbf{N}} \mu_i(\prod_{i \in \mathbf{N}} A_i)$ for every elementary rectangular set $A = \prod_{i \in \mathbf{N}} A_i$ in $\prod_{i \in \mathbf{N}} E_i$. By Charatheodory theorem and the latter relation we claim that the measures $\overline{\nu}^{ordinary}$ and $\prod_{i \in \mathbf{N}} \mu_i$ coincide on the cylindrical σ-algebra $C(\prod_{i \in \mathbf{N}} E_i)$. Since $\overline{\lambda}_{\prod_{i \in \mathbf{N}} E_i}^{ordinary}$ coincides with the completion of $\prod_{i \in \mathbf{N}} \mu_i$ and same times, with the $\overline{\lambda}^{ordinary}$, we claim that $\overline{\lambda}^{ordinary} = \overline{\prod_{i \in \mathbf{N}} \mu_i}$. This ends the proof of the validity of the Item (b).

For $R \in \mathcal{R}(O)^+$, we denote by μ_R a probability cylindrical measure defined by

$$(\forall X)\left(X \in C\Big(\prod_{i \in \mathbf{N}} E_i\Big)\right) \to \mu_R(X) = \frac{\lambda_R^{ordinary}(X)}{\lambda_R^{ordinary}(R)}\Big).$$

It is obvious that, for $R \in \mathcal{R}(O)^0$, the following equality

$$dom(\overline{\lambda}_R^{ordinary}) \cup \mathcal{P}\Big(\prod_{i\in\mathbf{N}} E_i \setminus R\Big) = \mathcal{P}\Big(\prod_{i\in\mathbf{N}} E_i\Big)$$

holds.

It is clear that, for $R \in \mathcal{R}(O)^+$ the following equality

$$dom(\overline{\mu}_R) = dom(\overline{\lambda}_R^{ordinary}) \cup \mathcal{P}\Big(\prod_{i\in\mathbf{N}} E_i \setminus R\Big)$$

holds. Hence, we have

$$U\Big(\prod_{i\in\mathbf{N}} E_i, \prod_{i\in\mathbf{N}} S_i\Big) = \cap_{\mu\in K}\Big(\prod_{i\in\mathbf{N}} S_i\Big)^{\mu} \subseteq \cap_{R\in\mathcal{R}(O)} dom(\overline{\mu}_R) \cup \mathcal{P}\Big(\prod_{i\in\mathbf{N}} E_i \setminus R\Big)$$

$$= \cap_{R\in\mathcal{R}(O)} dom(\overline{\lambda}_R^{ordinary}) \cup \mathcal{P}\Big(\prod_{i\in\mathbf{N}} E_i \setminus R\Big)\Big) = S^{ordinary}.$$

This ends the proof of the validity of the Item (d). □

Definition 8.3.7. The measure $\overline{\nu}^{ordinary}$ is called a strict ordinary product of the family of measures $(\mu_i)_{i\in\mathbf{N}}$ and is denoted by $(sO)\prod_{i\in N}\mu_i$.

Remark 8.3.2. In [98] has been constructed an ordinary product $(O)\prod_{i\in N}\mu_i$ of measures $(\mu_i)_{i\in\mathbf{N}}$. We do not know whether the following equality

$$dom\Big(\overline{(O)\prod_{i\in N}\mu_i}\Big) = dom\Big((sO)\prod_{i\in N}\mu_i\Big)$$

holds. In other words, we do not know whether the completion of the ordinary product of measures is a strict ordinary product.

Theorem 8.3.2. *Let $D \in U(\prod_{i\in\mathbf{N}} E_i, \prod_{i\in\mathbf{N}} S_i)$. For every $(a_{n+1}, a_{n+2}, \ldots) \in R^{N\setminus\{1,\ldots,n\}}$, we denote by $D_{(a_{n+1},a_{n+2},\ldots)}$ $(a_{n+1}, a_{n+2}, \ldots)$-section of the D, i.e.,*

$$D_{(a_{n+1},a_{n+2},\ldots)} = \{(x_1,\ldots,x_n) \in R^{\{1,\ldots,n\}} : (x_1,\ldots,x_n,a_{n+1},a_{n+2},\ldots) \in D\}.$$

Analogously, for $(a_1,a_2,\ldots,a_n) \in R^{\{1,\ldots,n\}}$, we denote by $D_{(a_1,a_2,\ldots,a_n)}$ $((a_1,a_2,\ldots,a_n))$-section of the D, i.e.,

$$D_{(a_1,a_2,\ldots,a_n)} = \{(x_{n+1},x_{n+2},\ldots) \in R^{N\setminus\{1,\ldots,n\}} : (a_1,\ldots,a_n,x_{n+1},x_{n+2},\ldots) \in D\}.$$

If D is covered by countable family of elements $(R_k)_{k\in N}$ of the class $\mathcal{R}(O)$, then the following formula

$$\overline{\nu}^{ordinary}(D) = \int_{R^{N\setminus\{1,\ldots,n\}}} (\overline{\nu}_{[1,\ldots,n]}^{ordinary}(D_{(a_{n+1},a_{n+2},\ldots)}) d\overline{\nu}_{[n+1,\ldots[}^{ordinary}(a_{n+1},a_{n+2},\ldots)$$

$$= \int_{R^{\{1,\ldots,n\}}} (\overline{\nu}_{[n+1,\ldots[}^{ordinary}(D_{(a_1,a_2,\ldots,a_n)}) d\overline{\nu}_{[1,\ldots,n]}^{ordinary}(a_1,a_2,\ldots,a_n)$$

is valid, where $\overline{\nu}_{[1,\ldots,n]}^{ordinary}$ denotes an ordinary complete product of measures $(\mu_i)_{1\le i\le n}$ and $\overline{\nu}_{[n+1,\ldots]}^{ordinary}$ denotes an ordinary complete product of measures $(\mu_i)_{i>n}$.

Proof. D is covered by countable family of elements $(R_k)_{k\in N}$ of the class $\mathcal{R}(O)$. We have $R_k=\prod_{i\in N}Y_i^{(k)}$ for $k\in N$.

We set $A_0=\prod_{i>n}Y_i^{(0)}$. If we have constructed a family $(A_s)_{0\le s\le m-1}$, then we set

$$A_m=\prod_{i>m}(Y_i^{(m)})\setminus\cup_{0\le s\le m-1}A_s.$$

Thus a disjoint family $(A_s)_{s\in N}$ is constructed. We set $A_{-1}=\prod_{i>n}E_i\setminus\cup_{s\in N}A_s$. We have

$$D=\Big(\sum_{s\in N}\prod_{0\le i\le n}E_i\times A_s\Big)\cap D=\sum_{s\in N\cup\{-1\}}\Big(\prod_{0\le i\le n}E_i\times A_s\cap D\Big)=\sum_{s\in N\cup\{-1\}}D^{(s)},$$

where $D^{(s)}=\big(\prod_{0\le i\le n}E_i\times A_s\cap D\big)$ for $s\in N\cup\{-1\}$.

Note, that for $s\in N\cup\{-1\}$, we have

$$\overline{\nu}^{ordinary}(D^{(s)})=\int_{R^{N\setminus\{1,\dots,n\}}}(\overline{\nu}^{ordinary}_{[1,\dots,n]}(D^{(s)}_{(a_{n+1},a_{n+2},\dots)})d\overline{\nu}^{ordinary}_{[n+1,\dots]}(a_{n+1},a_{n+2},\dots)$$

Since $\overline{\nu}^{ordinary}(D)<+\infty$, we have

$$\overline{\nu}^{ordinary}(D)=\sum_{s\in N\cup\{-1\}}\overline{\nu}^{ordinary}(D^{(s)})$$

$$=\sum_{s\in N\cup\{-1\}}\int_{R^{N\setminus\{1,\dots,n\}}}(\overline{\nu}^{ordinary}_{[1,\dots,n]}(D^{(s)}_{(a_{n+1},a_{n+2},\dots)})d\overline{\nu}^{ordinary}_{[n+1,\dots[}(a_{n+1},a_{n+2},\dots)$$

$$=\int_{R^{N\setminus\{1,\dots,n\}}}\sum_{s\in N\cup\{-1\}}\overline{\nu}^{ordinary}_{[1,\dots,n]}(D^{(s)}_{(a_{n+1},a_{n+2},\dots)})d\overline{\nu}^{ordinary}_{[n+1,\dots[}(a_{n+1},a_{n+2},\dots)$$

$$=\int_{R^{N\setminus\{1,\dots,n\}}}(\overline{\nu}^{ordinary}_{[1,\dots,n]}(D_{(a_{n+1},a_{n+2},\dots)})d\overline{\nu}^{ordinary}_{[n+1,\dots[}(a_{n+1},a_{n+2},\dots).$$

On the other hand, we have

$$\overline{\nu}^{ordinary}(D)=\sum_{s\in N\cup\{-1\}}\overline{\nu}^{ordinary}(D^{(s)})$$

$$=\sum_{s\in N\cup\{-1\}}\int_{R^{\{1,\dots,n\}}}(\overline{\nu}^{ordinary}_{[n+1,\dots[}(D^{(s)}_{(a_1,a_2,\dots,a_n)})d\overline{\nu}^{ordinary}_{[1,\dots,n]}(a_1,a_2,\dots,a_n)$$

$$=\int_{R^{\{1,\dots,n\}}}\sum_{s\in N\cup\{-1\}}\overline{\nu}^{ordinary}_{[n+1,\dots[}(D^{(s)}_{(a_1,a_2,\dots,a_n)})d\overline{\nu}^{ordinary}_{[1,\dots,n]}(a_1,a_2,\dots,a_n)$$

$$=\int_{R^{\{1,\dots,n\}}}(\overline{\nu}^{ordinary}_{[n+1,\dots[}(D_{(a_1,a_2,\dots,a_n)})d\overline{\nu}^{ordinary}_{[1,\dots,n]}(a_1,a_2,\dots,a_n).$$

This ends the proof of Theorem 8.3.2. □

Corollary 8.3.1 (A construction of the Loev-Ross strict ordinary measure). *Let $(X_i;\mathcal{B}_i;\mu_i)(i\in\mathbf{N})$ be a sequence of regular Borel measure spaces, where X_i is a Hausdorff topological space. Let $\mathcal{K}(O)$ be a class of rectangles of the form $\prod_{i\in\mathbf{N}}K_i\in\mathcal{R}(O)$, such that K_i is compact in X_i for $i\in\mathbf{N}$. Then, by Lemma* 8.3.1, *the family of measures $(\lambda_R^{ordinary})_{R\in\mathcal{K}(O)}$ is consistent as a subfamily of the consistent family of measures $(\lambda_R^{ordinary})_{R\in\mathcal{R}(O)}$. Application of Lemma* 8.2.2 *yields that the measure $\mu=\mu_{\mathcal{K}(O)}$ satisfies the following conditions:*

(a) for every $R=\prod_{i\in\mathbf{N}}Y_i\in\mathcal{K}(O)$ we have

$$\mu(\prod_{i\in\mathbf{N}}Y_i)=(\mathbf{O})\prod_{i\in\mathbf{N}}\mu_i(Y_i);$$

(b) if each μ_i is a probability measure on X_i, then $\overline{\nu}^{ordinary}$ coincides with the completion of the $\prod_{i\in\mathbf{N}}\mu_i$;

(c) if cardinality of indices i, for which X_i contains disjoint measurable subsets A_i,B_i such that $\mu_i(A_i)=\mu_i(B_i)=1$, is not finite then the measure $\overline{\nu}^{ordinary}$ is not σ-finite;

(d) the domain of the $\overline{\nu}^{ordinary}$ contains the class of all universally measurable subsets of the $\prod_{i\in\mathbf{N}}X_i$.

Corollary 8.3.2 (A construction of the Baker's strict ordinary measure). *Let $(X_i;\mathcal{B}_i;\mu_i)$ $(i\in\mathbf{N})$ be a sequence of regular Borel measure spaces, where X_i is a Hausdorff topological space. Then, by Lemma* 8.3.1, *the family of measures $(\lambda_R^{ordinary})_{R\in\mathcal{R}(O)}$ is consistent. Application of Lemma* 8.2.2 *yields that the measure $\lambda=\mu_{\mathcal{R}(O)}$ satisfies the following conditions:*

(a) for every $R=\prod_{i\in\mathbf{N}}Y_i\in\mathcal{R}(O)$ we have

$$\lambda\Big(\prod_{i\in\mathbf{N}}Y_i\Big)=(\mathbf{O})\prod_{i\in\mathbf{N}}\mu_i(Y_i);$$

(b) if each μ_i is a probability measure on X_i, then $\overline{\nu}^{ordinary}$ coincides with the completion of the $\prod_{i\in\mathbf{N}}\mu_i$;

(c) if cardinality of indices i, for which X_i contains disjoint measurable subsets A_i,B_i such that $\mu_i(A_i)=\mu_i(B_i)=1$, is not finite then the measure $\overline{\nu}^{ordinary}$ is not σ-finite;

(d) the domain of the $\overline{\nu}^{ordinary}$ contains the class of all universally measurable subsets of the $\prod_{i\in\mathbf{N}}X_i$.

By the scheme used in the proof of Theorem 8.3.1, one can prove the validity of the following assertion.

Theorem 8.3.3. *For $i\in\mathbf{N}$, let μ_i be a measure defined on a measure space (E_i,S_i). Assume, that there exists $R_0=\prod_{i\in\mathbf{N}}Z_i\in\mathcal{R}(S)$ with*

$$0<(\mathbf{S})\prod_{i\in\mathbf{N}}\mu_i(Z_i)<+\infty.$$

Then there exists a measure $\overline{\nu}^{standard}$ on $C(\prod_{i\in\mathbf{N}}E_i)$ such that:
(a) for every $R=\prod_{i\in\mathbf{N}}Y_i\in\mathcal{R}(S)$ we have

$$\overline{\nu}^{standard}\Big(\prod_{i\in\mathbf{N}}Y_i\Big)=(\mathbf{S})\prod_{i\in\mathbf{N}}\mu_i(Y_i);$$

(b) if each μ_i is a probability measure on X_i, then $\overline{\nu}^{standard}$ coincides with the completion of the $\prod_{i\in\mathbf{N}}\mu_i$;

(c) if cardinality of indices i, for which X_i contains disjoint measurable subsets A_i, B_i such that $\mu_i(A_i)=\mu_i(B_i)=1$, is not finite then the measure $\overline{\nu}^{standard}$ is not σ-finite;

(d) the domain of the $\overline{\nu}^{standard}$ contains the class of all universally measurable subsets of the $\prod_{i\in\mathbf{N}}X_i$.

Definition 8.3.8. The measure $\overline{\nu}^{standard}$ is called a strict standard product of the family of measures $(\mu_i)_{i\in\mathbf{N}}$ and is denoted by $(sS)\prod_{i\in N}\mu_i$.

Corollary 8.3.3 (A construction of the Loev-Ross strict standard measure). *Let $(X_i;\mathcal{B}_i;\mu_i)(i\in\mathbf{N})$ be a sequence of regular Borel measure spaces, where X_i is a Hausdorff topological space. Let $\mathcal{K}(S)$ be a class of rectangles of the form $\prod_{i\in\mathbf{N}}K_i\in\mathcal{R}(S)$, such that K_i is compact in X_i for $i\in\mathbf{N}$. Then, by Lemma* 8.3.1*, the family of measures $(\lambda_R^{standard})_{R\in\mathcal{K}(S)}$ is consistent as a subfamily of the consistent family of measures $(\lambda_R^{standard})_{R\in\mathcal{R}(S)}$. Application of Lemma* 8.2.2 *yields that the measure $\mu=\mu_{\mathcal{K}(S)}$ satisfies the following conditions:*

(a) for every $R=\prod_{i\in\mathbf{N}}Y_i\in\mathcal{K}(S)$ we have

$$\mu\Big(\prod_{i\in\mathbf{N}}Y_i\Big)=(\mathbf{S})\prod_{i\in\mathbf{N}}\mu_i(Y_i);$$

(b) if each μ_i is a probability measure on E_i, then μ coincides with the completion of the $\prod_{i\in\mathbf{N}}\mu_i$;

(c) if cardinality of indices i, for which E_i contains disjoint measurable subsets A_i, B_i such that $\mu_i(A_i)=\mu_i(B_i)=1$, is not finite then the measure μ is not σ-finite;

(d) the domain of the μ contains the class of all universally measurable subsets of the $\prod_{i\in\mathbf{N}}X_i$.

Corollary 8.3.4 (A construction of the Baker's strict standard measure). *Let $(X_i;\mathcal{B}_i;\mu_i)$ $(i\in\mathbf{N})$ be a sequence of regular Borel measure spaces, where X_i is a Hausdorff topological space. Then, by Lemma* 8.3.2*, the family of measures $(\lambda_R^{standard})_{R\in\mathcal{R}(S)}$ is consistent. Application of Lemma* 8.2.1 *yields that the measure $\lambda=\mu_{\mathcal{R}(S)}$ satisfies the following conditions:*

(a) for every $R=\prod_{i\in\mathbf{N}}Y_i\in\mathcal{R}(S)$ we have

$$\lambda\Big(\prod_{i\in\mathbf{N}}Y_i\Big)=(\mathbf{S})\prod_{i\in\mathbf{N}}\mu_i(Y_i);$$

(b) if each μ_i is a probability measure on E_i, then λ coincides with the completion of the $\prod_{i\in\mathbf{N}}\mu_i$;

(c) if cardinality of indices i, for which E_i contains disjoint measurable subsets A_i, B_i such that $\mu_i(A_i)=\mu_i(B_i)=1$, is not finite then the measure λ is not σ-finite;

(d) the domain of the λ contains the class of all universally measurable subsets of the $\prod_{i\in\mathbf{N}}X_i$.

The next statements are immediate consequences of Theorems 8.3.1–8.3.2.

Theorem 8.3.4. *Let G_i be a group of measurable transformations of E_i for $i \in \mathbf{N}$. Under assumptions of Theorem 8.3.1, if each measure μ_i is G_i-invariant for $i \in \mathbf{N}$, then the measure $(\mathbf{sO})\prod_{i\in\mathbf{N}}\mu_i$ is $\prod_{i\in\mathbf{N}} G_i$-invariant.*

Proof. Setting $G = \prod_{i\in\mathbf{N}} G_i$, by Lemma 8.2.1 (see, condition (iii)) we deduce that the measure $(\mathbf{sO})\prod_{i\in\mathbf{N}}\mu_i$ is G-invariant. □

By the scheme of the proof of Theorem 8.3.3 we can prove the following

Theorem 8.3.5. *Let G_i be a group of measurable transformations of E_i for $i \in \mathbf{N}$. Under assumptions of Theorem 8.3.2, if each measure μ_i is G_i-invariant for $i \in \mathbf{N}$, then the measure $(\mathbf{sS})\prod_{i\in\mathbf{N}}\mu_i$ is $\prod_{i\in\mathbf{N}} G_i$-invariant.*

Remark 8.3.3. For $i \in \mathbf{N}$, we set $E_i = \mathbf{R}$ and $\mu_i = l_i$, where l_i denotes a linear Lebesgue measure on $\mathbf{R}$.

The measure $(\mathbf{sO})\prod_{i\in\mathbf{N}} l_i$ is called a "strict ordinary Lebesgue measure" on $\mathbf{R}^\infty$.

The measure $(\mathbf{sS})\prod_{i\in\mathbf{N}} l_i$ is called a "strict standard Lebesgue measure" on $\mathbf{R}^\infty$.

Note that strict ordinary and strict standard Lebesgue measures on $\mathbf{R}^\infty$, like ordinary and standard Lebesgue measures on the same space, are different.

8.4 On a Strict Standard Product of an Arbitrary Family of σ-Finite Borel Measures with Domain in Polish Spaces

The purpose of the present section is consider a new modification of the standard product of an arbitrary family of σ-finite Borel measures with domain in Polish spaces constructed [101].

Let α be again an arbitrary infinite parameter set and let $(\alpha_i)_{i\in I}$ be its any partition such that α_i is a non-empty finite subset of the α for every $i \in I$. Let μ_j be a σ-finite continuous Borel measure defined on a Polish space (E_j, ρ_j) for $j \in \alpha$.

We denote by $\mathcal{R}_{(\alpha_i)_{i\in I}}$ the family of all measurable rectangles $R \subseteq \prod_{j\in\alpha} E_j$ of the form $\prod_{i\in I} R_i$ with the property $0 \le (\mathbf{S})\prod_{i\in I}\tau_i(R_i) < \infty$ such that at most $\aleph_0$ of them are noncompact.

We suppose that there exists $R_0 = \prod_{i\in I} R_i^{(0)} \in \mathcal{R}_{(\alpha_i)_{i\in I}}$ such that

$$0 < (\mathbf{S})\prod_{i\in I}\tau_i(R_i^{(0)}) < \infty.$$

We say that a Borel measure $\nu_{(\alpha_i)_{i\in I}}$ defined on $\mathcal{B}(\prod_{j\in\alpha} E_j)$ is a strict standard $(\alpha_i)_{i\in I}$-product of the family of σ-finite continuous Borel measures $(\mu_j)_{j\in\alpha}$ if :

(i) for every

$$R = \prod_{i\in I} R_i \in \mathcal{R}_{(\alpha_i)_{i\in I}}$$

we have

$$\nu_{(\alpha_i)_{i\in I}}(R) = (\mathbf{S})\prod_{i\in I}\tau_i(R_i),$$

where $\tau_i = \prod_{j\in\alpha_i}\mu_j$ for $i \in I$.

(ii) for every $X \in \mathcal{B}(\prod_{i\in I} E_i)$ we have

$$(\nu_{(\alpha_i)_{i\in I}}(X) = 0 \Leftrightarrow (\forall R)(R \in \mathcal{R}_{(\alpha_i)_{i\in I}} \to \nu_{(\alpha_i)_{i\in I}}(X \cap R) = 0)).$$

Theorem 8.4.1. *Let μ_j be a quasi-finite Borel measure defined on a Polish space (E_j, ρ_j) for $j \in \alpha$. Let α be again an arbitrary infinite parameter set and let $(\alpha_i)_{i\in I}$ be its any partition such that α_i is a non-empty finite subset of the α for every $i \in I$, and let us suppose that there exists $R_0 = \prod_{i\in I} R_i^{(0)} \in \mathcal{R}_{(\alpha_i)_{i\in I}}$ such that*

$$0 < (\mathbf{S})\prod_{i\in I} \tau_i(R_i^{(0)}) < \infty.$$

Then there exists a strict standard $(\alpha_i)_{i\in I}$-product of the family $(\mu_j)_{j\in\alpha}$.

Proof. We denote by $\mathcal{R}_{(\alpha_i)_{i\in I}}$ the family of all measurable rectangles $R \subseteq \prod_{j\in\alpha} E_j$ of the form $\prod_{i\in I} R_i$ with the property $0 \le (\mathbf{S})\prod_{i\in I} \tau_i(R_i) < \infty$ such that at most $\aleph_0$ of them are noncompact.

For $X \in \mathcal{B}(R)$, we set $\mu_R(X) = 0$ if

$$(\mathbf{S})\prod_{i\in I} \tau_i(R_i) = 0,$$

and

$$\mu_R(X) = (\mathbf{S})\prod_{i\in I} \tau(R_i) \times \Big(\prod_{i\in I} \frac{\tau_{iR_i}}{\tau_i(R_i)}\Big)(X)$$

otherwise, where $\frac{\tau_{iR_i}}{\tau_i(R_i)}$ is a Borel probability measure defined on R_i as follows

$$(\forall X)\Big(X \in \mathcal{B}(R_i) \to \frac{\tau_{iR_i}}{\tau_i(R_i)}(X) = \frac{\tau_i(Y \cap R_i)}{\tau_i(R_i)}\Big).$$

By Lemma 7.7, the family of measures $(\mu_R)_{R\in\mathcal{R}}$ is consistent. We set

$$\nu_{(\alpha_i)_{i\in I}} = \lambda^{(2)}_{\mathcal{R}_{(\alpha_i)_{i\in I}}},$$

where the measure $\lambda^{(2)}_{\mathcal{R}_{(\alpha_i)_{i\in I}}}$ is defined by Lemma 8.2.2. Application of Lemma 8.2.2 ends the proof of Theorem 8.4.2. □

We denote a strict standard $(\alpha_i)_{i\in I}$-product of the family $(\mu_j)_{j\in\alpha}$ by

$$(\mathbf{sS}, (\alpha_i)_{i\in I})\prod_{j\in\alpha} \mu_j.$$

Here we present a certain example of the family of σ-finite continuous Borel measures $(\mu_j)_{j\in N}$ defined on the real axis $\mathbf{R}$ and of two different partitions $(\alpha_i)_{i\in N}$ and $(\beta_i)_{i\in N}$ of N for which

$$(\mathbf{sS}, (\alpha_i)_{i\in N})\prod_{j\in N} \mu_j \neq (\mathbf{sS}, (\beta_i)_{i\in N})\prod_{j\in N} \mu_j.$$

Example 8.4.1. We set $\alpha = N$. For $j \in N$, let l_j be a linear Lebesgue measure on $\mathbf{R}$. Let $\alpha_i = \{i\}$ and $\beta_i = \{2i+1, 2(i+1)\}$ for $i \in N$.

We set

$$Y_i = \left[0, \frac{1}{2}\right] \times [0,2].$$

It is obvious that

$$\Big((\mathbf{sS}, (\beta_i)_{i\in N}) \prod_{j\in N} l_j\Big)\Big(\prod_{i\in N} Y_i\Big) = 1$$

and

$$\Big((\mathbf{sS}, (\alpha_i)_{i\in N}) \prod_{j\in N} l_j\Big)\Big(\prod_{i\in N} Y_i\Big) = 0.$$

In connection with Theorems 8.4.1 and Example 8.4.1 we state the following

Problem 8.4.1. Under assumptions of Theorem 8.4.1, describe all pairs of partitions $(\alpha_i)_{i\in I}$ and $(\beta_i)_{i\in I}$ of the α for which $(\mathbf{sS}, (\alpha_i)_{i\in I}) \prod_{j\in\alpha} \mu_j = (\mathbf{sS}, (\beta_i)_{i\in I}) \prod_{j\in\alpha} \mu_j$.

The next two statements are immediate consequences of Theorem 8.4.2 and Lemma 8.2.2.

Theorem 8.4.2. *Under assumptions of Theorem* 8.4.2, *if each measure* μ_j *is* G_j*-left-and-right-invariant, where* G_j *denotes a group of Borel transformations of the* E_j *for* $j \in \alpha$, *then the measure* $(\mathbf{sS}, (\alpha_i)_{i\in I}) \prod_{j\in\alpha} \mu_j$ *is* $\prod_{j\in\alpha} G_j$*-left-and-right-invariant.*

Theorem 8.4.3. *Under assumption of Theorem* 8.4.1, *if each measure* μ_j *is* G_j*-left-invariant, where* G_j *denotes a group of Borel transformations of the* E_j *for* $j \in \alpha$, *then the measure* $(\mathbf{sS}, (\alpha_i)_{i\in I}) \prod_{j\in\alpha} \mu_j$ *is a* $\prod_{j\in\alpha} G_j$*-left-invariant.*

Observation 8.4.1. *Under conditions of Theorem* 8.4.2, *the measure* $(\mathbf{sS}, (\alpha_i)_{i\in I}) \prod_{j\in\alpha} \mu_j$ *is Radon.*

Proof. Let $0 < \big((\mathbf{sS}, (\alpha_i)_{i\in I}) \prod_{j\in\alpha} \mu_j\big)(X) < \infty$. It means that $X \in \mathcal{B}(\prod_{j\in\alpha} E_j)$ is presented as follows

$$X = (\cup_{n\in\{0\}\cup N} A_n \cap X) \cup (X \setminus \cup_{n\in\{0\}\cup N} A_n),$$

where $(A_n)_{n\in\{0\}\cup N}$ is a countable family of elements of the $\mathcal{R}_{(\alpha_i)_{i\in I}}$ with $A_{-1} = A_0 = \emptyset$ and

$$(\mathbf{sS}, (\alpha_i)_{i\in I}) \prod_{j\in\alpha} \mu_j(X \setminus \cup_{n\in\{0\}\cup N} A_n) = 0.$$

Since the measure μ_{A_n} is Radon, we can choose a compact set

$$F_n \subseteq (A_n \setminus \cup_{k=1}^{n-1} A_k) \cap X$$

such that

$$\mu_{A_n}\big(((A_n \setminus \cup_{k=1}^{n-1} A_k) \cap X) \setminus F_n\big) < \frac{\varepsilon}{2^{n+1}}$$

for $n \in N$.

Also, we can choose a natural number n_ε such that

$$\sum_{n=n_\varepsilon+1}^{\infty} \lambda_{A_n}((A_n \setminus \cup_{k=1}^{n-1} A_k) \cap X) < \frac{\varepsilon}{2}.$$

Finally, we get

$$\Big((\mathbf{S}, (\alpha_i)_{i\in I}) \prod_{j\in\alpha} \mu_j\Big)(X \setminus \cup_{s=1}^{n_\varepsilon} F_s)$$

$$= \sum_{n=1}^{\infty} \lambda_{A_n}((A_n \setminus \cup_{k=1}^{n-1} A_k) \cap (X \setminus \cup_{s=1}^{n_\varepsilon} F_s))$$

$$= \sum_{n=1}^{n_\varepsilon} \lambda_{A_n}((A_n \setminus \cup_{k=1}^{n-1} A_k) \cap (X \setminus \cup_{s=1}^{n_\varepsilon} F_s))$$

$$+ \sum_{n=n_\varepsilon+1}^{\infty} \lambda_{A_n}((A_n \setminus \cup_{k=1}^{n-1} A_k) \cap (X \setminus \cup_{s=1}^{n_\varepsilon} F_s))$$

$$\leq \sum_{n=1}^{n_\varepsilon} \lambda_{A_n}(((A_n \setminus \cup_{k=1}^{n-1} A_k) \cap X) \setminus F_n)$$

$$+ \sum_{n=n_\varepsilon+1}^{\infty} \lambda_{A_n}((A_n \setminus \cup_{k=1}^{n-1} A_k) \cap X) \leq \frac{\varepsilon}{2} + \frac{\varepsilon}{2} = \varepsilon. \qquad \square$$

Remark 8.4.1. For $j \in \alpha$, we set $E_j = \mathbf{R}$ and $\mu_j = m$, where m denotes a linear Lebesgue measure on $\mathbf{R}$.

Let $(\alpha_i)_{i\in I}$ be any partition of α such that α_i is a non-empty finite for every $i \in I$.

It is clear that $\prod_{j\in\alpha}[a_j, b_j] \in \mathcal{R}_{(\alpha_i)_{i\in I}}$ if $0 \leq (\mathbf{S})\prod_{i\in I} m^{\alpha_i}(\prod_{j\in\alpha_i}[a_j, b_j]) < \infty$, where m^{α_i} is a Lebesgue measure on $\mathbf{R}^{\alpha_i}$.

Then the measure $(\mathbf{sS}, (\alpha_i)_{i\in I})\prod_{j\in\alpha}\mu_j$ has the following properties:

$$(i)\ \Big((\mathbf{sS}, (\alpha_i)_{i\in I}) \prod_{j\in\alpha} \mu_j\Big)\Big(\prod_{j\in\alpha}[a_j, b_j]\Big) = (\mathbf{S}) \prod_{j\in\alpha} (b_i - a_i);$$

$$(ii)\ \Big((\mathbf{sS}, (\alpha_i)_{i\in I}) \prod_{j\in\alpha} \mu_j\Big)(X) = 0 \Leftrightarrow (\forall R)\Big(R \in \mathcal{R}_{(\alpha_i)_{i\in I}} \to (\mathbf{sS}, (\alpha_i)_{i\in I}) \prod_{j\in\alpha} \mu_j\Big)(X \cap R) = 0)$$

for every $X \in \mathcal{B}(\mathbf{R}^\alpha)$.

The measure $(\mathbf{sS}, (\alpha_i)_{i\in I})\prod_{j\in\alpha}\mu_i$ is called a strict standard "$(\alpha_i)_{i\in I}$-Lebesgue measure" on $\mathbf{R}^\alpha$.

When $card(\alpha_i) = 1$ for every $i \in I$, then $(\mathbf{sS}, (\alpha_i)_{i\in I})\prod_{j\in\alpha}\mu_i$ is called a strict standard Lebesgue measure on $\mathbf{R}^\alpha$ and is denoted by ℓ_α.

Let f be any permutation of α. A mapping $A_f : \mathbf{R}^\alpha \to \mathbf{R}^\alpha$, defined by $A_f((x_i)_{i\in\alpha}) = (x_{f(i)})_{i\in\alpha}$ for $(x_i)_{i\in\alpha} \in \mathbf{R}^\alpha$, is called a canonical permutation of $\mathbf{R}^\alpha$.

Note that in our situation $\mathcal{R}_{(\alpha_i)_{i\in I}}$ is the family of all measurable rectangles $R \subseteq \mathcal{B}(\mathbf{R}^\alpha)$ of the form $\prod_{i\in\alpha} Y_i$ with the property $0 \leq (\mathbf{S})\prod_{i\in\alpha} m(Y_i) < \infty$ such that at most $\aleph_0$ of them are noncompact (i.e., $card\{i : i \in I \ \&\ Y_i$ is not compact in $\prod_{i\in\alpha_i} E_i\} \leq \aleph_0$.). It is obvious to see that a measure ℓ_α is invariant under a group $\mathcal{P}(\mathbf{R}^\alpha)$ generated by shifts and canonical permutations of $\mathbf{R}^\alpha$ and

$$\ell_\alpha\Big(\prod_{i\in\alpha} Y_i\Big) = (\mathbf{S}) \prod_{i\in\alpha} m(Y_i).$$

Remark 8.4.2. It can be shown that a standard Lebesgue measure m^{α} [101] and a strict standard Lebesgue measure ℓ_{α} are different when $card(\alpha) \geq \omega$. Such a difference between the measures m^{α} and ℓ_{α} is caused by the phenomena that every set of finite m^{α}-measure can be covered by countable elements of the family $\mathcal{R}_{(\alpha_i)_{i \in I}}$ which we can not say for the measure ℓ_{α}.

8.5 Some Auxiliary Propositions in Solovay Model

Solovay model (cf. [132]) is the following system of axioms:

$$(ZF)\&(DC)\&(\text{every subset of } \mathbf{R} \text{ is measurable in the Lebesgue sense}),$$

where (ZF) denotes the Zermelo-Fraenkel set theory and (DC) denotes the axiom of Dependent Choices (cf. [20]).

Now, let (E_1, S_1, μ_1) and (E_2, S_2, μ_2) be two measure spaces. The measures μ_1 and μ_2 are called isomorphic if there exists a measurable isomorphism from E_1 onto E_2 such that

$$(\forall X)(X \in S_1 \rightarrow \mu_1(X) = \mu_2(f(X))).$$

Lemma 8.5.1. *Let E_1 and E_2 be any two Polish topological spaces. Let μ_1 be a probability diffused Borel measure on E_1 and let μ_2 be a probability diffused Borel measure on E_2. Then there exists a Borel isomorphism $\varphi : (E_1, B(E_1)) \rightarrow (E_2, B(E_2))$ such that*

$$\mu_1(X) = \mu_2(\varphi(X))$$

for every $X \in B(E_1)$.

The proof of Lemma 8.5.1 can be found in [20].

Lemma 8.5.2 ([92], Lemma 7.2, p. 118)**.** *Let E be a Polish space and let μ be a probability diffused Borel measure on E. Then, in Solovay's model the completion $\overline{\mu}$ of μ is defined on the power set of E.*

We have the following simple consequence of Lemma 8.5.2.

Corollary 8.5.1. *Let E be a Polish space. Then the class of universally measurable subsets of E coincides with the power set of the E in Solovay model.*

8.6 On Infinite Versions of Rademacher Theorem for R^{∞} in Solovay Model

Let $(R^{\infty}, +)$ be an abelian Polish group with metric ρ defined by

$$\rho((x_k)_{k \in N}, (y_k)_{k \in N}) = \sum_{k \in N} \frac{|x_k - y_k|}{2^k(1 + |x_k - y_k|)}$$

for $(x_k)_{k \in N}, (y_k)_{k \in N} \in R^{\infty}$.

We say that a function $f: R^\infty \to R$ is differentiable at the point $a \in R^\infty$ if there exists a continuous strict linear mapping $L: \mathbf{R}^\infty \to R$ such that the following condition

$$\lim_{t\to 0}\frac{f(a+tv)-f(a)}{t} = L(v)$$

for every $v \in \mathbf{R}^\infty$.

Under a continuous linear mapping $L: \mathbf{R}^\infty \to R$ we understood such a mapping for which the following two conditions holds:

(i) if $\sum_{k\in N} c_k$ tends to the c in the metric ρ then $\sum_{k\in N} L(c_k)$ tends to the $L(c)$ in the usual sense on R.

(ii) for every $x,y \in \mathbf{R}^\infty$ and $\alpha,\beta \in R$ we have $L(\alpha x+\beta y) = \alpha L(x)+\beta L(y)$.

It is obvious that if $L: \mathbf{R}^\infty \to R$ is a continuous linear mapping then

$$(\forall (x_k)_{k\in N})\Big((x_k)_{k\in N} \in \mathbf{R}^\infty \to L\Big(\sum_{k\in N} x_k e_k\Big) = \sum_{k\in N} x_k L(e_k)\Big).$$

Hence, a continuous linear mapping is uniquely defined by its values on the family $\{e_k : k \in N\}$. The latter relation means that $card(\{L(e_k) : L(e_k) \neq 0\}) < \omega$ because for every infinite sequence of non-zero real numbers $(L(e_k))_{k\in N}$ we can found such a sequence of real numbers $(x_k)_{k\in N} \in R^\infty$ that the sum $\sum_{k\in N} x_k L(e_k)$ will be divergent.

Let G be any subset of R^N. We say that a function $f: R^\infty \to R$ is called an G-Gâteaux differentiable at the point $a \in \mathbf{R}^\infty$ if there exists a continuous linear operator $A_a: R^N \to R$ such that for each $v \in G$ one has

$$f(a+tv)-f(a)-tA_a(v) = r(v,t), \lim_{t\to 0}|t^{-1}r(v,t)| = 0.$$

We say that a function $f: R^\infty \to R$ is called an G-Fréchet differentiable at the point $a \in \mathbf{R}^\infty$ if there exists a continuous linear operator $A_a: R^N \to R$ such that for each $v \in G$ one has

$$f(a+v)-f(a)-\rho(0,v)A_a(v) = r(v), \quad \lim_{\rho(0,v)\to 0}\frac{r(v)}{\rho(0,v)} = 0.$$

We say that f is Lipschitz function on $\mathbf{R}^\infty$ if there exists a positive number L such that

$$(\forall x,y)(x,y \in \mathbf{R}^\infty \to |f(x)-f(y)| < L\rho(x,y)).$$

Lemma 8.6.1. *Let $f: R^n \to R$ be a Lipschitz function. Then f is ℓ_n-almost everywhere R^n-Gâteaux differentiable on R^n.*

Lemma 8.6.2. *Let $f: \mathbf{R}^\infty \to R$ be a Lipschitz function. For $n \in N$, we set $G_n = R^n \times \{0\}^{N\setminus\{1,\dots,n\}}$.*

Let $a \in R^{(N)}$. Then, for $n \in N$, we have

$$\ell_n(\{(x_1,\dots,x_n) : f \textit{ is not } G_n - \textit{Gâteaux differentiable at the point}$$

$$(x_1,\dots,x_n,a_{n+1},a_{n+2},\dots)\ \}) = 0.$$

Proof. Let us define $g : R^n \to R$ by

$$g(x_1,\dots,x_n) = f(x_1,\dots,x_n,a_{n+1},a_{n+2},\dots).$$

Let $(v_1,\dots,v_n,0,\dots) \in G_n$.
We have

$$\lim_{t\to 0} \frac{f((x_1,\dots,x_n,a_{n+1},a_{n+2},\dots)+t(v_1,\dots,v_n,0,\dots)) - f((x_1,\dots,x_n,a_{n+1},a_{n+2},\dots))}{t}$$

$$= \lim_{t\to 0} \frac{g((x_1,\dots,x_n)+t(v_1,\dots,v_n)) - g(x_1,\dots,x_n)}{t}.$$

Let us show that g is Lipschitz function on R^n. Indeed, we have

$$|g(x_1,\dots,x_n) - g(y_1,\dots,y_n)| = |f(x_1,\dots,x_n,a_{n+1},a_{n+2},\dots) - f(y_1,\dots,y_n,a_{n+1},a_{n+2},\dots)|$$

$$\le L\sum_{k=1}^{n} \frac{|x_k - y_k|}{2^k(1+|x_k-y_k|)} \le L\sum_{k=1}^{n} |x_k - y_k| \le nL \sup_{1\le k\le n} |x_k - y_k|.$$

By Lemma 8.6.1 we claim that g is ℓ_n-almost everywhere R^n-Gâteaux differentiable on R^n. This ends the proof of Lemma 8.6.2. □

Lemma 8.6.3. *Let $f : \mathbf{R}^\infty \to R$ be a Lipschitz function. For $n \in N$, we set $G_n = R^n \times \{0\}^{N\setminus\{1,\dots,n\}}$. Then, in Solovay model f is λ-almost everywhere G_n-Gâteaux differentiable on R^N, where the λ denotes the strict ordinary Lebesgue measure on R^N.*

Proof. Let λ_n be a Lebesgue measure on $R^{\{1,\dots,n\}}$ and $\lambda_{N\setminus\{1,\dots,n\}}$ be a strict standard Lebesgue measure on $R^{N\setminus\{1,\dots,n\}}$. It is clear that λ is a strict standard product of measures λ_n and $\lambda_{N\setminus\{1,\dots,n\}}$.

We denote by D a collection of points in R^N for which f is not G_n-Gâteaux differentiable.

For every $(a_{n+1},a_{n+2},\dots) \in R^{N\setminus\{1,\dots,n\}}$, we denote by $D_{(a_{n+1},a_{n+2},\dots)}$ a $(a_{n+1},a_{n+2},\dots)$-section of the D, i.e.,

$$D_{(a_{n+1},a_{n+2},\dots)} = \{(x_1,\dots,x_n) \in R^{\{1,\dots,n\}} : (x_1,\dots,x_n,a_{n+1},a_{n+2},\dots) \in D\}.$$

By Lemma 8.6.2, for every $(a_{n+1},a_{n+2},\dots) \in R^{N\setminus\{1,\dots,n\}}$ we have that

$$\ell_n(\{(x_1,\dots,x_n) : f \text{ is not } G_n- \text{ differentiable at the point}$$

$$(x_1,\dots,x_n,a_{n+1},a_{n+2},\dots)\ \}) = 0.$$

For every $E = \prod_{k\in N} E_k \in \mathcal{R}(O)^+$, by Fubini theorem we get

$$\mu_R(D\cap E) = \int_{\prod_{k\in N\setminus\{1,\dots,n\}} E_k} \lambda_n((D\cap E)_{(a_{n+1},a_{n+2},\dots)})d\lambda_{N\setminus\{1,\dots,n\}}(a_{n+1},a_{n+2},\dots)$$

$$\le \int_{\prod_{k\in N\setminus\{1,\dots,n\}} E_k} \lambda_n(D_{(a_{n+1},a_{n+2},\dots)})d\lambda_{N\setminus\{1,\dots,n\}}(a_{n+1},a_{n+2},\dots) = 0.$$

By Lemma 8.6.2, we claim that $\lambda(D) = 0$. This ends the proof of Lemma 8.6.3. □

Theorem 8.6.1. *Let $f : \mathbf{R}^\infty \to R$ be a Lipschitz function and $R^{(N)}$ be a group of all eventually zero sequences. Then, in Solovay model, f is λ-almost everywhere $R^{(N)}$-Gâteaux differentiable on R^N.*

Proof. Let D_n be a set of points in R^N where f is not G_n-Gâteaux differentiable. Let D be a set of points in R^N where f is not $R^{(N)}$-Gâteaux differentiable. Since $R^{(N)} = \cup_{n \subset N} G_n$ we claim that $D = \cup_{n \in N} D_n$. By Lemma 8.6.3 we have that $\lambda(D_n) = 0$. Hence, $\lambda(D) \leq \sum_{n \in N} \lambda(D_n) = 0$. This ends the proof of Theorem 8.6.1. □

8.7 On Uniform Measures on ℓ^α

We begin this section by the following definition.

Definition 8.7.1. Let $\mathcal{S}$ be a translation-invariant σ-algebra of subsets of B such that

$$\mathcal{F}(B) \subseteq \mathcal{S},$$

where $\mathcal{F}(B)$ denotes a minimal σ-algebra of subsets of B generated by all open balls.

We say that a measure u defined on $\mathcal{S}$ is uniformly distributed (or, shortly is uniform measure) if the u measure of a ball depends only of the radius and not of the center.

It is clear that an arbitrary translation-invariant measure on B with $\mathcal{F}(B) \subseteq dom(\mu)$ is uniformly distributed.

In [103], has been investigated the following

Problem 8.7.1. Let μ_1 and μ_2 are two uniform measures on Banach space B such that

$$\mathcal{F}(B) \subseteq (dom)(\mu_1) = (dom)(\mu_2)$$

and the values of measures μ_1 and μ_2 coincide on an arbitrary ball.

Can we assert that the measures μ_1 and μ_2 coincide?

In [103], has been obtained the following results.

Theorem 8.7.1 ([103], Theorem 4.1). *There exist two translation-invariant uniform measures μ_1 and μ_2 defined on the σ-algebra $\mathcal{B}(\ell^\infty) \cap \mathcal{B}(\mathcal{R}^\infty)$ such that*

(i) for every closed ball $B(x,r)$ with center at x and radius r we have $\mu_1(B(x,r)) = \mu_2(B(x,r))$;

(ii) the measure μ_1 is absolutely continuous with respect to μ_2, and μ_1 and μ_2 are not equivalent.

Theorem 8.7.2 ([103], Theorem 4.2). *For an arbitrary non-empty parameter set α, we denote by the ℓ^α a Banach space of all real-valued bounded functions defined on the α and equipped with a norm $||\cdot||_\alpha$ defined by*

$$||(x_i)_{i\in\alpha}||_\alpha = \sup_{i\in\alpha} |x_i|$$

for $(x_i)_{i\in\alpha} \in \ell^\alpha$.

Let $\mathbb{F}(\ell^\alpha)$ be a minimal σ-algebra of subsets of the ℓ^α generated by the family of all open balls in ℓ^α.

Further, let B_0 be a closed unite ball defined by

$$B_0 = \{(x_i)_{i\in\alpha} : (x_i)_{i\in\alpha} \in \ell^\alpha \ \&\ ||(x_i)_{i\in\alpha}||_\alpha \leq 1\}.$$

Then there exists a translation-invariant measure μ defined on the measure space $(\ell^\infty, \mathbb{F}(\ell^\alpha))$ such that $\mu(B_0) = 1$.

Theorem 8.7.3 ([103], Theorem 4.3)**.** *For an arbitrary infinite parameter set α, there exist two uniform measures λ_1 and λ_2 defined on the Banach space ℓ^α such that:*

(i) $dom(\lambda_1) = dom(\lambda_2) = \mathcal{F}(\ell^\alpha)$;

(ii) the values of measures λ_1 and λ_2 coincide on an arbitrary ball such that $\lambda_1(B_0) = \lambda_2(B_0) = 1$;

(iii) the measures λ_1 and λ_2 are not equivalent.

The validity of Theorems 2.7.1-2.7.3 has been obtained in [**?**] by the technique of ordinary and standard products of linear Lebesgue measures. Note here that same results can be obtained by the technique of strict ordinary and strict standard products of linear Lebesgue measures.

8.8 On a Completion of the $\mathcal{B}(R^\infty)$ by the Baker Measure

Suppose that $\mathcal{R}_1$ be the class of all infinite dimensional rectangles $R \in B(\mathbb{R}^\mathbb{N})$ of a form

$$R = \prod_{k=1}^{\infty}(a_i, b_i), \quad -\infty < a_i \leq b_i < \infty$$

such that $0 \leq \prod_{k=1}^{\infty}(b_k - a_k) < \infty$.

Following [92] (see, Definition 15.4.1), a translation-invariant Borel measure λ in $\mathbb{R}^\mathbb{N}$ is called Baker measure if

$$\lambda(R) = \prod_{k=1}^{\infty}(b_k - a_k)$$

for $R \in \mathcal{R}_1$.

In [106] has been considered the following

Question 8.8.1. *Is it valid that*

$$\mathcal{B}_{||\cdot||_\infty}(\ell^\infty) \subseteq \overline{\mathcal{B}}_\lambda(\mathbf{R}^\infty),$$

where $\overline{\mathcal{B}}_\lambda(\mathbf{R}^\infty)$ denotes the completion of the $\mathcal{B}(\mathbf{R}^\infty)$ by the Baker's measure λ?

This question has been solved negatively by Peter Holický and Jaroslav Tišer for Baker's measures [2] and [3] in the theory ZFC (see, [106]).

The purpose of our final section is to show that Question 2.8.1 is independent from the theory $ZF + DC$. In this direction it is sufficient to construct such a consistent extension of the theory $ZF + DC$ that Question 8.8.1 is solved negatively. Indeed, we have the following

Theorem 8.8.1. *In Solovay model there exists a Baker measure in $\mathbf{R}^\infty$ for which Question 8.8.1 is solved positively.*

Proof. Let λ be Baker's strict ordinary measure. Following condition (a) of Corollary 8.3.2, the λ is Baker's measure. Following condition (d) of Lemma 8.3.2, the domain of the λ contains the class of all universally measurable subsets of the $\mathbf{R}^\infty$. Following Corollary 8.5.1, the class of universally measurable subsets of the $\mathbf{R}^\infty$ coincides with the power set $\mathcal{P}(\mathbf{R}^\infty)$ of the $\mathbf{R}^\infty$ in Solovay model. Thus, in Solovay model we have

$$\mathcal{B}_{||\cdot||_\infty}(\ell^\infty) \subseteq \mathcal{P}(\mathbf{R}^\infty) = \overline{\mathcal{B}}_\lambda(\mathbf{R}^\infty).$$

This ends the proof of Theorem 8.8.1. □

Chapter 9

On an Explicit Representation of a Particular Solution of the Non-Homogeneous Differential Equation of the Higher Order with Real Constant Coefficients

9.1 Introduction

Following [73], to obtain the solution to the non-homogeneous (sometimes called inhomogeneous) ordinary differential equations with constant coefficients

$$\sum_{n=0}^{m} A_n \frac{d^n}{dx^n}\Psi = f, \tag{9.1.1}$$

where f is any continuous function on $[-l,l]$, $m \geq 1$ and $(A_n)_{0\leq n\leq m} \in \mathbf{C}^{m+1}$, we find a particular integral Ψ_p by either the method of undetermined coefficients or the method of variation of parameters; the general solution Ψ to the linear differential equation (9.1.1) is the sum of the general solution Ψ_c of the related homogeneous equation

$$\sum_{n=0}^{m} A_n \frac{d^n}{dx^n}\Psi = 0 \tag{9.1.2}$$

and the particular integral Ψ_p. Or, when the initial conditions are set, we use Laplace transform to obtain the particular solution directly.

The first method of solving of (9.1.1) is due to Euler, who realized that solutions have the form e^{zx}, for possibly-complex values of z. The exponential function is one of the few functions that keep its shape after differentiation. In order for the sum of multiple derivatives of a function to sum up to zero, the derivatives must cancel each other out and the only way for them to do so is for the derivatives to have the same form as the initial function (see, for example [11], [124]). Thus, to solve (9.1.2), we set $\Theta(x) = e^{zx}$, leading to

$$\sum_{n=0}^{m} A_n z^n e^{zx} = 0. \tag{9.1.3}$$

Division by e^{zx} gives the nth-order polynomial equation

$$F(z) = \sum_{n=0}^{m} A_n z^n = 0. \tag{9.1.4}$$

This algebraic equation $F(z) = 0$, is the characteristic equation considered later by Gaspard Monge and Augustin-Louis Cauchy.

Formally, the terms $\frac{d^n}{dx^n}\Theta$ of the original differential equation are replaced by z^n. Solving the polynomial gives m values of z, $z_1, \cdots, z_m$. Substitution of any of those values for z_i into $e^{z_i x}$ gives a solution $\Theta_i(x) = e^{z_i x}$. Since homogeneous linear differential equations obey the superposition principle, any linear combination of these functions also satisfies the differential equation. When these roots are all distinct, we have m distinct solutions to the differential equation. It can be shown that these are linearly independent, by applying the Vandermonde determinant, and together they form a basis $(\Theta_i)_{1 \le i \le m}$ of the space of all solutions of the differential equation.

The preceding gave a solution for the case when all zeros are distinct, that is, each has multiplicity 1. For the general case, if z is a (possibly complex) zero (or root) of $F(z)$ having multiplicity k, then, for $i \in \{0, 1, \cdots, k-1\}$, $x^i e^{zx}$ is a solution of (9.1.3). Applying this to all roots gives a collection of m distinct and linearly independent functions, where m is the degree of $F(z)$. As before, these functions make up a basis of the solution space.

If the coefficients $(A_n)_{0 \le n \le m}$ of the differential equation are real, then real-valued solutions are generally preferable. Since non-real roots z then come in conjugate pairs, so do their corresponding basis functions $x^i e^{zx}$, and the desired result is obtained by replacing each pair with their real-valued linear combinations $Re(\Theta)$ and $Im(\Theta)$, where Θ is one of the pair.

Let $(\Theta_n)_{1 \le n \le m}$ be a basis of the space of all solutions of the differential equation (9.1.2).

We now seek a particular integral Ψ_p by the variation of parameters method. Let the coefficients of the linear combination be functions of x:

$$\Psi_p(x) = \sum_{n=1}^{m} U_n(x)\Theta_n(x). \tag{9.1.5}$$

For ease of notation we will drop the dependency on x (i.e. the various (x)). Using the "operator" notation $D = \frac{d}{dx}$ and a broad-minded use of notation, the ODE in question is $P(D)\Psi = f$; so

$$P(D)\Psi_p(x) = f(x) = \sum_{n=1}^{m} P(D)(U_n(x)\Theta_n(x)). \tag{9.1.6}$$

With the constraints

$$0 = \sum_{n=1}^{m} U_n'(x)\Theta_n(x), 0 = \sum_{n=1}^{m} U_n'(x)\Theta_n'(x), \cdots, 0 = \sum_{n=1}^{m} U_n'(x)\Theta_n^{(m-2)}(x), >$$

the parameters commute out, with a little "dirt":

$$f(x) = \sum_{n=1}^{m} U_n(x)P(D)(\Theta_n(x)) + \sum_{n=1}^{m} U_n^{'}(x)\Theta_n^{(m-1)}(x). \tag{9.1.7}$$

But $P(D)(\Theta_n(x)) = 0$ for $1 \leq n \leq m$, therefore

$$f(x) = \sum_{n=1}^{m} U_n^{'}(x)\Theta_n^{(m-1)}(x). \tag{9.1.8}$$

This, with the constraints, gives a linear system in the $U_n^{'}(1 \leq n \leq m)$. This much can always be solved; in fact, combining Cramer's rule with the Wronskian,

$$U_n^{'}(x) = \frac{W(\Theta_1, \cdots, \Theta_{n-1}, \Theta_{n+1}, \cdots, \Theta_m)\begin{pmatrix} 0 \\ \vdots \\ 0 \\ f \end{pmatrix}}{W(\Theta_1, \cdots, \Theta_m)}. \tag{9.1.9}$$

The rest is a matter of integrating $U_n^{'}$.

Thus, the above described approach does not allow us to represent in an explicit form a particular solution of non-homogeneous differential equation of the higher order with constant coefficients. This phenomena is caused by the fact that for a sufficiently large natural number m, we are not able to solve in an explicit form the characteristic equation (9.1.4), without speaking nothing about difficulties of performance remained components of such an approach.

The aim of the present chapter is twofold. First, under some general restrictions, we describe a new and essentially different approach for a construction in an explicit form of a particular solution of non-homogeneous differential equation of the higher order with real constant coefficients. Second, we specify some formulas obtained in [107].

The chapter is organized as follows.

In Section 9.2, by using a structure of the "Fourier differential operator" in R^∞ (cf. [107]), we describe a certain approach for a solution of some operator equations. In Section 9.2, we present our main results.

9.2 Some Auxiliary Notions and Facts

Definition 9.1. "Fourier differential operator" $(\mathcal{F})\frac{d}{dx}$ in R^∞ is defined as follows :

$$(\mathcal{F})\frac{d}{dx}\begin{pmatrix}\frac{a_0}{2}\\ a_1\\ b_1\\ a_2\\ b_2\\ a_3\\ b_3\\ \vdots\end{pmatrix}=\begin{pmatrix}0&0&0&0&0&0&0&\dots\\ 0&0&\frac{1\pi}{l}&0&0&0&0&\dots\\ 0&-\frac{1\pi}{l}&0&0&0&0&0&\dots\\ 0&0&0&0&\frac{2\pi}{l}&0&0&\dots\\ 0&0&0&-\frac{2\pi}{l}&0&0&0&\dots\\ 0&0&0&0&0&0&\frac{3\pi}{l}&\ddots\\ 0&0&0&0&0&-\frac{3\pi}{l}&0&\ddots\\ \vdots&\vdots&\vdots&\vdots&\vdots&\vdots&\ddots&\ddots\end{pmatrix}\times\begin{pmatrix}\frac{a_0}{2}\\ a_1\\ b_1\\ a_2\\ b_2\\ a_3\\ b_2\\ \vdots\end{pmatrix}, \qquad (9.2.1)$$

where $l > 0$.

For $n \in N$, we denote by $FD^{(n)}[-l,l]$ a vector space of all n-times differentiable functions Ψ on $[-l,l]$ such that a series obtained by k-times differentiation term by term of the Fourier trigonometric series of Ψ pointwise converges to $\Psi^{(k)}$ for all $x \in [-l,l]$ and $0 \le k \le n$.

Lemma 9.2.1. *Let G_M be an embedding of the $FD^{(0)}[-l,l]$ in to R^∞ which sends a function to a sequence of real numbers consisting from its Fourier coefficients. i.e., if*

$$\Psi(x) = \frac{c_0}{2} + \sum_{k=1}^{\infty} c_k \cos(\frac{k\pi x}{l}) + d_k \sin(\frac{k\pi x}{l}) \ (x \in [-l,l]),$$

then $G_F(\Psi) = (\frac{c_0}{2}, c_1, d_1, c_2, d_2, \dots)$. Then, for $\Psi \in FD^{(1)}[-l,l]$, the following equality

$$\left(G_F^{-1} \circ (\mathcal{F})\frac{d}{dx} \circ G_F\right)(\Psi) = \frac{d}{dx}(\Psi) \qquad (9.2.2)$$

holds.

Proof. Assume that for $\Psi \in FD^{(1)}[-l,l]$, we have the following representation

$$\Psi(x) = \frac{c_0}{2} + \sum_{k=1}^{\infty} c_k \cos\left(\frac{k\pi x}{l}\right) + d_k \sin\left(\frac{k\pi x}{l}\right) \ (x \in [-l,l]).$$

By the definition of the class $FD^{(1)}[-l,l]$, we have

$$\frac{d}{dx}(\Psi) = \frac{d}{dx}(\frac{c_0}{2} + \sum_{k=1}^{\infty} c_k \cos\left(\frac{k\pi x}{l}\right) + d_k \sin\left(\frac{k\pi x}{l}\right)) =$$

$$\sum_{k=1}^{\infty} c_k \frac{d}{dx}(\cos\left(\frac{k\pi x}{l}\right)) + d_k \frac{d}{dx}(\sin\left(\frac{k\pi x}{l}\right)) =$$

$$\sum_{k=1}^{\infty} -c_k \frac{k\pi}{l} \sin\left(\frac{k\pi x}{l}\right) + d_k \frac{k\pi}{l} \cos\left(\frac{k\pi x}{l}\right) =$$

$$\sum_{k=1}^{\infty} \frac{k\pi d_k}{l} \cos\left(\frac{k\pi x}{l}\right) - \frac{k\pi c_k}{l} \sin\left(\frac{k\pi x}{l}\right).$$

By the definition of the composition of mappings, we have

$$\Big(G_F^{-1}\circ(\mathcal{F})\frac{d}{dx}\circ G_F\Big)(\Psi)=G_F^{-1}((\mathcal{F})\frac{d}{dx}((G_F(\Psi))))=G_F^{-1}((\mathcal{F})\frac{d}{dx}\begin{pmatrix}\frac{c_0}{2}\\ c_1\\ d_1\\ c_2\\ d_2\\ c_3\\ d_3\\ \vdots\end{pmatrix})=$$

$$G_F^{-1}(\begin{pmatrix}0&0&0&0&0&0&0&\cdots\\ 0&0&\frac{1\pi}{l}&0&0&0&0&\cdots\\ 0&-\frac{1\pi}{l}&0&0&0&0&0&\cdots\\ 0&0&0&0&\frac{2\pi}{l}&0&0&\cdots\\ 0&0&0&-\frac{2\pi}{l}&0&0&0&\cdots\\ 0&0&0&0&0&0&\frac{3\pi}{l}&\ddots\\ 0&0&0&0&0&-\frac{3\pi}{l}&0&\ddots\\ \vdots&\vdots&\vdots&\vdots&\vdots&\vdots&\ddots&\ddots\end{pmatrix}\times\begin{pmatrix}\frac{c_0}{2}\\ c_1\\ d_1\\ c_2\\ d_2\\ c_3\\ d_3\\ \vdots\end{pmatrix})=$$

$$G_F^{-1}(\begin{pmatrix}0\\ \frac{1\pi d_1}{l}\\ -\frac{1\pi c_1}{l}\\ \frac{2\pi d_2}{l}\\ -\frac{2\pi c_2}{l}\\ \frac{3\pi d_3}{l}\\ -\frac{3\pi c_3}{l}\\ \vdots\end{pmatrix})=\sum_{k=1}^{\infty}\frac{k\pi d_k}{l}\cos\left(\frac{k\pi x}{l}\right)-\frac{k\pi c_k}{l}\sin\left(\frac{k\pi x}{l}\right). \tag{9.2.3}$$

□

By using the principle of mathematical induction and the result of Lemma 9.2.2, we can get the validity of the following assertion.

Lemma 9.2.2. *For $\Psi\in FD^{(n)}[-l,l]$ and $A_k\in R(0\leq k\leq n)$, the following equality*

$$\Big(G_F^{-1}\circ\Big(\sum_{k=0}^{n}A_k((\mathcal{F})\frac{d}{dx})^k\Big)\circ G_F\Big)(\Psi)=\sum_{k=0}^{n}A_k\frac{\partial^k}{\partial x^k}(\Psi) \tag{9.2.4}$$

holds.

Lemma 9.2.3. *For $m\geq 1$, let us consider an operator equation in R^∞*

$$\sum_{n=0}^{2m}A_n\Big((\mathcal{F})\frac{d}{dx}\Big)^n\Big(\frac{a_0}{2},a_1,b_1,a_2,b_2,\dots\Big)=(\frac{c_0}{2},c_1,d_1,c_2,d_2,\dots), \tag{9.2.5}$$

where $A_n\in R$ for $0\leq n\leq 2m$ and $(\frac{c_0}{2},c_1,d_1,c_2,d_2,\dots)\in R^\infty$.

Then (9.2.5) has a unique solution if and only if the following two conditions hold:
a) $A_0 \neq 0$;
b) $\sigma_k^2 + \omega_k^2 \neq 0$ *for each* $k \geq 1$, *where*

$$\sigma_k = \sum_{n=0}^{m} (-1)^n A_{2n} \left(\frac{k\pi}{l}\right)^{2n}, \tag{9.2.6}$$

$$\omega_k = \sum_{n=0}^{m-1} (-1)^n A_{2n+1} \left(\frac{k\pi}{l}\right)^{2n+1}. \tag{9.2.7}$$

If conditions a)-b) hold, then the solution of (9.2.5) *is given by*

$$\begin{pmatrix} \frac{a_0}{2} \\ a_1 \\ b_1 \\ a_2 \\ b_2 \\ a_3 \\ b_3 \\ \vdots \end{pmatrix} = \begin{pmatrix} \frac{c_0}{2A_0} \\ \frac{c_1\sigma_1 - d_1\omega_1}{\sigma_1^2+\omega_1^2} \\ \frac{c_1\omega_1 + d_1\sigma_1}{\sigma_1^2+\omega_1^2} \\ \frac{c_2\sigma_2 - d_2\omega_2}{\sigma_2^2+\omega_2^2} \\ \frac{c_2\omega_2 + d_2\sigma_2}{\sigma_2^2+\omega_2^2} \\ \vdots \end{pmatrix}. \tag{9.2.8}$$

Proof. Note that the matrix $((\mathcal{F})\frac{d}{dx})^{2n}$ is a $(1,2,2,\dots)$-cellular matrix B with cells $(B_k)_{k\in N}$ for which $B_0 = (0)$ and B_k is 2×2-diagonal matrix whose diagonal elements are $\frac{(-1)^n k^{2n}\pi^{2n}}{l^{2n}}$. Hence, the matrix $\sum_{n=0}^{m} A_{2n}((\mathcal{F})\frac{d}{dx})^{2n}$ has the form

$$\begin{pmatrix}
0 & 0 & 0 & 0 & 0 & 0 & \dots \\
0 & \sum_{n=0}^{m} A_{2n}\frac{(-1)^n 1^{2n}\pi^{2n}}{l^{2n}} & 0 & 0 & 0 & 0 & \dots \\
0 & 0 & \sum_{n=0}^{m} A_{2n}\frac{(-1)^n 1^{2n}\pi^{2n}}{l^{2n}} & 0 & 0 & 0 & \dots \\
0 & 0 & 0 & \sum_{n=0}^{m} A_{2n}\frac{(-1)^n 2^{2n}\pi^{2n}}{l^{2n}} & 0 & 0 & \dots \\
0 & 0 & 0 & 0 & \sum_{n=0}^{m} A_{2n}\frac{(-1)^n 2^{2n}\pi^{2n}}{l^{2n}} & 0 & \dots \\
\vdots & \vdots & \vdots & \vdots & \vdots & \ddots & \ddots
\end{pmatrix}. \tag{9.2.9}$$

Note that the matrix $((\mathcal{F})\frac{d}{dx})^{2n+1}$ has the form

$$\begin{pmatrix}
0 & 0 & 0 & 0 & 0 & 0 & \dots \\
0 & 0 & (-1)^n(\frac{1\pi}{l})^{2n+1} & 0 & 0 & 0 & \dots \\
0 & (-1)^{n+1}(\frac{1\pi}{l})^{2n+1} & 0 & 0 & 0 & 0 & \dots \\
0 & 0 & 0 & 0 & (-1)^n(\frac{2\pi}{l})^{2n+1} & 0 & \dots \\
0 & 0 & 0 & (-1)^{n+1}(\frac{2\pi}{l})^{2n+1} & & 0 & \dots \\
\vdots & \vdots & \vdots & \vdots & \vdots & \ddots & \ddots
\end{pmatrix}. \tag{9.2.10}$$

The matrix $\sum_{n=0}^{m-1} A_{2n+1}((\mathcal{F})\frac{d}{dx})^{2n+1}$ is $(1,2,\ldots)$-cellular matrix D with cells $(D_k)_{k\in N}$ for which $D_0=(0)$ and

$$D_k = \begin{pmatrix} 0 & \omega_k \\ -\omega_k & 0 \end{pmatrix} \tag{9.2.11}$$

for $k \geq 1$.

The matrix $\sum_{n=0}^{2m} A_n((\mathcal{F})\frac{d}{dx})^n$ is $(1,2,\ldots)$-cellular matrix E with cells $(E_k)_{k\in N}$ for which $E_0=(A_0)$ and

$$E_k = \begin{pmatrix} \sigma_k & \omega_k \\ -\omega_k & \sigma_k \end{pmatrix} \tag{9.2.12}$$

for $k \geq 1$, where σ_k and ω_k are defined by (9.2.6) and (9.2.7), respectively.

Necessity. If (9.2.5) has a unique solution then we claim that E is invertible. The latter relation means that all cells are invertible which follows that $A_0 \neq 0$ and $det(E_k) \neq 0$ for $k \geq 1$.

Sufficiency. Conditions a)-b) follow that each cell E_k is invertible for $k \in N$. Obviously, E is invertible and hence an operator equation (9.2.5) has a unique solution.

Now, if conditions a)-b) are satisfied then $E_0^{-1} = (\frac{1}{A_0})$ and

$$E_k^{-1} = \begin{pmatrix} \frac{\sigma_k}{\sigma_k^2+\omega_k^2} & -\frac{\omega_k}{\sigma_k^2+\omega_k^2} \\ \frac{\omega_k}{\sigma_k^2+\omega_k^2} & \frac{\sigma_k}{\sigma_k^2+\omega_k^2} \end{pmatrix} \tag{9.2.13}$$

for $k \geq 1$. Hence the solution of (9.2.5) has the following form

$$\begin{pmatrix} \frac{a_0}{2} \\ a_1 \\ b_1 \\ a_2 \\ b_2 \\ a_3 \\ b_3 \\ \vdots \end{pmatrix} = \Big(\sum_{n=0}^{2m} A_n((\mathcal{F})\frac{d}{dx})^n\Big)^{-1} \times \begin{pmatrix} \frac{c_0}{2} \\ c_1 \\ d_1 \\ c_2 \\ d_2 \\ c_3 \\ d_3 \\ \vdots \end{pmatrix} = \begin{pmatrix} \frac{c_0}{2A_0} \\ \frac{c_1\sigma_1-d_1\omega_1}{\sigma_1^2+\omega_1^2} \\ \frac{c_1\omega_1+d_1\sigma_1}{\sigma_1^2+\omega_1^2} \\ \frac{c_2\sigma_2-d_2\omega_2}{\sigma_2^2+\omega_2^2} \\ \frac{c_2\omega_2+d_2\sigma_2}{\sigma_2^2+\omega_2^2} \\ \vdots \end{pmatrix} \tag{9.2.14}$$

□

Remark 9.2.1. Let $\frac{\pi}{l} \in \mathbb{Q}$, where $\mathbb{Q}$ denotes a set of all rational numbers. For $m \geq 1$, let $(A_n)_{0\leq n\leq 2m}$ be a sequence of real numbers which are linearly independent over $\mathbb{Q}$. Then for a sequence of real numbers $(A_n)_{0\leq n\leq 2m}$ the condition b) in Lemma 9.2.4 is fulfilled.

Remark 9.2.2. Let us consider an operator equation (9.2.5). We know that, for $m \geq 1$, the matrix $\sum_{n=0}^{2m} A_n((\mathcal{F})\frac{d}{dx})^n$ is $(1,2,\cdots)$-cellular matrix E with cells $(E_k)_{k\in\mathbb{N}}$ for which $E_0 = (A_0)$ and

$$E_k = \begin{pmatrix} \sigma_k & \omega_k \\ -\omega_k & \sigma_k \end{pmatrix} \tag{9.2.15}$$

for $k \geq 1$.

We denote by $\hat{E}_0 : \mathbb{R} \to \mathbb{R}$ a linear operator defined by $\hat{E}_0(x) = A_0 x$.

We denote by $\hat{E}_k : \mathbb{R}\times\mathbb{R} \to \mathbb{R}\times\mathbb{R}$ a linear operator whose matrix in a standard basis of $\mathbb{R}^2$ coincides with the call E_k for $k \geq 1$.

Then we can do the following conclusions:

(i) if $\sigma_k^2+\omega_k^2 \neq 0$ for $k \geq 1$ and $A_0 \neq 0$, then for each $(\frac{c_0}{2}, c_1, d_1, \cdots) \in \mathbb{R}^\infty$, the operator equation (9.2.5) has a unique solution defined by

$$\Big(\frac{c_0}{2A_0}, \frac{c_1\sigma_1 - d_1\omega_1}{\sigma_1^2+\omega_1^2}, \frac{c_1\omega_1 + d_1\sigma_1}{\sigma_1^2+\omega_1^2}, \frac{c_2\sigma_2 - d_2\omega_2}{\sigma_2^2+\omega_2^2}, \frac{c_2\omega_2 + d_2\sigma_2}{\sigma_2^2+\omega_2^2}, \cdots\Big). \tag{9.2.16}$$

(ii) if $\sigma_k^2+\omega_k^2 \neq 0$ for $k \geq 1$ and $A_0 = 0$, then for each $(0, c_1, d_1, \cdots) \in \mathbb{R}^\infty$, the operator equation (9.2.5) has infinitely many solutions and the general solution has a form

$$\Big(\frac{c}{2}, \frac{c_1\sigma_1 - d_1\omega_1}{\sigma_1^2+\omega_1^2}, \frac{c_1\omega_1 + d_1\sigma_1}{\sigma_1^2+\omega_1^2}, \frac{c_2\sigma_2 - d_2\omega_2}{\sigma_2^2+\omega_2^2}, \frac{c_2\omega_2 + d_2\sigma_2}{\sigma_2^2+\omega_2^2}, \cdots\Big), \tag{9.2.17}$$

where $c \in \mathbb{R}$ is a free coefficient.

(iii) if $\sigma_k^2+\omega_k^2 \neq 0$ for $k \geq 1$ and $A_0 = 0$, then for each $(\frac{c_0}{2}, c_1, d_1, \cdots) \in \mathbb{R}^\infty$ with $\frac{c_0}{2} \neq 0$, the operator equation (9.2.5) has no any solution.

(iv) if $J = \{k : k \in \mathbb{N}\ \&\ \sigma_k^2+\omega_k^2 = 0\} \neq \emptyset$ and $A_0 = 0$, then for each $(0, c_1, d_1, \cdots) \in \mathbb{R}^\infty$ with $(c_k, d_k) \in \hat{E}_k(\mathbb{R}\times\mathbb{R})$ for $k \geq 1$ the operator equation (9.2.5) has infinitely many solutions and its general solution has a form

$$\Big(\frac{\psi_0}{2}, \psi_1, \xi_1, \psi_2, \xi_2, \cdots\Big), \tag{9.2.18}$$

where $\frac{\psi_0}{2} \in \mathbb{R}$ is a free coefficient, $\psi_k = \frac{c_k\sigma_k - d_1\omega_k}{\sigma_k^2+\omega_k^2}$ and $\xi_k = \frac{c_k\omega_k + d_k\sigma_k}{\sigma_k^2+\omega_k^2}$ for $k \in \mathbb{N}\setminus J$, and (ψ_k, ξ_k) are free coefficients belonging to the set $\{(x,y) : \hat{E}_k(x,y) = (c_k, d_k)\}$) for $k \in J$.

(v) if $J = \{k : k \in \mathbb{N}\ \&\ \sigma_k^2+\omega_k^2 = 0\} \neq \emptyset$ and $A_0 \neq 0$, then for each $(\frac{c_0}{2}, c_1, d_1, \cdots) \in \mathbb{R}^\infty$ with $(c_k, d_k) \in \hat{E}_k(\mathbb{R}\times\mathbb{R})$ for $k \geq 1$, the operator equation (9.2.5) has infinitely many solutions and its general solution has a form

$$\Big(\frac{\psi_0}{2}, \psi_1, \xi_1, \psi_2, \xi_2, \cdots\Big), \tag{9.2.19}$$

where

$\psi_0 = \frac{c_0}{A_0}$, $\psi_k = \frac{c_k\sigma_k - d_1\omega_k}{\sigma_k^2+\omega_k^2}$ and $\xi_k = \frac{c_k\omega_k + d_k\sigma_k}{\sigma_k^2+\omega_k^2}$ for $k \in \mathbb{N}\setminus J$, and (ψ_k, ξ_k) are free coefficients belonging to the set $\{(x,y) : \hat{E}_k(x,y) = (c_k, d_k)\}$) for $k \in J$.

(vi) if $J = \{k : k \in \mathbb{N}\ \&\ \sigma_k^2+\omega_k^2 = 0\} \neq \emptyset$, then for each $(\frac{c_0}{2}, c_1, d_1, \cdots) \in \mathbb{R}^\infty$, for which exists $k_0 \in J$ with $(c_{k_0}, d_{k_0}) \notin \hat{E}_{k_0}(\mathbb{R}\times\mathbb{R})$ or $c_0 \notin \hat{E}_0(\mathbb{R})$, the operator equation (9.2.5) has no any solution.

9.3 Main Results

The following assertion is a simple consequence of Lemmas 9.2.3-9.2.4.

Theorem 9.1. *For $m \geq 1$, let us consider an ordinary differential equation*

$$\sum_{n=0}^{2m} A_n \frac{d^n}{dx^n} \Psi = f, \tag{9.3.1}$$

where

$$f(x) = \frac{c_0}{2} + \sum_{k=1}^{\infty} c_k \cos\left(\frac{k\pi x}{l}\right) + d_k \sin\left(\frac{k\pi x}{l}\right) \in FD^{(0)}[-l,l] \tag{9.3.2}$$

and $A_n \in R$ for $0 \leq n \leq 2m$.

Suppose that $A_0 \neq 0$ and $\sigma_k^2 + \omega_k^2 \neq 0$ for $k \geq 1$, where σ_k and ω_k are defined by (9.2.6) and (9.2.7), respectively.

If $(\frac{c_0}{2}, c_1, d_1, c_2, d_2, \dots)$ is such a sequence of real numbers that the series Ψ_p, defined by

$$\Psi_p(x) = \frac{c_0}{2A_0} + \sum_{k=1}^{\infty} \left(\frac{c_k\sigma_k - d_k\omega_k}{\sigma_k^2 + \omega_k^2}\right) \cos\left(\frac{k\pi x}{l}\right) + \left(\frac{c_k\omega_k + d_k\sigma_k}{\sigma_k^2 + \omega_k^2}\right) \sin\left(\frac{k\pi x}{l}\right), \tag{9.3.3}$$

belongs to the class $FD^{(2m)}[-l,l]$, then Ψ_p is a particular solution of (9.3.1).

Theorem 9.2. *For $m \geq 1$, let us consider an ordinary differential equation (9.3.1), where*

$$f(x) \in C[-l,l] \tag{9.3.4}$$

and $A_n \in R$ for $0 \leq n \leq 2m$.

Suppose that $A_0 \neq 0$ and $\sigma_k^2 + \omega_k^2 \neq 0$ for $k \geq 1$, where σ_k and ω_k are defined by (9.2.6) and (9.2.7), respectively. Let $(\frac{c_0}{2}, c_1, d_1, c_2, d_2, \dots)$ be Fourier coefficients of f and $(\frac{c_0}{2}, c_1, d_1, c_2, d_2, \dots) \in \ell_1$.

Then the series Ψ_p, defined by

$$\Psi_p(x) = \frac{c_0}{2A_0} + \sum_{k=1}^{\infty} \left(\frac{c_k\sigma_k - d_k\omega_k}{\sigma_k^2 + \omega_k^2}\right) \cos\left(\frac{k\pi x}{l}\right) + \left(\frac{c_k\omega_k + d_k\sigma_k}{\sigma_k^2 + \omega_k^2}\right) \sin\left(\frac{k\pi x}{l}\right), \tag{9.3.5}$$

is a particular solution of (9.3.1).

Proof. For $N \geq 1$, let us consider an ordinary differential equation

$$\sum_{n=0}^{2m} A_n \frac{d^n}{dx^n} \Psi = f_N$$

where

$$f_N(x) = \frac{c_0}{2} + \sum_{k=1}^{N} c_k \cos\left(\frac{k\pi x}{l}\right) + d_k \sin\left(\frac{k\pi x}{l}\right). \tag{9.3.6}$$

Since $(\frac{c_0}{2}, c_1, d_1, c_2, d_2, \dots, c_N, d_N, 0, 0, \dots)$ is such a sequence of real numbers that a partial sum $\Psi_p^{(N)}$, defined by

$$\Psi_p^{(N)}(x) = \frac{c_0}{2A_0} + \sum_{k=1}^{N} \left(\frac{c_k\sigma_k - d_k\omega_k}{\sigma_k^2 + \omega_k^2}\right) \cos\left(\frac{k\pi x}{l}\right) +$$

$$\left(\frac{c_k\omega_k + d_k\sigma_k}{\sigma_k^2 + \omega_k^2}\right) \sin\left(\frac{k\pi x}{l}\right), \tag{9.3.7}$$

belongs to the class $FD^{(2m)}[-l,l]$, by Theorem 9.3.1 we deduce that $\Psi_p^{(N)}$ is a particular solution of

$$\sum_{n=0}^{2m} A_n \frac{d^n}{dx^n} \Psi = f_N, \tag{9.3.8}$$

where

$$f_N(x) = \frac{c_0}{2} + \sum_{k=1}^{N} c_k \cos\left(\frac{k\pi x}{l}\right) + d_k \sin\left(\frac{k\pi x}{l}\right) \in FD^{(0)}[-l,l]. \tag{9.3.9}$$

Note that $(\frac{c_0}{2}, c_1, d_1, c_2, d_2, \dots) \in \ell_1$ implies that

$$\left(\frac{c_0}{2A_0}, \frac{c_k\sigma_k - d_k\omega_k}{\sigma_k^2 + \omega_k^2}, \frac{c_k\omega_k + d_k\sigma_k}{\sigma_k^2 + \omega_k^2}, \dots, \frac{c_k\sigma_k - d_k\omega_k}{\sigma_k^2 + \omega_k^2}, \frac{c_k\omega_k + d_k\sigma_k}{\sigma_k^2 + \omega_k^2}, \dots\right) \in \ell_1,$$

because

$$\lim_{k\to\infty} \frac{\sigma_k}{\sigma_k^2 + \omega_k^2} = 0$$

and

$$\lim_{k\to\infty} \frac{\omega_k}{\sigma_k^2 + \omega_k^2} = 0.$$

Hence, the trigonometric series

$$\Psi_p(x) = \frac{c_0}{2A_0} + \sum_{k=1}^{\infty} \left(\frac{c_k\sigma_k - d_k\omega_k}{\sigma_k^2 + \omega_k^2}\right) \cos\left(\frac{k\pi x}{l}\right) +$$

$$\left(\frac{c_k\omega_k + d_k\sigma_k}{\sigma_k^2 + \omega_k^2}\right) \sin\left(\frac{k\pi x}{l}\right) \tag{9.3.10}$$

converges to Ψ_p uniformly (and hence also pointwise.) On the one hand, $\sum_{n=0}^{2m} A_n \frac{d^n}{dx^n} \Psi_p^{(N)}$ converges to $\sum_{n=0}^{2m} A_n \frac{d^n}{dx^n} \Psi_p$ uniformly (and hence also pointwise). On the other hand, $\sum_{n=0}^{2m} A_n \frac{d^n}{dx^n} \Psi_p^{(N)} = f_N$ converges to f uniformly (and hence also pointwise). The later relations mean that $\sum_{n=0}^{2m} A_n \frac{d^n}{dx^n} \Psi_p = f$. □

Remark 9.3.1. If for $m \geq 1$ we have an ordinary differential equation

$$\sum_{n=0}^{2m+1} A_n \frac{d^n}{dx^n} \Psi = f, \tag{9.3.11}$$

then setting $A_{2m+2} = 0$, we can rewrite (9.3.11) as follows

$$\sum_{n=0}^{2(m+1)} A_n \frac{d^n}{dx^n} \Psi = f. \tag{9.3.12}$$

This discussion shows us, that in general, we can will be limited to consideration only ordinary differential equations of the form (9.3.11).

Remark 9.3.2. Under some general restrictions, one can use Remark 9.2.6 for an investigation of a problem about existence and uniqueness of solutions of the non-homogeneous differential equation of the higher order with real constant coefficients (9.3.1).

Finally, we state the following

Problem 9.3.1. Describe the class of all real-valued continuous functions f on $[-l, l]$ for which the result of Theorem 9.3.2 remains valid.

Chapter 10

An Invariant Measure for the Non-Homogeneous Ordinary Differential Equation of Infinite Order with Real Constant Coefficients

10.1 Introduction

Let us consider a linear differential equation of infinite order

$$A_0 y + A_1 y' + A_2 y'' + \cdots = \phi(x), \tag{10.1.1}$$

where the coefficients A_k are constants, $\phi(x)$ is a given function and $y(x)$ is the function to be determined.

The problem of a solution of (10.1.1) itself is interesting and it is so important that has been studied for more than one century ago by many people in various manuscripts(cf. [116] - [61]). After the pioneer work of S. Pincherle [116], several investigators have contributed to the development of the theory of linear differential equations of infinite order with constant coefficients and of systems of such equations. The most elegant known theorem concerning solutions of equation (10.1.1) is perhaps the following assertion.

Theorem 10.1. *([17], Theorem 1, p. 202) In the linear differential equation of infinite order (10.1.1) let the coefficients $(A_n)_{n\in\mathbb{N}}$ be constants such that the function $F(z)$,*

$$F(z) = A_0 + A_1 z + A_2 z^2 + \cdots, \tag{10.1.2}$$

is analytic in the region $|z| \leq q$, where q is a given positive constant or zero, and let $\Phi(x)$ be a function of exponential type not exceeding q. If $F(z)$ vanishes at least once in the region $|z| \leq q$, let m be the number of its zeros in this region (each counted according to its multiplicity) and let $P(z)$ be the polynomial of degree m with leading coefficient unity such *that* $\frac{F(z)}{P(z)}$ *does not vanish in the region. If* $F(z)$ *does not vanish in the region, let* $P(z)$ *be identically equal to* 1*. When* $P(z) \equiv 1$*, let* $P_{m-1}(z)$ *be identically equal to zero; otherwise let it be an arbitrary polynomial of degree* $m-1$ *(including the case of an arbitrary constant when* $m = 1$*). Then the general solution* $y(x)$ *of (10.1.1), subject to the condition that it shall be a function of exponential type not exceeding* q*, may be written in the form*

$$y(x) = \frac{1}{2\pi i}\int_{C_p} \frac{\Psi(s)}{F(s)} e^{xs} ds + \frac{1}{2\pi i}\int_{C_p} \frac{P_{m-1}(s)}{P(s)} e^{xs} ds, \tag{10.1.3}$$

where

$$\Psi(s) = \sum_{n=0}^{\infty} \frac{\phi^{(n)}(0)}{s^{n+1}}, \tag{10.1.4}$$

and where C_p *is a circle of radius* p *about* 0 *as center,* p *being greater than* q *and such that* $F(z)$ *is analytic in the region* $q > |z| \geq p$ *and does not vanish there. If* $\phi(x)$ *is precisely of exponential type* q*, then the named solution* $y(x)$ *is also of exponential type* q*. If we take* $P_{m-1} \equiv 0$*, we have in (10.1.3) a particular solution of equation (10.1.1).*

If for a solution of non-homogeneous differential equations of the infinite order with real constant coefficients the technique of the theory of power series and of the theory of complex numbers was used everywhere in papers [116] - [61], our approach unlike their, uses the technique of the theory of Fourier trigonometric series and the theory of cellular matrices of infinite order.

If we consider a set of all functions f for which (10.1.1) has a unique solution then such a class of functions constitutes a vector space of functions on which an inverse of the differential operator $\sum_{k=0}^{\infty} A_k \frac{d^k}{dx^k}$ is defined. Hence, there naturally arises a question asking whether there exists a measure on that class which will be invariant under the action of the above-mentioned operator. If such a measure exists then we say that it is an invariant measure for the corresponding differential equation. Note that a problem of the existence and uniqueness of an invariant measure for various differential equations has been studied by many authors(see, for example, [146], [133], [1], [10], [?], [?]).

The aim of the present chapter is the twofold: first, by using a structure of the "Fourier differential operator" in $\mathbb{R}^\infty$, we plan to express in an explicit form the general solutions of (10.1.1); second, under some restriction, we are going to construct an invariant measure for the differential equation (10.1.1)

The chapter is organized as follows.

In Section 10.2, we consider some applications of methods of the theory of cellular matrices of infinite order. In particular, by using a structure of the "Fourier differential operator" in $\mathbb{R}^\infty$, we describe a new approach for a solution of some operator equations in R^∞ defined by (10.1.1).

In Section 10.3, under some restrictions, we give explicit expressions of general solutions of the non-homogeneous ordinary differential equation of infinite order with real constant coefficients.

In Section 10.4 we construct an invariant measure for the differential equation (10.1.1).

10.2 Some Auxiliary Notions and Facts from the Theory of Cellular Matrices of the Infinite Order

Definition 10.2.1. Let $(\mathcal{A}_n)_{n\in\mathbb{N}} = \left((a_{ij}^{(n)})_{i,j\in\mathbb{N}}\right)_{n\in\mathbb{N}}$ be a sequence of infinite-dimensional matrices. We say that $\lim_{n\to\infty}\mathcal{A}_n = (c_{ij})_{i,j\in\mathbb{N}}$ iff $\lim_{n\to\infty} a_{ij}^{(n)} = c_{ij}$ for each $i,j\in\mathbb{N}$.

Definition 10.2.2. Let $(\mathcal{A}_n)_{n\in\mathbb{N}} = \left((a_{ij}^{(n)})_{i,j\in\mathbb{N}}\right)_{n\in\mathbb{N}}$ be a sequence of infinite-dimensional matrices. We say that $\sum_{n=1}^{\infty}\mathcal{A}_n = (c_{ij})_{i,j\in\mathbb{N}}$ iff $\lim_{n\to\infty}\sum_{i=1}^{n}\mathcal{A}_i = (c_{ij})_{i,j\in\mathbb{N}}$.

Further, we suppose that $\mathbb{R}^\infty$ is equipped with Tychonoff metric.

Lemma 10.2.1. *Suppose that $(A_n)_{n\in\mathbb{N}}\in\mathbb{R}^{\mathbb{N}}$ be such a sequence of real numbers that*

$$\sigma_k = \sum_{n=0}^{\infty}(-1)^n A_{2n}(\frac{k\pi}{l})^{2n} \tag{10.2.1}$$

and

$$\omega_k = \sum_{n=0}^{\infty}(-1)^n A_{2n+1}(\frac{k\pi}{l})^{2n+1} \tag{10.2.2}$$

are convergent for each $k\geq 1$.

Then the following equality

$$\Big(\sum_{n=0}^{\infty}A_n\Big((\mathcal{F})\frac{d}{dx}\Big)^n\Big)\times\Big(\frac{a_0}{2},a_1,b_1,a_2,b_2,\dots\Big) = \sum_{n=0}^{\infty}A_n\Big((\mathcal{F})\frac{d}{dx}\Big)^n\Big)\times\Big(\frac{a_0}{2},a_1,b_1,a_2,b_2,\dots\Big) \tag{10.2.3}$$

holds for each $\left(\frac{a_0}{2},a_1,b_1,a_2,b_2,\dots\right)\in\mathbb{R}^\infty$.

Lemma 10.2.2. *Suppose that $(A_n)_{n\in\mathbb{N}}\in\mathbb{R}^{\mathbb{N}}$ be such a sequence of real numbers that σ_k and ω_k, defined (10.2.1) and (10.2.2) respectively, are convergent for each $k\geq 1$.*

Let us consider an operator equation in $\mathbb{R}^\infty$

$$\Big(\sum_{n=0}^{\infty}A_n\Big((\mathcal{F})\frac{d}{dx}\Big)^n\Big)\Big(\frac{a_0}{2},a_1,b_1,a_2,b_2,\dots\Big) = (\frac{c_0}{2},c_1,d_1,c_2,d_2,\dots), \tag{10.2.4}$$

where $(\frac{c_0}{2},c_1,d_1,c_2,d_2,\dots)\in R^\infty$.

Then (10.2.4) has a unique solution if and only if the following two conditions

a) $A_0\neq 0$;

b) $\sigma_k^2+\omega_k^2\neq 0$ for each $k\geq 1$;

hold.

If conditions a)-b) hold, then the solution of (10.2.4) *is given by*

$$\begin{pmatrix} \frac{a_0}{2} \\ a_1 \\ b_1 \\ a_2 \\ b_2 \\ a_3 \\ b_3 \\ \vdots \end{pmatrix} = \begin{pmatrix} \frac{c_0}{2A_0} \\ \frac{c_1\sigma_1 - d_1\omega_1}{\sigma_1^2+\omega_1^2} \\ \frac{c_1\sigma_1 + d_1\omega_1}{\sigma_1^2+\omega_1^2} \\ \frac{c_2\sigma_2 - d_2\omega_2}{\sigma_2^2+\omega_2^2} \\ \frac{c_2\sigma_2 + d_2\omega_2}{\sigma_2^2+\omega_2^2} \\ \vdots \end{pmatrix}. \tag{10.2.5}$$

Proof. Note that the matrix B corresponding to the linear operator $((\mathcal{F})\frac{d}{dx})^{2n}$ is a $(1,2,2,\dots)$-cellular matrix with cells $(B_k)_{k\in\mathbb{N}}$ for which $B_0 = (0)$ and B_k is 2×2-diagonal matrix whose diagonal elements are $\frac{(-1)^n k^{2n}\pi^{2n}}{l^{2n}}$. Hence, the matrix $\sum_{n=0}^{\infty} A_{2n}((\mathcal{F})\frac{d}{dx})^{2n}$ has the form

$$\begin{pmatrix} A_0 & 0 & 0 & 0 & 0 & 0 & \dots \\ 0 & \sum_{n=0}^{\infty} A_{2n}\frac{(-1)^n 1^{2n}\pi^{2n}}{l^{2n}} & 0 & 0 & 0 & 0 & \dots \\ 0 & 0 & \sum_{n=0}^{\infty} A_{2n}\frac{(-1)^n 1^{2n}\pi^{2n}}{l^{2n}} & 0 & 0 & 0 & \dots \\ 0 & 0 & 0 & \sum_{n=0}^{\infty} A_{2n}\frac{(-1)^n 2^{2n}\pi^{2n}}{l^{2n}} & 0 & 0 & \dots \\ 0 & 0 & 0 & 0 & \sum_{n=0}^{\infty} A_{2n}\frac{(-1)^n 2^{2n}\pi^{2n}}{l^{2n}} & 0 & \dots \\ \vdots & \vdots & \vdots & \vdots & \vdots & \ddots & \ddots \end{pmatrix}. \tag{10.2.6}$$

Note that the matrix $((\mathcal{F})\frac{d}{dx})^{2n+1}$ has the form

$$\begin{pmatrix} 0 & 0 & 0 & 0 & 0 & 0 & \dots \\ 0 & 0 & (-1)^n(\frac{1\pi}{l})^{2n+1} & 0 & 0 & 0 & \dots \\ 0 & (-1)^{n+1}(\frac{1\pi}{l})^{2n+1} & 0 & 0 & 0 & 0 & \dots \\ 0 & 0 & 0 & 0 & (-1)^n(\frac{2\pi}{l})^{2n+1} & 0 & \dots \\ 0 & 0 & 0 & (-1)^{n+1}(\frac{2\pi}{l})^{2n+1} & 0 & 0 & \dots \\ \vdots & \vdots & \vdots & \vdots & \vdots & \ddots & \ddots \end{pmatrix}. \tag{10.2.7}$$

The matrix $\sum_{n=0}^{\infty} A_{2n+1}((\mathcal{F})\frac{d}{dx})^{2n+1}$ is $(1,2,\dots)$-cellular matrix D with cells $(D_k)_{k\in N}$ for which $D_0 = (0)$ and

$$D_k = \begin{pmatrix} 0 & \omega_k \\ -\omega_k & 0 \end{pmatrix} \tag{10.2.8}$$

for $k \geq 1$.

The matrix $\sum_{n=0}^{\infty} A_n((\mathcal{F})\frac{d}{dx})^n$ is $(1,2,\ldots)$-cellular matrix E with cells $(E_k)_{k\in N}$ for which $E_0 = (A_0)$ and

$$E_k = \begin{pmatrix} \sigma_k & \omega_k \\ -\omega_k & \sigma_k \end{pmatrix} \tag{10.2.9}$$

for $k \geq 1$, where σ_k and ω_k are defined by $(10.2.1)$ and $(10.2.2)$, respectively.

Necessity. If $(10.2.4)$ has a unique solution then we claim that E is invertible. The latter relation means that all cellular matrices are invertible which follows that $A_0 \neq 0$ and $det(E_k) \neq 0$ for all $k \geq 1$.

Sufficiency. Conditions a)-b) follow that each cell E_k is invertible for $k \in \mathbb{N}$. Obviously, E is invertible and hence an operator equation $(10.2.4)$ has a unique solution.

Now, if conditions a)-b) are satisfied then $E_0^{-1} = (\frac{1}{A_0})$ and

$$E_k^{-1} = \begin{pmatrix} \frac{\sigma_k}{\sigma_k^2+\omega_k^2} & -\frac{\omega_k}{\sigma_k^2+\omega_k^2} \\ \frac{\omega_k}{\sigma_k^2+\omega_k^2} & \frac{\sigma_k}{\sigma_k^2+\omega_k^2} \end{pmatrix} \tag{10.2.10}$$

for $k \geq 1$. Hence the solution of $(10.2.4)$ has the following form

$$\begin{pmatrix} \frac{a_0}{2} \\ a_1 \\ b_1 \\ a_2 \\ b_2 \\ a_3 \\ b_3 \\ \vdots \end{pmatrix} = \Big(\sum_{n=0}^{\infty} A_n((\mathcal{F})\frac{d}{dx})^n\Big)^{-1} \times \begin{pmatrix} \frac{c_0}{2} \\ c_1 \\ d_1 \\ c_2 \\ d_2 \\ c_3 \\ d_3 \\ \vdots \end{pmatrix} = \begin{pmatrix} \frac{c_0}{2A_0} \\ \frac{c_1\sigma_1-d_1\omega_1}{\sigma_1^2+\omega_1^2} \\ \frac{c_1\omega_1+d_1\sigma_1}{\sigma_1^2+\omega_1^2} \\ \frac{c_2\sigma_2-d_2\omega_2}{\sigma_2^2+\omega_2^2} \\ \frac{c_2\omega_2+d_2\sigma_2}{\sigma_2^2+\omega_2^2} \\ \vdots \end{pmatrix} \tag{10.2.11}$$

□

Remark 10.2.1. Let us consider an operator equation $(10.2.4)$ and suppose that the sequence of real numbers $(\sigma_k)_{k\in\mathbb{N}}$ and $(\omega_k)_{k\in\mathbb{N}}$ defined by $(10.2.1)-(10.2.2)$ are convergent. We know that the matrix $\sum_{n=0}^{\infty} A_n((\mathcal{F})\frac{d}{dx})^n$ is $(1,2,\cdots)$-cellular matrix E with cells $(E_k)_{k\in\mathbb{N}}$ for which $E_0 = (A_0)$ and

$$E_k = \begin{pmatrix} \sigma_k & \omega_k \\ -\omega_k & \sigma_k \end{pmatrix} \tag{10.2.12}$$

for $k \geq 1$.

We denote by $\hat{E}_0 : \mathbb{R} \to \mathbb{R}$ a linear operator defined by $\hat{E}_0(x) = A_0 x$.

We denote by $\hat{E}_k : \mathbb{R}\times\mathbb{R} \to \mathbb{R}\times\mathbb{R}$ a linear operator whose matrix in a standard basis of $\mathbb{R}^2$ coincides with the call E_k for $k \geq 1$.

Then we can do the following conclusions:

(i) if $\sigma_k^2+\omega_k^2 \neq 0$ for $k \geq 1$ and $A_0 \neq 0$, then for each $(\frac{c_0}{2}, c_1, d_1, \cdots) \in \mathbb{R}^\infty$, the operator equation (10.2.4) has a unique solution defined by

$$(\frac{c_0}{2A_0}, \frac{c_1\sigma_1 - d_1\omega_1}{\sigma_1^2+\omega_1^2}, \frac{c_1\omega_1 + d_1\sigma_1}{\sigma_1^2+\omega_1^2}, \frac{c_2\sigma_2 - d_2\omega_2}{\sigma_2^2+\omega_2^2}, \frac{c_2\omega_2 + d_2\sigma_2}{\sigma_2^2+\omega_2^2}, \cdots). \tag{10.2.13}$$

(ii) if $\sigma_k^2+\omega_k^2 \neq 0$ for $k \geq 1$ and $A_0 = 0$, then for each $(0, c_1, d_1, \cdots) \in \mathbb{R}^\infty$, the operator equation (10.2.4) has an infinite number of solutions and the general solution has a form

$$(\frac{c}{2}, \frac{c_1\sigma_1 - d_1\omega_1}{\sigma_1^2+\omega_1^2}, \frac{c_1\omega_1 + d_1\sigma_1}{\sigma_1^2+\omega_1^2}, \frac{c_2\sigma_2 - d_2\omega_2}{\sigma_2^2+\omega_2^2}, \frac{c_2\omega_2 + d_2\sigma_2}{\sigma_2^2+\omega_2^2}, \cdots), \tag{10.2.14}$$

where $c \in \mathbb{R}$ is a free coefficient.

(iii) if $\sigma_k^2+\omega_k^2 \neq 0$ for $k \geq 1$ and $A_0 = 0$, then for each $(\frac{c_0}{2}, c_1, d_1, \cdots) \in \mathbb{R}^\infty$ with $\frac{c_0}{2} \neq 0$, the operator equation (10.2.4) has no any solution.

(iv) if $J = \{k : k \in \mathbb{N} \; \& \; \sigma_k^2+\omega_k^2 = 0\} \neq \emptyset$ and $A_0 = 0$, then for each $(0, c_1, d_1, \cdots) \in \mathbb{R}^\infty$ with $(c_k, d_k) \in \hat{E}_k(\mathbb{R} \times \mathbb{R})$ for $k \geq 1$ the operator equation (10.2.4) has infinitely many solutions and its general solution has a form

$$(\frac{\psi_0}{2}, \psi_1, \xi_1, \psi_2, \xi_2, \cdots), \tag{10.2.15}$$

where $\frac{\psi_0}{2} \in \mathbb{R}$ is a free coefficient, $\psi_k = \frac{c_k\sigma_k - d_1\omega_k}{\sigma_k^2+\omega_k^2}$ and $\xi_k = \frac{c_k\omega_k + d_k\sigma_k}{\sigma_k^2+\omega_k^2}$ for $k \in \mathbb{N} \setminus J$, and (ψ_k, ξ_k) are free coefficients belonging to the set $\{(x,y) : \hat{E}_k(x,y) = (c_k, d_k)\}$ for $k \in J$.

(v) if $J = \{k : k \in \mathbb{N} \; \& \; \sigma_k^2+\omega_k^2 = 0\} \neq \emptyset$ and $A_0 \neq 0$, then for each $(\frac{c_0}{2}, c_1, d_1, \cdots) \in \mathbb{R}^\infty$ with $(c_k, d_k) \in \hat{E}_k(\mathbb{R} \times \mathbb{R})$ for $k \geq 1$, the operator equation (10.2.4) has infinitely many solutions and its general solution has a form

$$(\frac{\psi_0}{2}, \psi_1, \xi_1, \psi_2, \xi_2, \cdots), \tag{10.2.15}$$

where

$\psi_0 = \frac{c_0}{A_0}$, $\psi_k = \frac{c_k\sigma_k - d_1\omega_k}{\sigma_k^2+\omega_k^2}$ and $\xi_k = \frac{c_k\omega_k + d_k\sigma_k}{\sigma_k^2+\omega_k^2}$ for $k \in \mathbb{N} \setminus J$, and (ψ_k, ξ_k) are free coefficients belonging to the set $\{(x,y) : \hat{E}_k(x,y) = (c_k, d_k)\}$ for $k \in J$.

(vi) if $J = \{k : k \in \mathbb{N} \; \& \; \sigma_k^2+\omega_k^2 = 0\} \neq \emptyset$, then for each $(\frac{c_0}{2}, c_1, d_1, \cdots) \in \mathbb{R}^\infty$, for which exists $k_0 \in J$ with $(c_{k_0}, d_{k_0}) \notin \hat{E}_{k_0}(\mathbb{R} \times \mathbb{R})$ or $c_0 \notin A_0\mathbb{R}$, the operator equation (10.2.4) has no any solution.

By using Remark 10.2.1 we can get the validity of the following assertions.

Theorem 10.2.1. *Let us consider an operator equation (10.2.4) and suppose that the sequence of real numbers* $(\sigma_k)_{k\in\mathbb{N}}$ *and* $(\omega_k)_{k\in\mathbb{N}}$ *defined by* (10.2.1) – (10.2.2) *are convergent. Then the operator equation* (10.2.4) *has a solution for* $(\frac{c_0}{2}, c_1, d_1, \cdots) \in \mathbb{R}^\infty$ *if and only if* $\frac{c_0}{2} \in A_0\mathbb{R}$ *and* $(c_k, d_k) \in \hat{E}_k(\mathbb{R} \times \mathbb{R})$ *for* $k \geq 1$.

Theorem 10.2.2. *Let us consider an operator equation (10.2.4) and suppose that the sequence of real numbers* $(\sigma_k)_{k\in\mathbb{N}}$ *and* $(\omega_k)_{k\in\mathbb{N}}$ *defined by* (10.2.1) – (10.2.2) *are convergent. Then the operator equation* (10.2.4) *has a unique solution for each* $(\frac{c_0}{2}, c_1, d_1, \cdots) \in \mathbb{R}^\infty$ *if and only if* $A_0 \neq 0$ *and* $\sigma_k^2+\omega_k^2 \neq 0$ *for* $k \geq 1$.

Theorem 10.2.3. *Let us consider an operator equation* $(10.2.4)$ *and suppose that the sequence of real numbers* $(\sigma_k)_{k\in\mathbb{N}}$ *and* $(\omega_k)_{k\in\mathbb{N}}$ *defined by* $(10.2.1)-(10.2.2)$ *are convergent. Then the operator equation* $(10.2.4)$ *has infinitely many solutions for* $(\frac{c_0}{2},c_1,d_1,\cdots)\in\mathbb{R}^\infty$ *if and only if the following conditions:*

(i) $\frac{c_0}{2}\in\hat{E}_0(\mathbb{R})$ *and* $(c_k,d_k)\in\hat{E}_k(\mathbb{R}\times\mathbb{R})$ *for* $k\geq 1$*;*

(ii) $A_0=0$ *or exists* $k_0\geq 1$ *such that* $\sigma_{k_0}^2+\omega_{k_0}^2=0$*;*

are fulfilled.

10.3 Explicit Expressions of General Solutions of the Non-Homogeneous Ordinary Differential Equation of the Infinite Order with Real Constant Coefficients

Suppose that $(A_n)_{n\in\mathbb{N}}\in\mathbb{R}^{\mathbb{N}}$. For $s\geq 1$ we set

$$\sigma_k^{(s)}=\sum_{n=0}^{s}(-1)^n A_{2n}(\frac{k\pi}{l})^{2n} \tag{10.3.1}$$

and

$$\omega_k^{(s)}=\sum_{n=0}^{s-1}(-1)^n A_{2n+1}(\frac{k\pi}{l})^{2n+1}. \tag{10.3.2}$$

Definition 10.1. We say that the sequence of real numbers $(A_n)_{n\in\mathbb{N}}$ satisfies the regularity property with respect to the sequence $(\frac{c_0}{2},c_1,d_1,c_2,d_2,\cdots)$, if the following conditions are fulfilled:

(i) for each $k\geq 1$, infinite series σ_k and ω_k, defined by $(10.2.1)$ and $(10.2.2)$ respectively, are convergent;

(ii)

$$\sigma_k^2+\omega_k^2\neq 0 \tag{10.3.3}$$

for each $k\geq 1$;

(iii)

$$\sum_{k=1}^{\infty}\Big(\frac{c_k\sigma_k-d_k\omega_k}{\sigma_k^2+\omega_k^2}\Big)\cos\Big(\frac{k\pi x}{l}\Big)+\Big(\frac{c_k\omega_k+d_k\sigma_k}{\sigma_k^2+\omega_k^2}\Big)\sin\Big(\frac{k\pi x}{l}\Big)\in FD^{(\infty)}[-l,l] \tag{10.3.4}$$

is infinitely times differentiable term by term on $[-l,l]$;

(iv)

$$\lim_{s\to\infty}\sum_{k=1}^{+\infty}\big(|c_k|+|d_k|\big)\Big(\Big|\frac{\sigma_k\sigma_k^{(s)}+\omega_k\omega_k^{(s)}}{\sigma_k^2+\omega_k^2}-1\Big|+\Big|\frac{\sigma_k\omega_k^{(s)}-\omega_k\sigma_k^{(s)}}{\sigma_k^2+\omega_k^2}\Big|\Big)=0. \tag{10.3.5}$$

The following assertion is a simple consequence of Lemmas 9.2.2 and 10.2.2.

Theorem 10.3.1. *Let us consider an ordinary differential equation*

$$\sum_{n=0}^{\infty}A_n\frac{d^n}{dx^n}\Psi=f \tag{10.3.6}$$

where

a) f is such a real valued function on $[-l,l]$ that its Fourier series

$$\frac{c_0}{2}+\sum_{k=1}^{+\infty} c_k\cos\Big(\frac{k\pi x}{l}\Big)+d_k\sin\Big(\frac{k\pi x}{l}\Big); \tag{10.3.7}$$

converges to f everywhere on $[-l,l]$.

b) the sequence $(A_n)_{n\in\mathbb{N}}\in\mathbb{R}^{\mathbb{N}}$ satisfies the regularity property with respect to the sequence $(\frac{c_0}{2},c_1,d_1,c_2,d_2,\cdots)$.

Then we have :

1) for $A_0\neq 0$, the trigonometric series Ψ_P, defined by

$$\Psi_P(x)=\frac{c_0}{2A_0}+\sum_{k=1}^{+\infty}\Big(\frac{c_k\sigma_k-d_k\omega_k}{\sigma_k^2+\omega_k^2}\Big)\cos\Big(\frac{k\pi x}{l}\Big)+$$

$$\Big(\frac{c_k\omega_k+d_k\sigma_k}{\sigma_k^2+\omega_k^2}\Big)\sin\Big(\frac{k\pi x}{l}\Big), \tag{10.3.8}$$

is a particular solution of (10.3.6)*;*

2) for $A_0=0$, (10.3.6) *has a particular solution iff $c_0=0$. If $A_0=0$ and $c_0=0$, then the trigonometric series Ψ_P, defined by*

$$\Psi_P(x)=\sum_{k=1}^{+\infty}\Big(\frac{c_k\sigma_k-d_k\omega_k}{\sigma_k^2+\omega_k^2}\Big)\cos\Big(\frac{k\pi x}{l}\Big)+$$

$$\Big(\frac{c_k\omega_k+d_k\sigma_k}{\sigma_k^2+\omega_k^2}\Big)\sin\Big(\frac{k\pi x}{l}\Big), \tag{10.3.9}$$

is a particular solution of (10.3.6)*.*

Proof. Let us prove that

$$\sum_{n=0}^{\infty}A_n\frac{d^n}{dx^n}\Psi_p=f \tag{10.3.10}$$

For $s>1$, by using the result of Lemma 9.2.2, we get

$$|\sum_{n=0}^{2s}A_n\frac{d^n}{dx^n}\Psi_p-f|=|\sum_{k=1}^{+\infty}[(\sum_{n=0}^{s}A_{2n}\frac{(-1)^nk^{2n}\pi^{2n}}{l^{2n}})\Big(\frac{c_k\sigma_k-d_k\omega_k}{\sigma_k^2+\omega_k^2}\Big)+$$

$$(\sum_{n=0}^{s-1}A_{2n+1}(-1)^n(\frac{k\pi}{l})^{2n+1})\Big(\frac{c_k\omega_k+d_k\sigma_k}{\sigma_k^2+\omega_k^2}\Big)-c_k]\cos\Big(\frac{k\pi x}{l}\Big)+$$

$$[\Big(-\sum_{n=0}^{s-1}A_{2n+1}(-1)^n(\frac{k\pi}{l})^{2n+1}\Big)\Big(\frac{c_k\sigma_k-d_k\omega_k}{\sigma_k^2+\omega_k^2}\Big)+$$

$$(\sum_{n=0}^{s}A_{2n}\frac{(-1)^nk^{2n}\pi^{2n}}{l^{2n}})\Big(\frac{c_k\omega_k+d_k\sigma_k}{\sigma_k^2+\omega_k^2}\Big)-d_k]\sin\Big(\frac{k\pi x}{l}\Big)|=$$

$$|\sum_{k=1}^{\infty}[\sigma_k^{(s)}\Big(\frac{c_k\sigma_k-d_k\omega_k}{\sigma_k^2+\omega_k^2}\Big)+\omega_k^{(s)}\Big(\frac{c_k\omega_k+d_k\sigma_k}{\sigma_k^2+\omega_k^2}\Big)-c_k]\cos\Big(\frac{k\pi x}{l}\Big)+$$

$$[-\omega_k^{(s)}\left(\frac{c_k\sigma_k-d_k\omega_k}{\sigma_k^2+\omega_k^2}\right)+\sigma_k^{(s)}\left(\frac{c_k\omega_k+d_k\sigma_k}{\sigma_k^2+\omega_k^2}\right)-d_k]\sin\left(\frac{k\pi x}{l}\right)|=$$

$$|\sum_{k=1}^{\infty}[\frac{c_k\sigma_k\sigma_k^{(s)}-d_k\omega_k\sigma_k^{(s)}+c_k\omega_k\omega_k^{(s)}+d_k\sigma_k\omega_k^{(s)}}{\sigma_k^2+\omega_k^2}-c_k]\cos\left(\frac{k\pi x}{l}\right)+$$

$$[\frac{-c_k\sigma_k\omega_k^{(s)}+d_k\omega_k^{(s)}\omega_k+c_k\sigma_k^{(s)}\omega_k+d_k\sigma_k\sigma_k^{(s)}}{\sigma_k^2+\omega_k^2}-d_k]\sin\left(\frac{k\pi x}{l}\right)|=$$

$$|\sum_{k=1}^{\infty}[\frac{c_k(\sigma_k\sigma_k^{(s)}+\omega_k\omega_k^{(s)})+d_k(\sigma_k\omega_k^{(s)}-\omega_k\sigma_k^{(s)})}{\sigma_k^2+\omega_k^2}-c_k]\cos\left(\frac{k\pi x}{l}\right)+$$

$$[\frac{c_k(\omega_k\sigma_k^{(s)}-\sigma_k\omega_k^{(s)})+d_k(\sigma_k\omega_k^{(s)}+\sigma_k^{(s)}\sigma_k)}{\sigma_k^2+\omega_k^2}-d_k]\sin\left(\frac{k\pi x}{l}\right)|\leq$$

$$\sum_{k=1}^{\infty}\left|c_k\frac{\sigma_k\sigma_k^{(s)}+\omega_k\omega_k^{(s)}}{\sigma_k^2+\omega_k^2}-c_k\right|+\left|d_k\frac{\sigma_k\omega_k^{(s)}-\omega_k\sigma_k^{(s)})}{\sigma_k^2+\omega_k^2}\right|+$$

$$\left|c_k\frac{\omega_k\sigma_k^{(s)}-\sigma_k\omega_k^{(s)})}{\sigma_k^2+\omega_k^2}\right|+\left|d_k\frac{\omega_k\omega_k^{(s)}+\sigma_k^{(s)}\sigma_k}{\sigma_k^2+\omega_k^2})-d_k\right|\leq$$

$$\sum_{k=1}^{\infty}\left|c_k\frac{\sigma_k\sigma_k^{(s)}+\omega_k\omega_k^{(s)}}{\sigma_k^2+\omega_k^2}-c_k\right|+$$

$$\left|d_k\frac{\sigma_k\omega_k^{(s)}-\omega_k\sigma_k^{(s)})}{\sigma_k^2+\omega_k^2}\right|+\left|c_k\frac{\omega_k\sigma_k^{(s)}-\sigma_k\omega_k^{(s)})}{\sigma_k^2+\omega_k^2}\right|+\left|d_k\frac{\omega_k\omega_k^{(s)}+\sigma_k^{(s)}\sigma_k}{\sigma_k^2+\omega_k^2})-d_k\right|\leq$$

$$\sum_{k=1}^{\infty}(|c_k|+|d_k|)\left|\frac{\sigma_k\sigma_k^{(s)}+\omega_k\omega_k^{(s)}}{\sigma_k^2+\omega_k^2}-1\right|+|c_k|\left|d_k\frac{\sigma_k\omega_k^{(s)}-\omega_k\sigma_k^{(s)})}{\sigma_k^2+\omega_k^2}\right|+|c_k|\left|\frac{\omega_k\sigma_k^{(s)}-\sigma_k\omega_k^{(s)})}{\sigma_k^2+\omega_k^2}\right|\leq$$

$$\sum_{k=1}^{\infty}(|c_k|+|d_k|)(\left|\frac{\sigma_k\sigma_k^{(s)}+\omega_k\omega_k^{(s)}}{\sigma_k^2+\omega_k^2}-1\right|+\left|\frac{\sigma_k\omega_k^{(s)}-\omega_k\sigma_k^{(s)})}{\sigma_k^2+\omega_k^2}\right|). \tag{10.3.11}$$

By using (10.3.5), we deduce that $\left|\sum_{n=0}^{2s}A_n\frac{d^n}{dx^n}\Psi_p-f\right|$ tends to zero when s tends to $+\infty$.

This ends the proof of theorem.

□

Corollary 10.3.1. *Let us consider an ordinary differential equation*

$$\sum_{n=0}^{\infty}A_n\frac{d^n}{dx^n}\Psi=f \tag{10.3.12}$$

where a) f is such a real valued function on $[-l,l]$ that

$$f(x)\quad\frac{c_0}{2}+\sum_{k=1}^{S}c_k\cos\left(\frac{k\pi x}{l}\right)+d_k\sin\left(\frac{k\pi x}{l}\right)(S\geq 1); \tag{10.3.13}$$

b) the sequence $(A_n)_{n\in\mathbb{N}}\in\mathbb{R}^{\mathbb{N}}$ satisfies conditions $(i)-(ii)$ participated in the Definition 10.3.1.

Then we have :

1) for $A_0 \neq 0$, the trigonometric series $\Psi_P^{(S)}$, defined by

$$\Psi_P^{(S)}(x) = \frac{c_0}{2A_0} + \sum_{k=1}^{S} \left(\frac{c_k\sigma_k - d_k\omega_k}{\sigma_k^2+\omega_k^2}\right)\cos\left(\frac{k\pi x}{l}\right) + \left(\frac{c_k\omega_k + d_k\sigma_k}{\sigma_k^2+\omega_k^2}\right)\sin\left(\frac{k\pi x}{l}\right), \quad (10.3.14)$$

is a particular solution of (10.3.12).

2) for $A_0 = 0$, (10.3.12) *has a particular solution iff $c_0 = 0$. If $A_0 = 0$ and $c_0 = 0$, then the trigonometric series $\Psi_P^{(S)}$, defined by*

$$\Psi_P^{(S)}(x) = \sum_{k=1}^{S} \left(\frac{c_k\sigma_k - d_k\omega_k}{\sigma_k^2+\omega_k^2}\right)\cos\left(\frac{k\pi x}{l}\right) + \left(\frac{c_k\omega_k + d_k\sigma_k}{\sigma_k^2+\omega_k^2}\right)\sin\left(\frac{k\pi x}{l}\right), \quad (10.3.15)$$

is a particular solution of (10.3.12).

Proof. It is not hard to check that the sequence of real numbers $(A_n)_{n\in\mathbb{N}}$ satisfies the regularity property with respect to the sequence $(\frac{c_0}{2}, c_1, d_1, \cdots, c_S, d_S, 0, 0, \cdots)$ (i.e., conditions (10.3.4) – (10.3.5) are fulfilled). For $A_0 \neq 0$, Ψ_P defined by (10.3.8) coincides with $\Psi_P^{(S)}$ defined by (10.3.14). For $A_0 = 0$, Ψ_P defined by (10.3.9) coincides with $\Psi_P^{(S)}$ defined by (10.3.15). By Theorem 10.3.2 we deduce that all conditions of Corollary 10.3.1 are satisfied. □

Example 10.3.1. Let us consider an ordinary differential equation

$$\sum_{n=0}^{\infty} \frac{(-1)^n}{n!}\frac{d^{2n}}{dx^{2n}}\Psi = f, \quad (10.3.16)$$

where f is defined by (10.3.13).

Then we have that $A_0 = 1$,

$$\sigma_k = \sum_{n=0}^{\infty} \frac{(-1)^n}{n!}(-1)^n\left(\frac{k\pi}{l}\right)^{2n} = \sum_{n=0}^{\infty} \frac{((\frac{k\pi}{l})^2)^n}{n!} = e^{(\frac{k\pi}{l})^2} \quad (10.3.17)$$

and

$$\omega_k = 0$$

for $k \geq 1$.

By Corollary 10.3.1 we deduce that $\Psi_P^{(S)}$ defined by

$$\Psi_P^{(S)}(x) = \frac{c_0}{2} + \sum_{k=1}^{S} e^{-\frac{k^2\pi^2}{l^2}}\left(c_k\cos\left(\frac{k\pi x}{l}\right) + d_k\sin\left(\frac{k\pi x}{l}\right)\right) \quad (10.3.18)$$

is a particular solution of (10.3.16).

Example 10.3.2. Let us consider an ordinary differential equation

$$\sum_{n=0}^{\infty} \frac{(-1)^n}{n!}\frac{d^{2n+1}}{dx^{2n+1}}\Psi = f, \quad (10.3.19)$$

where f is defined by (10.3.13).

Then we have that $A_0 = 0$,

$$\omega_k = \sum_{n=0}^{\infty} \frac{(-1)^n}{n!}(-1)^n\left(\frac{k\pi}{l}\right)^{2n+1} = \frac{k\pi}{l}\sum_{n=0}^{\infty}\frac{((\frac{k\pi}{l})^2)^n}{n!} = \frac{k\pi}{l}e^{(\frac{k\pi}{l})^2} \tag{10.3.20}$$

and $\sigma_k = 0$ for $k \geq 1$.

By Corollary 10.3.3 we deduce that $\Psi_P^{(S)}$ defined by

$$\Psi_P^{(S)}(x) = \sum_{k=1}^{S} \frac{l}{k\pi} e^{-\frac{k^2\pi^2}{l^2}}\left(-d_k \cos\left(\frac{k\pi x}{l}\right) + c_k \sin\left(\frac{k\pi x}{l}\right)\right) \tag{4.21}$$

is a particular solution of (10.3.19).

The following corollary of Theorem 10.3.1, under some natural restrictions, helps us to express in an explicit form a particular solution of non-homogeneous differential equation of finite order with real constant coefficients.

Corollary 10.3.2. *Let us consider an ordinary differential equation*

$$\sum_{n=0}^{2m} A_n \frac{d^n}{dx^n}\Psi = f \tag{10.3.22}$$

where $m \geq 1$ and f is defined by (10.3.7).

We have $\sigma_k = \sum_{n=0}^{m}(-1)^n A_{2n}(\frac{k\pi}{l})^{2n}$ *and* $\omega_k = \sum_{n=0}^{m-1}(-1)^n A_{2n+1}(\frac{k\pi}{l})^{2n+1}$ *for* $k \geq 1$. *Suppose that the sequence* $(A_n)_{0 \leq n \leq 2m}$ *satisfies the condition* (10.3.3), *i.e., the condition* $\sigma_k^2 + \omega_k^2 \neq 0$ *for* $k \geq 1$ [1]

Then we have :

1) for $A_0 \neq 0$, *the trigonometric series* Ψ_P, *defined by*

$$\Psi_P(x) = \frac{c_0}{2A_0} + \sum_{k=1}^{\infty}\left(\frac{c_k\sigma_k - d_k\omega_k}{\sigma_k^2+\omega_k^2}\right)\cos\left(\frac{k\pi x}{l}\right) + \left(\frac{c_k\omega_k + d_k\sigma_k}{\sigma_k^2+\omega_k^2}\right)\sin\left(\frac{k\pi x}{l}\right), \tag{10.3.23}$$

is a particular solution of (10.3.22).

2) for $A_0 = 0$, (10.3.22) *has a particular solution iff* $c_0 = 0$. *If* $A_0 = 0$ *and* $c_0 = 0$, *then the trigonometric series* Ψ_P, *defined by*

$$\Psi_P(x) = \sum_{k=1}^{\infty}\left(\frac{c_k\sigma_k - d_k\omega_k}{\sigma_k^2+\omega_k^2}\right)\cos\left(\frac{k\pi x}{l}\right) + \left(\frac{c_k\omega_k + d_k\sigma_k}{\sigma_k^2+\omega_k^2}\right)\sin\left(\frac{k\pi x}{l}\right), \tag{10.3.24}$$

is a particular solution of (10.3.22).

Proof. It is not hard to check that the sequence of real numbers $(A_n)_{1 \leq n \leq 2m}$ satisfies the regularity property with respect to the sequence $(\frac{c_0}{2}, c_1, d_1, \cdots)$. For $A_0 \neq 0$, Ψ_P defined by (9.3.5) coincides with Ψ_P defined by (10.3.23). For $A_0 = 0$, Ψ_P defined by (10.3.9) coincides with Ψ_P defined by (10.3.24). By Theorem 10.3.1 we deduce that all conditions of Corollary 10.3.2 are satisfied. □

[1]Note that if $\frac{\pi}{l} \in \mathbb{Q}$ and a family of real numbers $(A_n)_{0 \leq n \leq 2m}$ are linearly independent over $\mathbb{Q}$, the condition (10.3.3) holds.

Since G_F is a linear embedding of the vector space $FD^{(0)}[-1,1]$ in to Tychonoff topological vector space $\mathbb{R}^\infty$, we can define a metric ρ_T on $FD^{(0)}[-1,1]$ as follows

$$\rho_T(v_1,v_2)=\rho(G_F(v_1),G_F(v_1))$$

for $v_1,v_2 \in FD^{(0)}[-1,1]$, where ρ is a metric on $\mathbb{R}^\infty$ defined by

$$(\forall z_1,z_2 \in \mathbb{R}^\infty \to$$

$$\rho(z_1,z_2)=\frac{|c_0^{(2)}-c_0^{(1)}|}{2(1+|c_0^{(2)}-c_0^{(1)}|)}+\sum_{k=1}^{\infty}\frac{1}{2^{k+1}}\Big(\frac{|c_k^{(2)}-c_k^{(1)}|}{1+|c_k^{(2)}-c_k^{(1)}|}+\frac{|d_k^{(2)}-d_k^{(1)}|}{1+|d_k^{(2)}-d_k^{(1)}|}\Big), \quad (10.3.25)$$

where $z_1=(c_0^{(1)},c_1^{(1)},d_1^{(1)},\cdots)$ and $z_2=(c_0^{(2)},c_1^{(2)},d_1^{(2)},\cdots)$.

It is obvious that convergence in the sense of Tychonoff metric in $\mathbb{R}^\infty$ is same as the convergence in the sense of the metric ρ.

Definition 10.3.1. We say that $\Psi \in FD^{(\infty)}[-l,l]$ is a particular solution of $(10.3.6)-(10.3.7)$ in the sense of Tychonoff if

$$\lim_{L\to\infty}\rho_T\Big(\sum_{n=0}^{L}A_n\frac{d^n}{dx^n}\Psi,f\Big)=0 \quad (10.3.26)$$

Theorem 10.3.2. *Let us consider an ordinary differential equation* (10.3.6), *where :*

a) $FD^{(0)}[-1,1]\ni f$ *is defined by* (10.3.7).

b) the sequence $(A_n)_{n\in\mathbb{N}}\in\mathbb{R}^{\mathbb{N}}$ *satisfies conditions* $(i)-(iii)$ *participated in the Definition 10.3.1.*

Then we have :

1) for $A_0\neq 0$, *the trigonometric series* Ψ_P, *defined by* (10.3.8) *is a particular solution of* (10.3.6) *in the sense of Tychonoff.*

2) for $A_0=0$, (10.3.6) *has a particular solution iff* $c_0=0$. *If* $A_0=0$ *and* $c_0=0$, *then the trigonometric series* Ψ_P, *defined by* (9.2.2) *is a particular solution of* (10.3.6) *in the sense of Tychonoff.*

Proof. We denote by $\mathbf{O}$ the zero element of the class $FD^{(0)}[-1,1]$.

Case 1. $A_0\neq 0$.

For $s>1$, by using the result of Lemma 9.2.2, we get

$$\rho_T\Big(\sum_{n=0}^{2s}A_n\frac{d^n}{dx^n}\Psi_p,f\Big)=\rho_T\Big(\sum_{k=1}^{\infty}\Big[\Big(\sum_{n=0}^{s}A_{2n}\frac{(-1)^n k^{2n}\pi^{2n}}{l^{2n}}\Big)\Big(\frac{c_k\sigma_k-d_k\omega_k}{\sigma_k^2+\omega_k^2}\Big)+$$

$$\Big(\sum_{n=0}^{s-1}A_{2n+1}(-1)^n\Big(\frac{k\pi}{l}\Big)^{2n+1}\Big)\Big(\frac{c_k\omega_k+d_k\sigma_k}{\sigma_k^2+\omega_k^2}\Big)-c_k\Big]\cos\Big(\frac{k\pi x}{l}\Big)+$$

$$\Big[\Big(-\sum_{n=0}^{s-1}A_{2n+1}(-1)^n\Big(\frac{k\pi}{l}\Big)^{2n+1}\Big)\Big(\frac{c_k\sigma_k-d_k\omega_k}{\sigma_k^2+\omega_k^2}\Big)+$$

$$\Big(\sum_{n=0}^{s}A_{2n}\frac{(-1)^n k^{2n}\pi^{2n}}{l^{2n}}\Big)\Big(\frac{c_k\omega_k+d_k\sigma_k}{\sigma_k^2+\omega_k^2}\Big)-d_k\Big]\sin\Big(\frac{k\pi x}{l}\Big),\mathbf{O}\Big)=$$

$$\rho_T\Big(\sum_{k=1}^{\infty}[\sigma_k^{(s)}\left(\frac{c_k\sigma_k-d_k\omega_k}{\sigma_k^2+\omega_k^2}\right)+\omega_k^{(s)}\left(\frac{c_k\omega_k+d_k\sigma_k}{\sigma_k^2+\omega_k^2}\right)-c_k]\cos\left(\frac{k\pi x}{l}\right)+$$

$$[-\omega_k^{(s)}\left(\frac{c_k\sigma_k-d_k\omega_k}{\sigma_k^2+\omega_k^2}\right)+\sigma_k^{(s)}\left(\frac{c_k\omega_k+d_k\sigma_k}{\sigma_k^2+\omega_k^2}\right)-d_k]\sin\left(\frac{k\pi x}{l}\right),\mathbf{O}\Big)=$$

$$\sum_{k=1}^{\infty}\frac{1}{2^{k+1}}\Bigg(\frac{\left|[\sigma_k^{(s)}\left(\frac{c_k\sigma_k-d_k\omega_k}{\sigma_k^2+\omega_k^2}\right)+\omega_k^{(s)}\left(\frac{c_k\omega_k+d_k\sigma_k}{\sigma_k^2+\omega_k^2}\right)-c_k]\right|}{1+\left|[\sigma_k^{(s)}\left(\frac{c_k\sigma_k-d_k\omega_k}{\sigma_k^2+\omega_k^2}\right)+\omega_k^{(s)}\left(\frac{c_k\omega_k+d_k\sigma_k}{\sigma_k^2+\omega_k^2}\right)-c_k]\right|}+$$

$$\frac{\left|[-\omega_k^{(s)}\left(\frac{c_k\sigma_k-d_k\omega_k}{\sigma_k^2+\omega_k^2}\right)+\sigma_k^{(s)}\left(\frac{c_k\omega_k+d_k\sigma_k}{\sigma_k^2+\omega_k^2}\right)-d_k]\right|}{1+\left|[-\omega_k^{(s)}\left(\frac{c_k\sigma_k-d_k\omega_k}{\sigma_k^2+\omega_k^2}\right)+\sigma_k^{(s)}\left(\frac{c_k\omega_k+d_k\sigma_k}{\sigma_k^2+\omega_k^2}\right)-d_k]\right|}\Bigg). \tag{10.3.27}$$

Let $\varepsilon>0$ be an arbitrary positive number. Let L_ε be such a natural number that $\sum_{k=L_\varepsilon}^{\infty}\frac{1}{2^k}<\frac{\varepsilon}{2}$.

For all fixed k with $1\le k<L_\varepsilon$, let M_ε be such a natural number that

$$\frac{\left|[\sigma_k^{(s)}\left(\frac{c_k\sigma_k-d_k\omega_k}{\sigma_k^2+\omega_k^2}\right)+\omega_k^{(s)}\left(\frac{c_k\omega_k+d_k\sigma_k}{\sigma_k^2+\omega_k^2}\right)-c_k]\right|}{1+\left|[\sigma_k^{(s)}\left(\frac{c_k\sigma_k-d_k\omega_k}{\sigma_k^2+\omega_k^2}\right)+\omega_k^{(s)}\left(\frac{c_k\omega_k+d_k\sigma_k}{\sigma_k^2+\omega_k^2}\right)-c_k]\right|}<2^{k+1}\frac{\varepsilon}{4L_\varepsilon} \tag{10.3.28}$$

and

$$\frac{\left|[-\omega_k^{(s)}\left(\frac{c_k\sigma_k-d_k\omega_k}{\sigma_k^2+\omega_k^2}\right)+\sigma_k^{(s)}\left(\frac{c_k\omega_k+d_k\sigma_k}{\sigma_k^2+\omega_k^2}\right)-d_k]\right|}{1+\left|[-\omega_k^{(s)}\left(\frac{c_k\sigma_k-d_k\omega_k}{\sigma_k^2+\omega_k^2}\right)+\sigma_k^{(s)}\left(\frac{c_k\omega_k+d_k\sigma_k}{\sigma_k^2+\omega_k^2}\right)-d_k]\right|}<2^{k+1}\frac{\varepsilon}{4L_\varepsilon} \tag{10.3.29}$$

for all $s\ge M_\varepsilon$.

Then we get, that

$$\rho_T\Big(\sum_{n=0}^{2s}A_n\frac{d^n}{dx^n}\Psi_p,f\Big)<\varepsilon \tag{10.3.30}$$

for $s\ge M_\varepsilon$. Now it is not hard to show that

$$\rho_T\Big(\sum_{n=0}^{m}A_n\frac{d^n}{dx^n}\Psi_p,f\Big)<\varepsilon \tag{10.3.31}$$

for $m\ge 2M_\varepsilon$

Case 2. For $A_0=0$, we can use the scheme applied in the Case 1.

□

Remark 10.3.1. Under some general restrictions, one can use Remark 10.2.1 for an investigation of the structure of solutions (in the sense of Tychonoff) of the non-homogeneous differential equation of infinite order with real constant coefficients (10.3.6).

Remark 10.3.2. Some applications of approaches introduced in [107] can be found in [109], [110], [111].

10.4 An Invariant Measure for the Non-Homogeneous Ordinary Differential Equation of the Infinite Order with Real Constant Coefficients

Suppose that for a sequence of real numbers $(A_n)_{n\in\mathbb{N}}$ the following four conditions

(i)

$$A_0 \neq 0, \tag{10.4.1}$$

(ii) for each $k \geq 1$, infinite series σ_k and ω_k, defined by (10.2.1) and (10.2.2) respectively, are convergent,

(5.2)

(iii)

$$\sigma_k \times \omega_k \neq 0 \tag{10.4.2}$$

for each $k \geq 1$,

(iv) for each $k \geq 1$

$$0 < \sup\{|\sigma_k|, |\omega_k| : k \in N\} := C < +\infty \tag{10.4.3}$$

are fulfilled.

Let us define an operator $T_F : FD^{(0)}[-l,l] \to FD^{(0)}[-l,l]$ by the following formula

$$T_F(f) = \frac{c_0}{2A_0} + \sum_{k=1}^{\infty} \left(\frac{c_k\sigma_k - d_k\omega_k}{\sigma_k^2+\omega_k^2}\right)\cos\left(\frac{k\pi x}{l}\right) + \left(\frac{c_k\omega_k + d_k\sigma_k}{\sigma_k^2+\omega_k^2}\right)\sin\left(\frac{k\pi x}{l}\right). \tag{10.4.4}$$

We consider a set $D(F) := T_F(FD^{(0)}[-l,l]) \cap FD^{(\infty)}[-l,l]$. It is obvious that $D(F)$ is a vector subspace of $FD^{(0)}[-l,l]$.

We begin our discussion by the construction of an invariant measure on the Borel σ-algebra of subsets of $D(F)$ induced by Tychonoff topology on R^∞.

For $f \in D(F)$ with

$$f(x) = \frac{c_0}{2} + \sum_{k=1}^{\infty} c_k \cos\left(\frac{k\pi x}{l}\right) + d_k \sin\left(\frac{k\pi x}{l}\right) \ (x \in [-l,l]) \tag{10.4.5}$$

we set

$$A(f) = (\frac{c_0}{2}, 1!c_1, 1!d_1, 2!c_2, 2!d_2, \cdots). \tag{10.4.6}$$

It is obvious that $A : D(F) \to R^\infty$ is a linear operator.

We denote by $\mathcal{B}(D(F))$ a σ-algebra of subsets of $D(F)$ defined by

$$\mathcal{B}(D(F)) = \{A^{-1}(X) : X \in \mathcal{B}(R^\infty). \tag{10.4.7}$$

Lemma 10.4.1. *Let $\mu_{(1,2,2,\cdots)}$ be $(1,2,2,\cdots)$-ordinary "Lebesgue measure" on R^∞. Then a measure λ defined by*

$$\lambda(X) = \mu_{(1,2,2,\cdots)}(A(X)) \tag{10.4.8}$$

for $X \in \mathcal{B}(D(F))$ is a translation-invariant Borel measure on $D(F)$ which gets a numerical value one on the set $A^{-1}([-1/2,1/2]^\infty)$.

Proof. For $X \in \mathcal{B}(D(F))$ and $f \in D(F)$, we have

$$\lambda(X+f) = \mu_{(1,2,2,\cdots)}(A(X+f)) = \mu_{(1,2,2,\cdots)}(A(X)+A(f)) = \mu_{(1,2,2,\cdots)}(A(X)) = \lambda(X). \tag{10.4.9}$$

Note that

$$A^{-1}([-\frac{1}{2},\frac{1}{2}]^{\infty}) = \{f : f(x) = \frac{c_0}{2} + \sum_{k=1}^{\infty} c_k \cos\left(\frac{k\pi x}{l}\right) + d_k \sin\left(\frac{k\pi x}{l}\right) \ (x \in [-l,l])\&$$

$$(\frac{c_0}{2}, 1!c_1, 1!d_1, 2!c_2, 2!d_2, \cdots) \in [-\frac{1}{2},\frac{1}{2}]^{\infty} =$$

$$\{f : f(x) = \frac{c_0}{2} + \sum_{k=1}^{\infty} c_k \cos\left(\frac{k\pi x}{l}\right) + d_k \sin\left(\frac{k\pi x}{l}\right) \ (x \in [-l,l])\&$$

$$|c_0| \le 1, |c_1| \le \frac{1}{1!2}, |d_1| \le \frac{1}{1!2}, |c_2| \le \frac{1}{2!2}, |d_2| \le \frac{1}{2!2}, \cdots\}. \tag{10.4.10}$$

Let us show that

$$\lambda(A^{-1}([-\frac{1}{2},\frac{1}{2}]^{\infty})) = 1. \tag{10.4.11}$$

Indeed, for each $(\frac{c_0}{2}, c_1, d_1, c_2, d_2, \cdots)$ with

$$|c_0| \le 1, |c_1| \le \frac{1}{1!2}, |d_1| \le \frac{1}{1!2}, |c_2| \le \frac{1}{2!2}, |d_2| \le \frac{1}{2!2} \tag{10.4.12}$$

we have that $f \in FD^{(\infty)}[-l,l]$, where

$$f(x) = \frac{c_0}{2} + \sum_{k=1}^{\infty} c_k \cos\left(\frac{k\pi x}{l}\right) + d_k \sin\left(\frac{k\pi x}{l}\right) \ (x \in [-l,l]) \tag{10.4.13}$$

On the other hand, for the same f we have the following representation

$$f(x) = T_F[\frac{A_0 c_0}{2} + \sum_{k=1}^{\infty} (c_k\sigma_k + \omega_k d_k)\cos\left(\frac{k\pi x}{l}\right) + (\sigma_k d_k - \omega_k c_k)\sin\left(\frac{k\pi x}{l}\right)]. \tag{10.4.14}$$

Indeed, we have

$$T_F[\frac{A_0 c_0}{2} + \sum_{k=1}^{\infty} (c_k\sigma_k + \omega_k d_k)\cos\left(\frac{k\pi x}{l}\right) + (\sigma_k d_k - \omega_k c_k)\sin\left(\frac{k\pi x}{l}\right)] =$$

$$\frac{A_0 c_0}{2A_0} + \sum_{k=1}^{\infty} \left(\frac{(c_k\sigma_k + \omega_k d_k)\sigma_k - (\sigma_k d_k - \omega_k c_k)\omega_k}{\sigma_k^2 + \omega_k^2}\right)\cos\left(\frac{k\pi x}{l}\right) +$$

$$\left(\frac{(c_k\sigma_k + \omega_k d_k)\omega_k + (\sigma_k d_k - \omega_k c_k)\sigma_k}{\sigma_k^2 + \omega_k^2}\right)\sin\left(\frac{k\pi x}{l}\right) =$$

$$\frac{c_0}{2} + \sum_{k=1}^{\infty} c_k \cos\left(\frac{k\pi x}{l}\right) + d_k \sin\left(\frac{k\pi x}{l}\right) \ (x \in [-l,l]). \tag{10.4.15}$$

These above mentioned relations follow that

$$\Delta := \{f : f(x) = \frac{c_0}{2} + \sum_{k=1}^{\infty} c_k \cos\left(\frac{k\pi x}{l}\right) + d_k \sin\left(\frac{k\pi x}{l}\right)\ (x \in [-l, l])\&$$

$$|c_0| \le 1, |c_1| \le \frac{1}{1!2}, |d_1| \le \frac{1}{1!2}, |c_2| \le \frac{1}{2!2}, |d_2| \le \frac{1}{2!2}, \cdots\} \subset D(F). \tag{10.4.16}$$

Now let us show that $\Delta = A^{-1}([-\frac{1}{2}, \frac{1}{2}]^{\infty})$.

From the uniqueness of the representation in Fourier's series of $f \in \Delta$, we deduce that a restriction of the mapping A on Δ is injective such that

$$A(\Delta) = [-\frac{1}{2}, \frac{1}{2}]^{\infty} \tag{10.4.17}$$

and

$$\Delta = A^{-1}([-\frac{1}{2}, \frac{1}{2}]^{\infty}). \tag{10.4.18}$$

The latter relation means that $\Delta \in \mathcal{B}(D(F))$.

Finally, by the definition of the measure λ, we get

$$\lambda(\Delta) = \lambda(A^{-1}([-\frac{1}{2}, \frac{1}{2}]^{\infty}) = \mu_{(1,2,2,\cdots)}([-\frac{1}{2}, \frac{1}{2}]^{\infty}) = 1. \tag{10.4.19}$$

□

Theorem 10.4.1. *Under restrictions (i)-(iv), For $X \in \mathcal{B}(D(F))$, we have*

$$\lambda(T_F(X)) = \frac{1}{|A_0|}\lambda((X)). \tag{10.4.20}$$

Proof. First, let us show that the following equality holds $A(T_F(f)) = T_F(A(f))$ for $f \in D(F)$. Indeed, for

$$f(x) = \frac{c_0}{2} + \sum_{k=1}^{\infty} c_k \cos\left(\frac{k\pi x}{l}\right) + d_k \sin\left(\frac{k\pi x}{l}\right)\ (x \in [-l, l]) \in D(F) \tag{10.4.21}$$

we have

$$A(T_F(f)) = A(T_F(\frac{c_0}{2} + \sum_{k=1}^{\infty} c_k \cos\left(\frac{k\pi x}{l}\right) + d_k \sin\left(\frac{k\pi x}{l}\right))) = A(\frac{c_0}{2A_0} +$$

$$\sum_{k=1}^{\infty}\left(\frac{c_k\sigma_k - d_k\omega_k}{\sigma_k^2 + \omega_k^2}\right)\cos\left(\frac{k\pi x}{l}\right) +$$

$$\left(\frac{c_k\omega_k + d_k\sigma_k}{\sigma_k^2 + \omega_k^2}\right)\sin\left(\frac{k\pi x}{l}\right)) =$$

$$\frac{c_0}{2A_0} + \sum_{k=1}^{\infty}\left(\frac{k!(c_k\sigma_k - d_k\omega_k)}{\sigma_k^2 + \omega_k^2}\right)\cos\left(\frac{k\pi x}{l}\right) +$$

$$\left(\frac{k!(c_k\omega_k + d_k\sigma_k)}{\sigma_k^2 + \omega_k^2}\right)\sin\left(\frac{k\pi x}{l}\right) =$$

$$\frac{c_0}{2A_0}+\sum_{k=1}^{\infty}\Big(\frac{(k!c_k)\sigma_k-(k!d_k)\omega_k)}{\sigma_k^2+\omega_k^2}\Big)\cos\Big(\frac{k\pi x}{l}\Big)+$$
$$\Big(\frac{(k!c_k)\omega_k+(k!d_k)\sigma_k)}{\sigma_k^2+\omega_k^2}\Big)\sin\Big(\frac{k\pi x}{l}\Big)=$$
$$T_F(\frac{c_0}{2}+\sum_{k=1}^{\infty}(k!c_k)\cos\Big(\frac{k\pi x}{l}\Big)+(k!d_k)\sin\Big(\frac{k\pi x}{l}\Big)))=T_F(A(f)). \qquad (10.4.22)$$

Take into account the later relation, for $X\in\mathcal{B}(D(F))$, we have

$$\lambda(T_F(X))=\mu_{(1,2,2,\cdots)}(A(T_F(X)))=\mu_{(1,2,2,\cdots)}(T_F(A(X)))=\mu_{(1,2,2,\cdots)}(T_F(A(X)))=$$
$$\mu_{(1,2,2,\cdots)}(\{E^{-1}\times x : x\in A(X)\}), \qquad (10.4.23)$$

where E^{-1} is $(1,2,2,\cdots)$ cellular matrix defined by the formula (10.2.10).

By Theorem 3.3, we get

$$\lambda(T_F(X))=\mu_{(1,2,2,\cdots)}(\{E^{-1}\times x : x\in A(X)\})=\frac{1}{|A_0|}\mu_{(1,2,2,\cdots)}(A(X))=\frac{1}{|A_0|}\lambda(X).$$

□

Corollary 10.4.1. *Under restrictions (i)-(iv), the measurable transformation $T_F : D(F)\to D(F)$ is :*

(i) stable with respect to the Borel measure λ if $|A_0|=1$ provided that $\lambda(T_F(X))=\lambda(X)$ for $X\in\mathcal{B}(D(F))$;

(ii) is extensible with respect to the Borel measure λ if $|A_0|<1$ provided that $\lambda(T_F(X))>\lambda(X)$ for $X\in\mathcal{B}(D(F))$ with $0<\lambda(X)<+\infty$;

(iii) is pressing with respect to the Borel measure λ if $|A_0|>1$ provided that $\lambda(T_F(X))<\lambda(X)$ for $X\in\mathcal{B}(D(F))$ with $0<\lambda(X)<+\infty$.

Remark 10.4.1. Some interesting applications of partial analogs of the Lebesgue measure in the theory of differential equations can be found in [92] and [43].

Chapter 11

Description of the Behaviour of von Foerster-Lasota Phase Motions in $\mathbf{R}^{\infty}$ in Terms of Ordinary and Standard "Lebesgue Measures"

11.1 Introduction

We begin our discussion by a short history described by Antony Leon Dawidowicz and Anna Poskrobko in [29]:

In 1926 McKendrick [81] proposed the first age-dependent model of the dynamics of a population. He assumed that the state of a population in time t is described by a function $u(t,\cdot)$. The number of individuals in age from the interval $[x_1;x_2]$ equals $\int_{x_1}^{x_2} u(t,x)dx$. From this paper the equation

$$\frac{\partial u}{\partial t} + \frac{\partial u}{\partial x} = \lambda(x)u \tag{11.1.1}$$

follows, called in the literature as McKendrick equation or more often as von Foerster equation.

McKendricks model was generalized in many ways, among others by Gurtin and MacCamy [45] or by the authors [28]. This equation is part of the mathematical description of a particular population, as the population of red blood cells is (see [142]). It is the model with a feedback, because a circulatory system controls the global number of erythrocytes to some quantity optimum which can change. It happens, for example, during mountain trips or in a case of any disease of the respiratory system. In their next paper [68] the authors of [142] introduced a new model of precursor cells. There the main assumption is the fact that cells mature with different intensity. The form of the equation is the following:

$$\frac{\partial u}{\partial t} + c(x)\frac{\partial u}{\partial x} = f(x,u), \tag{11.1.2}$$

where $c : [0;1] \to \mathbf{R}$ and $f : [0;1] \times [0;+\infty) \to \mathbf{R}$ are the given functions fulfilling suitable conditions. In this model x denotes the degree of cell differentiation (maturity) and

$\int_{x_0}^{x_1} u(t_0,x)dx$ is the number of cells having at time $t = t_0$ the value x in the interval $[x_0;x_1]$. The coefficient c is the velocity of cell differentiation. Because of biological application the above equation has been the matter of interest for many mathematicians: Lasota and Pianigiani [67], Rudnicki [126], Joskot [52], and Lasota and Szarek [69]. In [29] has been considered a simpler case of the equation, that is

$$\frac{\partial u}{\partial t} + x\frac{\partial u}{\partial x} = \gamma u \ (t \geq 0, 0 \leq x \leq 1) \tag{11.1.3}$$

with the initial condition

$$u(0,x) = v(x), 0 \leq x \leq 1, \tag{11.1.4}$$

where v belongs to some normed vector space V of functions defined on $[0,1]$. Then the semidynamical system $(\overline{T}_t)_{t\geq 0}$ is given by the formula

$$\overline{T}_t(v) = \overline{u}(t,x) = e^{\gamma t} v(xe^{-t}), \ \ x \in [0,1] \tag{11.1.5}$$

where $\overline{u}$ is the unique solution of (11.1.3) and (11.1.4).

In the present chapter we consider the following partial differential equation

$$\frac{\partial u}{\partial t} = \sum_{n=0}^{m} A_n x^n \frac{\partial^n u}{\partial x^n} \tag{11.1.6}$$

with initial condition $u(0,x) = f(x)$, where $A_n \in \mathbf{R}$ for $0 \leq n \leq m$ $(m \in N)$ and $f : [0;1] \to \mathbf{R}$ is the given function fulfilling suitable conditions.

It is obvious that (11.1.6) is a certain generalization of the von Foerster-Lasota equation (11.1.3).

We propose a new approach for a solution of (11.1.6) which is essentially different from the approach developed for a solution of (11.1.3) in [29](In this context, see, for example [107]). We show that a semidynamical system defined by (11.1.6), under some restrictions, is stable in the sense of [29]. Finally, in terms of ordinary and standard "Lebesgue measures" we study the behavior of semidynamical systems in $\mathbf{R}^\infty$ defined by generalized von Foerster-Lasota operator equations.

The chapter is organized as follows.

In Section 11.2 we give some auxiliary notions and facts from the measure theory and linear algebra. In Section 11.3 we give a solution of some operator equation in R^∞ and use its structure to construct a solution of (11.1.6) in an explicit form. In Section 11.4 we present observation of some semidynamical systems in terms of ordinary and standard "Lebesgue measures" in $\mathbf{R}^\infty$.

11.2 Auxiliary Facts from Linear Algebra

In the sequel we will need the following auxiliary proposition from linear algebra.

Lemma 11.2.1. *([40], §6, Section 1) Let* $\alpha = (n_i)_{i\in\mathbb{N}} \in (\mathbb{N} \setminus \{0\})^{\mathbb{N}}$ *and, let* $A = (a_{ij})_{i,j\in N}$ *be an infinite-dimensional real-valued* α*-cellular matrix. Let us consider a linear partial autonomous differential equation*

$$\frac{\partial}{\partial t}((a_k)_{k\in N}) = A \times ((a_k)_{k\in N}) \tag{11.2.1}$$

with an initial condition

$$(a_k(0))_{k\in N} = (c_k)_{k\in N} \in \mathbf{R}^\infty. \tag{11.2.2}$$

Then the solution of the $(11.2.1)-(11.2.2)$ *is given by*

$$(a_k(t))_{k\in N} = \exp(tA) \times (c_k)_{k\in N}. \tag{11.2.3}$$

Proof. Let us present the column $(a_k(t))_{k\in N}$ in the Maclaurin's series as follows:

$$(a_k(t))_{k\in N} = \sum_{m=0}^{\infty} \frac{(a_k^{(m)}(0))_{k\in N}}{m!} t^m. \tag{11.2.4}$$

Take into account the validity of the formula

$$(a_k^{(m)}(0))_{k\in N} = (\frac{d^m a_k(t)}{dt^m}|_{t=0})_{k\in N} = A^m \times (a_k(0))_{k\in N}, \tag{11.2.5}$$

we get

$$(a_k(t))_{k\in N} = \sum_{m=0}^{\infty} \frac{(a_k^{(m)}(0)t^m)_{k\in N}}{m!} = \sum_{m=0}^{\infty} \frac{(tA)^m}{m!} \times (a_k(0))_{k\in N} = \exp(tA) \times (c_k)_{k\in N} \tag{11.2.6}$$

□

In the sequel we will need some notions characterizing the behavior of some phase motions $(\Phi_t)_{t\geq 0}$ in $\mathbf{R}^\infty$.

Let ν be any " Lebesque measure" in $\mathbf{R}^\infty$ (see, for example, [91], [98]).

Definition 11.2.1. We say that the phase motion $(\Phi_t)_{t\geq 0}$ is stable in the sense of a " Lebesque measure" ν if it preserves the measure ν, i.e.

$$(\forall t)(\forall D)(0 < t < \infty \,\&\, 0 < \nu(D) < \infty \rightarrow \nu(\Phi_t(D)) = \nu(D).$$

Definition 11.2.2. We say that the phase motion $(\Phi_t)_{t\geq 0}$ is expansible in the sense of a "Lebesque measure" ν if

$$(\forall t_1)(\forall t_2)(\forall D)(0 < t_1 < t_2 < \infty \,\&\, 0 < \nu(D) < \infty \rightarrow \nu(\Phi_{t_1}(D)) < \nu(\Phi_{t_2}(D))).$$

Definition 11.2.3. We say that the phase motion $(\Phi_t)_{t\geq 0}$ is pressing in the sense of a "Lebesque measure" ν if

$$(\forall t_1)(\forall t_2)(\forall D)(0 < t_1 < t_2 < \infty \,\&\, 0 < \nu(D) < \infty \rightarrow \nu(\Phi_{t_1}(D)) > \nu(\Phi_{t_2}(D))).$$

Definition 11.2.4. We say that the phase motion $(\Phi_t)_{t\geq 0}$ is totally expansible in the sense of a "Lebesque measure" ν if

$$(\forall t)(\forall D)(0 < t < \infty \,\&\, 0 < \nu(D) < \infty \rightarrow \nu(\Phi_t(D)) = +\infty).$$

Definition 11.2.5. We say that the phase motion $(\Phi_t)_{t\geq 0}$ is totally pressing in the sense of a "Lebesque measure" ν if

$$(\forall t)(\forall D)(0<t<\infty \,\&\, 0<\nu(D)<\infty \rightarrow \nu(\Phi_t(D))=0).$$

Definition 11.2.6. Let (V,ρ) be a topological vector space and $(T_t)_{t\geq 0}$ be any semidynamical system. Following [27], $(T_t)_{t\geq 0}$ is called strongly stable in V iff for every $v\in V$ the following condition

$$\lim_{t\rightarrow\infty}\rho(T_t(v),\mathbf{0})=0$$

holds, where $\mathbf{0}$ denotes the zero of V.

In context to Definition 11.2.6, the following result is of some interest.

Lemma 11.2.2. *([27], Theorem 3, p.809) Let $(\overline{T}_t)_{t\geq 0}$ be the semidynamical system defined by (1.3)–(1.4) and acting on the Banach space $\widetilde{L^p}$ (see, [27], Proposition 3, p.806). Then $(\overline{T}_t)_{t\geq 0}$ is strongly stable in $\widetilde{L^p}$ for $\gamma\leq 0$.*

11.3 Maclaurin's Differential Operators in $\mathbf{R}^\infty$

Definition 11.1. Maclaurin's differential operator $(\mathcal{M})\frac{\partial}{\partial x}$ in $\mathbf{R}^\infty$ is defined as follows :

$$(\mathcal{M})\frac{\partial}{\partial x}(\begin{pmatrix} a_0\\ a_1\\ a_2\\ \vdots\end{pmatrix})=\begin{pmatrix} 0 & 1 & 0 & 0 & \dots\\ 0 & 0 & 2 & 0 & \ddots\\ 0 & 0 & 0 & 3 & \ddots\\ 0 & 0 & 0 & 0 & \ddots\\ \vdots & \vdots & \vdots & \vdots & \ddots\end{pmatrix}\times\begin{pmatrix} a_0\\ a_1\\ a_2\\ \vdots\end{pmatrix}. \tag{11.3.1}$$

Definition 11.2. An operator $(\mathcal{M})x\frac{\partial}{\partial x}$ in $\mathbf{R}^\infty$ is defined as follows:

$$(\mathcal{M})x\frac{\partial}{\partial x}((a_k)_{k\in N})=\begin{pmatrix} 0 & 0 & 0 & 0 & \dots\\ 0 & 1 & 0 & 0 & \ddots\\ 0 & 0 & 2 & 0 & \ddots\\ 0 & 0 & 0 & 3 & \ddots\\ \vdots & \vdots & \vdots & \ddots & \ddots\end{pmatrix}\times\begin{pmatrix} a_0\\ a_1\\ a_2\\ \vdots\end{pmatrix}. \tag{11.3.2}$$

Definition 11.3. An operator $(\mathcal{M})\S^{\in}\frac{\partial^{\in}}{\partial\S^{\in}}$ in $\mathbf{R}^\infty$ is defined as follows:

$$(\mathcal{M})\S^{\in}\frac{\partial^{\in}}{\partial\S^{\in}}((a_k)_{k\in N})=\begin{pmatrix} 0 & 0 & 0 & 0 & \dots\\ 0 & 0 & 0 & 0 & \ddots\\ 0 & 0 & 1\times 2 & 0 & \ddots\\ 0 & 0 & 0 & 2\times 3 & \ddots\\ \vdots & \vdots & \vdots & \ddots & \ddots\end{pmatrix}\times\begin{pmatrix} a_0\\ a_1\\ a_2\\ \vdots\end{pmatrix}. \tag{11.3.3}$$

Definition 11.4. Formally, we set that factorial of all negative integer number is equal to $+\infty$. Then an operator $(\mathcal{M})\S\backslash\frac{\partial\backslash}{\partial\S\backslash}$ in $\mathbf{R}^{\infty}$ is defined as follows:

$$(\mathcal{M})\S\backslash\frac{\partial\backslash}{\partial\S\backslash}((a_k)_{k\in N}) = \begin{pmatrix} \frac{0!}{(0-n)!} & 0 & 0 & 0 & \cdots \\ 0 & \frac{1!}{(1-n)!} & 0 & 0 & \ddots \\ 0 & 0 & \frac{2!}{(2-n)!} & 0 & \ddots \\ 0 & 0 & 0 & \frac{3!}{(3-n)!} & \ddots \\ \vdots & \vdots & \vdots & \ddots & \ddots \end{pmatrix} \times \begin{pmatrix} a_0 \\ a_1 \\ a_2 \\ \vdots \end{pmatrix}. \quad (11.3.4)$$

We set

$$MS[-1,1] = \{f : f \in \mathbf{R}^{[-1,1]} \ \& \ f(x) = \sum_{k=0}^{\infty} a_k x^k \ \& \ [-1,1] \subset (-R_f, R_f)\},$$

where R_f denotes a radius of convergence of f.

It is obvious that every element of $MS[-1,1]$ is k-times differentiable term by term for $k \in N$.

Theorem 11.1. *Let G_M be an embedding of the $MS[-1,1]$ in to $\mathbf{R}^{\infty}$ defined by:*

$$MS[-1,1] \ni f(x) = \sum_{k\in N} a_k x^k \to G_M(f) = (a_k)_{k\in N} \in \mathbf{R}^{\infty}. \quad (11.3.5)$$

Then, for $f \in MS[-1,1]$, the following equalities

$$(G_M^{-1} \circ (\mathcal{M})\frac{\partial}{\partial x} \circ G_M)(f) = \frac{\partial}{\partial x}(f) \quad (11.3.6)$$

and

$$(G_M^{-1} \circ (\mathcal{M})x\frac{\partial}{\partial x} \circ G_M)(f) = x\frac{\partial}{\partial x}(f) \quad (11.3.7)$$

hold, where $\frac{\partial}{\partial x}$ and $x\frac{\partial}{\partial x}$ denote usual differential operators on $MS[-1,1]$.

Proof. We have

$$(G_M^{-1} \circ (\mathcal{M})\frac{\partial}{\partial x} \circ G_M)(f) =$$

$$G_M^{-1}((\mathcal{M})\frac{\partial}{\partial x} \times G_M(f)) =$$

$$G_M^{-1}(\begin{pmatrix} 0 & 1 & 0 & 0 & \cdots \\ 0 & 0 & 2 & 0 & \ddots \\ 0 & 0 & 0 & 3 & \ddots \\ 0 & 0 & 0 & 0 & \ddots \\ \vdots & \vdots & \vdots & \vdots & \ddots \end{pmatrix} \times \begin{pmatrix} a_0 \\ a_1 \\ a_2 \\ \vdots \end{pmatrix}) =$$

$$G_M^{-1}(\begin{pmatrix} 1a_1 \\ 2a_2 \\ 3a_3 \\ \vdots \end{pmatrix}) = \sum_{k \in N} (k+1)a_{k+1}x^k = \frac{\partial}{\partial x}(\sum_{k \in N} a_k x^k).$$

Similarly, we have

$$(G_M^{-1} \circ (\mathcal{M})x\frac{\partial}{\partial x} \circ G_M)(f) =$$

$$G_M^{-1}((\mathcal{M})x\frac{\partial}{\partial x} \times G_M(f)) =$$

$$G_M^{-1}(\begin{pmatrix} 0 & 0 & 0 & 0 & \dots \\ 0 & 1 & 0 & 0 & \ddots \\ 0 & 0 & 2 & 0 & \ddots \\ 0 & 0 & 0 & 3 & \ddots \\ \vdots & \vdots & \vdots & \vdots & \ddots \end{pmatrix} \times \begin{pmatrix} a_0 \\ a_1 \\ a_2 \\ \vdots \end{pmatrix}) =$$

$$G_M^{-1}(\begin{pmatrix} 0a_0 \\ 1a_1 \\ 2a_2 \\ \vdots \end{pmatrix}) = \sum_{k \in N} k a_k x^k = x\frac{\partial}{\partial x}(\sum_{k \in N} a_k x^k).$$

□

We have the following consequence of Theorem 11.3.1.

Theorem 11.3.1. *Let G_M be an embedding of the $MS[-1,1]$ in to $\mathbf{R}^\infty$ defined by (11.3.5). Let $(A_n)_{0 \le n \le m} (m \in N)$ be a sequence of real numbers. Then the following equality*

$$(G_M^{-1} \circ (\sum_{n=0}^{m} A_n(\mathcal{M})(x^n \frac{\partial^n}{\partial x^n})) \circ G_M)(f) = \sum_{n=0}^{m} A_n x^n \frac{\partial^n}{\partial x^n}(f) \tag{11.3.8}$$

holds.

Proof. We have

$$(G_M^{-1} \circ (\sum_{n=0}^{m} A_n(\mathcal{M})(x^n \frac{\partial^n}{\partial x^n})) \circ G_M)(f) =$$

$$G_M^{-1}(\sum_{n=0}^{m} A_n \begin{pmatrix} \frac{0!}{(0-n)!} & 0 & 0 & 0 & \dots \\ 0 & \frac{1!}{(1-n)!} & 0 & 0 & \ddots \\ 0 & 0 & \frac{2!}{(2-n)!} & 0 & \ddots \\ 0 & 0 & 0 & \frac{3!}{(3-n)!} & \ddots \\ \vdots & \vdots & \vdots & \ddots & \ddots \end{pmatrix} \times G_M(f)) =$$

$$G_M^{-1}\left(\begin{pmatrix} \sum_{n=0}^m A_n \frac{0!}{(0-n)!} & 0 & 0 & 0 & \cdots \\ 0 & \sum_{n=0}^m A_n \frac{1!}{(1-n)!} & 0 & 0 & \ddots \\ 0 & 0 & \sum_{n=0}^m A_n \frac{2!}{(2-n)!} & 0 & \ddots \\ 0 & 0 & 0 & \sum_{n=0}^m A_n \frac{3!}{(3-n)!} & \ddots \\ \vdots & \vdots & \vdots & \ddots & \ddots \end{pmatrix} \times \begin{pmatrix} a_0 \\ a_1 \\ a_2 \\ \vdots \end{pmatrix}\right) =$$

$$G_M^{-1}\left(\begin{pmatrix} a_0 \sum_{n=0}^m A_n \frac{0!}{(0-n)!} \\ a_1 \sum_{n=0}^m A_n \frac{1!}{(1-n)!} \\ a_2 \sum_{n=0}^m A_n \frac{2!}{(2-n)!} \\ \vdots \end{pmatrix}\right) = \sum_{i\in N} (a_i \sum_{n=0}^m A_n \frac{i!}{(i-n)!})x^i = \sum_{n=0}^m A_n \sum_{i\in N} a_i \frac{i!}{(i-n)!} x^i =$$

$$\sum_{n=0}^m A_n x^n \sum_{i\in N} a_i \frac{i!}{(i-n)!} x^{i-n} = \sum_{n=0}^m A_n x^n \frac{\partial^n}{\partial x^n} (\sum_{i\in N} a_i x^i) = \sum_{n=0}^m A_n x^n \frac{\partial^n}{\partial x^n} (f(x)).$$

□

Theorem 11.2. *Let* $(A_n)_{0\le n\le m} (m \in N)$ *be a sequence of real numbers. Let consider a nonhomogeneous PDE*

$$\frac{\partial}{\partial t}((a_k)_{k\in N}) = (\sum_{n=0}^m A_n((\mathcal{M})(x^n \frac{\partial^n}{\partial x^n})((a_k)_{k\in N}) + (f_k(t))_{k\in N} \qquad (11.3.9)$$

with initial condition

$$(a_k(0))_{k\in N} = (C_k)_{k\in N} \qquad (11.3.10),$$

where

(i) $(C_k)_{k\in N} \in \mathbf{R}^\infty$;

(ii) $(f_k(t))_{k\in N}$ *is the sequence of continuous functions on* $\mathbf{R}$.

Then

$$(a_k(t))_{k\in N} = \Big(e^{t\sum_{n=0}^m A_n \frac{k!}{(k-n)!}} C_k + \int_0^t e^{(t-\tau)\sum_{n=0}^m A_n \frac{k!}{(k-n)!}} f_k(\tau) d\tau\Big)_{k\in N}.$$

Proof. Note that

$$(a_k(t))_{k\in N} = e^{t\left(\left(\sum_{n=0}^m A_n((\mathcal{M})\left(x^n \frac{\partial^n}{\partial x^n}\right)\right)\right)} \times (C_k)_{k\in N} + \int_0^t e^{(t-\tau)\left(\left(\sum_{n=0}^m A_n((\mathcal{M})\left(x^n \frac{\partial^n}{\partial x^n}\right)\right)\right)} f(\tau) d\tau, \qquad (11.3.11)$$

where

$$e^{t\left(\sum_{n=0}^m A_n (\mathcal{M}) x^n \frac{\partial^n}{\partial x^n}\right)} = \begin{pmatrix} e^{t\sum_{n=0}^m A_n \frac{0!}{(0-n)!}} & 0 & 0 & 0 & \cdots \\ 0 & e^{t\sum_{n=0}^m A_n \frac{1!}{(1-n)!}} & 0 & 0 & \ddots \\ 0 & 0 & e^{t\sum_{n=0}^m A_n \frac{2!}{(2-n)!}} & 0 & \ddots \\ 0 & 0 & 0 & e^{t\sum_{n=0}^m A_n \frac{3!}{(3-n)!}} & \ddots \\ \vdots & \vdots & \vdots & \ddots & \ddots \end{pmatrix}.$$

and

$$\int_0^t e^{(t-\tau)\left(\sum_{n=0}^m A_n(\mathcal{M})x^n \frac{\partial^n}{\partial x^n}\right)} f(\tau)d\tau = \begin{pmatrix} \int_0^t e^{(t-\tau)\sum_{n=0}^m A_n \frac{0!}{(0-n)!}} f_0(\tau)d\tau \\ \int_0^t e^{(t-\tau)\sum_{n=0}^m A_n \frac{1!}{(1-n)!}} f_1(\tau)d\tau \\ \int_0^t e^{(t-\tau)\sum_{n=0}^m A_n \frac{2!}{(2-n)!}} f_2(\tau)d\tau \\ \int_0^t e^{(t-\tau)\sum_{n=0}^m A_n \frac{3!}{(3-n)!}} f_2(\tau)d\tau \\ \vdots \end{pmatrix}.$$

Finally we get that

$$(a_k(t))_{k\in N} = \left(e^{t\sum_{n=0}^m A_n \frac{k!}{(k-n)!}} C_k + \int_0^t e^{(t-\tau)\sum_{n=0}^m A_n \frac{k!}{(k-n)!}} f_k(\tau)d\tau\right)_{k\in N}.$$

□

Theorem 11.3.2. *Let consider a nonhomogeneous PDE*

$$\frac{\partial}{\partial t}(\Psi(t,x)) = \sum_{n=0}^m A_n x^n \frac{\partial^n}{\partial x^n}(\Psi(t,x)) + \sum_{k\in N} f_k(t)x^k \tag{11.3.12}$$

with initial condition

$$\Psi(t,x) = \sum_{k\in N} C_k x^k, \tag{11.3.13}$$

where

(i) $A_m < 0$;

(ii) $(C_k)_{k\in N} \in G_M(MS[-1,1])$;

(iii) $f(t) = (f_k(t))_{k\in N}$ *is the sequence of continuous functions on* $\mathbf{R}$ *such that*

$$\int_0^t e^{(t-\tau)\sum_{n=0}^m A_n(\mathcal{M})x^n \frac{\partial^k}{\partial x^k}} f(\tau)d\tau \in G_M(MS[-1,1]) \tag{11.3.14}$$

for every $t \geq 0$.

Then the solution of (11.3.9)-(11.3.10) has the following form

$$\Psi(t,x) = \sum_{k\in N} \left(e^{t\sum_{n=0}^m A_n \frac{k!}{(k-n)!}} C_k + \int_0^t e^{(t-\tau)\sum_{n=0}^m A_n \frac{k!}{(k-n)!}} f_k(\tau)d\tau\right)x^k. \tag{11.3.15}$$

Proof. First, we have to show that $\Psi(t,x) \in MS[-1,1]$ for each $t \geq 0$. Without loss of generality, we can assume that $t > 0$. Following (3.11.14), we have

$$\sum_{k\in N} \left(\int_0^t e^{(t-\tau)\sum_{n=0}^m A_n \frac{k!}{(k-n)!}} f_k(\tau)d\tau\right)x^k \in MS[-1,1].$$

Since $MS[-1,1]$ is the vector space, we have only to show that

$$\sum_{k\in N} \left(e^{t\sum_{n=0}^m A_n \frac{k!}{(k-n)!}} C_k\right)x^k \in MS[-1,1].$$

Note that

$$\lim_{k\to\infty} t\sum_{n=0}^{m} A_n \frac{k!}{(k-n)!} = \lim_{k\to\infty} t(A_0 + kA_1 + k(k-1)A_2 + \cdots + k(k-1)\cdots(k-m+1)A_m) =$$

$$= \lim_{k\to\infty} tk(k-1)\cdots(k-m+1)\Big(\frac{A_0}{k(k-1)\cdots(k-m+1)} + \frac{kA_1}{k(k-1)\cdots(k-m+1)} +$$

$$\frac{k(k-1)A_2}{k(k-1)\cdots(k-m+1)} + \cdots + A_m\Big) = \lim_{k\to\infty} tk(k-1)\cdots(k-m+1)A_m = -\infty.$$

This means that there exists $k_0(t) \in N$ such that $t\sum_{n=0}^{m} A_n \frac{k!}{(k-n)!} < 0$ for each $k \geq k_0(t)$. We set $C = \max_{k<k_0(t)}\{e^{t\sum_{n=0}^{m} A_n \frac{k!}{(k-n)!}}\} + 1$. Then we get

$$\left|e^{t\sum_{n=0}^{m} A_n \frac{k!}{(k-n)!}} C_k\right| \leq C|C_k|$$

for all $k \in N$. The latter relation means that

$$\sum_{k\in N} (e^{t\sum_{n=0}^{m} A_n \frac{k!}{(k-n)!}} C_k) x^k \in MS[-1,1].$$

Note that $G_M(\Psi)$, as a vector-function of a parameter t, is differentiable w.r.t. same parameter.

Following Theorem 11.3.3, $G_M(\Psi)$ is a solution of (11.3.9), i.e.,

$$\frac{\partial}{\partial t}(G_M(\Psi)) = \Big(\sum_{n=0}^{m} A_n((\mathcal{M})\big(x^n \frac{\partial^n}{\partial x^n}\big)(G_M(\Psi)) + (f_k(t))_{k\in N}$$

with initial condition

$$G_M(\Psi)(0) = (C_k)_{k\in N}.$$

This means that

$$\Big(\sum_{n=0}^{m} A_n((\mathcal{M})\big(x^n \frac{\partial^n}{\partial x^n}\big)(G_M(\Psi)) = \frac{\partial}{\partial t}(G_M(\Psi)) - (f_k(t))_{k\in N}.$$

By Theorem 11.3.2, we have

$$G_M^{-1} \circ \Big(\sum_{n=0}^{m} A_n((\mathcal{M})\big(x^n \frac{\partial^n}{\partial x^n}\big) \circ G_M(\Psi) = \sum_{n=0}^{m} A_n x^n \frac{\partial^n}{\partial x^n}\Psi.$$

But

$$G_M^{-1} \circ \Big(\frac{\partial}{\partial t}(G_M(\Psi)) - (f_k(t))_{k\in N}\Big) = \frac{\partial}{\partial t}\Psi - \sum_{k\in N} f_k(t)x^k.$$

Thus we obtain

$$\sum_{n=0}^{m} A_n x^n \frac{\partial^n}{\partial x^n}\Psi = \frac{\partial}{\partial t}\Psi - \sum_{k\in N} f_k(t)x^k.$$

The latter relation means that Ψ is a solution of (11.3.12).

To check the validity of (11.3.13) is obvious.

□

We have the following consequences of Theorem 11.3.4.

Corollary 11.3.1. *Let us consider PDE*

$$\frac{\partial}{\partial t}(\Psi(t,x)) = \sum_{n=0}^{m} A_n x^n \frac{\partial^n}{\partial x^n}(\Psi(t,x)) \qquad (11.3.16)$$

with initial condition

$$\Psi(t,x) = \sum_{k\in N} C_k x^k \qquad (11.3.17),$$

where

(i) $A_m < 0$;

(ii) $(C_k)_{k\in N} \in G_M(MS[-1,1])$.

Then the solution of (11.3.16)-(11.3.17) has the following form

$$\Psi(t,x) = \sum_{k\in N} e^{t\sum_{n=0}^{m} A_n \frac{k!}{(k-n)!}} C_k x^k. \qquad (11.3.18)$$

The following assertion is a simple consequence of Corollary 11.3.1

Corollary 11.3.2. *Let consider the von Foerster-Lasota differential equation*

$$\frac{\partial \Psi(t,x)}{\partial t} = -x\frac{\partial \Psi(t,x)}{\partial x} + \gamma\Psi(t,x). \qquad (11.3.19)$$

with initial condition

$$\Psi(0,x) = \sum_{k\in N} C_k x^k \qquad (11.3.20),$$

where $(C_k)_{k\in N} \in G_M(MS[-1,1])$.

Then the solution of the von Foerster-Lasota differential equation (11.3.19)-(11.3.20) can be given in an explicit form as follows

$$\Psi(t,x) = \sum_{k\in N} e^{t(\gamma\frac{k!}{(k-0)!} - \frac{k!}{(k-1)!})} C_k x^k = C_0 e^{\gamma t} + \sum_{k=1}^{\infty} e^{t(\gamma-k)} C_k x^k. \qquad (11.3.21)$$

Remark 11.3.1. Following [27](cf. p. 805), there exists a connection between equation (11.1.1) and (11.1.3). If ψ and Ψ are the solutions of the equations (11.1.1) and (11.1.3) with same initial conditions (11.1.4), respectively, then we have the equality

$$\psi(t,x) = e^{\int_0^x \frac{\lambda(0)-\lambda(s)}{s} ds}\Psi(t,x).$$

Now we claim that the solution ψ of (11.1.1) with initial condition

$$\psi(0,x) = \sum_{k\in N} C_k x^k \in MS[-1,1] \qquad (11.3.22)$$

can be given in an explicit form as follows

$$\psi(t,x) = e^{\int_0^x \frac{\lambda(0)-\lambda(s)}{s} ds}\Big(C_0 e^{\gamma t} + \sum_{k=1}^{\infty} e^{t(\gamma-k)} C_k x^k\Big). \qquad (11.3.23)$$

Remark 11.3.2. Note that the von Foerster-Lasota differential equation (11.3.19) is a partial case of the differential equation (11.3.16) and it can be obtained from (11.3.16) when $n = 2, A_0 = \gamma$ and $A_1 = -1$. Since in this model x denotes the degree of cell differentiation (maturity) and $\int_{x_0}^{x_1} \Psi(t_0;x)dx$ is the number of cells having at time $t = t_0$ the value x in the interval $[x_0;x_1]$, we get

$$\int_{x_0}^{x_1} \Psi(x;t_0)dx = \int_{x_0}^{x_1} (C_0 e^{\gamma t_0} + \sum_{k=1}^{\infty} e^{t_0(\gamma-k)} C_k x^k)dx =$$

$$C_0 e^{\gamma t_0}(x_1 - x_0) + \sum_{k=1}^{\infty} e^{t_0(\gamma-k)} C_k \frac{x_1^{k+1} - x_0^{k+1}}{k+1}. \tag{11.3.24}$$

By using (11.3.18), we can define semidynamical system $(\overline{T}_t)_{t \geq 0}$ induced by (11.3.16)–(11.3.17) as follows

$$(\forall v)(v \in MS[-1,1] \to \overline{T}_t(v) = \sum_{k \in N} e^{t \sum_{n=0}^{m} A_n \frac{k!}{(k-n)!}} \frac{v^{(k)}(0)}{k!} x^k). \tag{11.3.25}$$

Since G_M is a linear embedding of the vector space $MS[-1,1]$ in to Tychonoff topological metric space $\mathbf{R}^\infty$, we can define a metric ρ on $MS[-1,1]$ as follows

$$\rho(v_1, v_2) = \rho_T(G_M(v_1), G_M(v_1)),$$

where ρ_T is a Tychonoff metric on $\mathbf{R}^\infty$ defined by

$$(\forall (x_i)_{i \in N}, (y_i)_{i \in N} \in \mathbf{R}^\infty \to \rho_T((x_i)_{i \in N}, (y_i)_{i \in N}) = \sum_{i \in N} \frac{|x_i - y_i|}{2^{i+1}(1 + |x_i - y_i|)}). \tag{11.3.26}$$

Theorem 11.3.3. *Let $(\overline{T}_t)_{t \geq 0}$ be the semidynamical system defined by (3.16) and acting on metric space $(MS[-1,1], \rho)$. If $A_m < 0, A_0 < 0$ and $A_i \leq 0$ for $1 \leq i < m$, then $(\overline{T}_t)_{t \geq 0}$ is strongly stable in $(MS[-1,1], \rho)$.*

Proof. Let $v \in MS[-1,1]$. By the definition of $MS[-1,1]$, we have that $\sum_{k \in N} \frac{v^{(k)}(0)}{k!} 1^k$ is absolutely convergent, which implies that the $\lim_{k \to \infty} |\frac{v^{(k)}(0)}{k!}| = 0$. The latter relation implies that there is a positive number C such that $|\frac{v^{(k)}(0)}{k!}| \leq C$ for $k \in N$.

Take into account this fact we get

$$\rho(\overline{T}_t(v), \mathbf{0}) = \sum_{k=0}^{\infty} \frac{|e^{t \sum_{n=0}^{m} A_n \frac{k!}{(k-n)!}} \frac{v^{(k)}(0)}{k!}|}{2^{k+1}(1 + |e^{t \sum_{n=0}^{m} A_n \frac{k!}{(k-n)!}} \frac{v^{(k)}(0)}{k!}|)} \leq$$

$$\sum_{k=0}^{\infty} \frac{e^{tA_0} C}{2^{k+1}} \leq e^{tA_0} C \sum_{k=0}^{\infty} \frac{1}{2^{k+1}} \leq e^{tA_0} C.$$

The latter relation follows that

$$\lim_{t \to \infty} \rho(\overline{T}_t(v), \mathbf{0}) = 0.$$

This ends the proof of Theorem 11.3.5. □

By using (11.3.21), we can define semidynamical system $(\overline{T}_t)_{t\geq 0}$ induced by (11.3.19)–(11.3.20) as follows

$$(\forall v)(v \in MS[-1,1] \rightarrow \overline{T}_t(v) = v(0)e^{\gamma t} + \sum_{k=1}^{\infty} e^{t(\gamma-k)} \frac{v^{(k)}(0)}{k!} x^k). \quad (11.3.27)$$

The following consequence of Theorem 11.3.5 is similar to the result of Lemma 11.2.7.

Theorem 11.3.4. *Let $(\overline{T}_t)_{t\geq 0}$ be the semidynamical system defined by (11.3.27) and acting on metric space $(MS[-1,1],\rho)$. Then $(\overline{T}_t)_{t\geq 0}$ is strongly stable in $(MS[-1,1],\rho)$ for $\gamma<0$.*

11.4 Observations

By Theorems 3.3 and 11.3.3, one can get the following observations.

Observation 11.4.1 *The semidynamical system defined by* (11.3.9) *in* $\mathbf{R}^\infty$ *is:*

(i) stable in the sense of $(1,1,\cdots)$*-ordinary "Lebesgue measure" iff* $A_n = 0$ *for* $0 \leq n \leq m$.

(ii) totally pressing in the sense of $(1,1,\cdots)$*-ordinary "Lebesgue measure" iff* $\sum_{k=0}^{\infty}\sum_{n=0}^{m} A_n \frac{k!}{(k-n)!} = -\infty$.

(iii) totally expansible in the sense of $(1,1,\cdots)$*-ordinary "Lebesgue measure" iff* $\sum_{k=0}^{\infty}\sum_{n=0}^{m} A_n \frac{k!}{(k-n)!} = +\infty$.

Observation 11.4.2 *Let consider the von Foerster-Lasota equation in* $\mathbf{R}^\infty$ *defined by*

$$\frac{\partial}{\partial t}((a_k)_{k\in N}) = -(\mathcal{M})(x\frac{\partial}{\partial x})((a_k)_{k\in N}) + \gamma(a_k)_{k\in N} \quad (11.4.1)$$

with initial condition

$$(a_k(0))_{k\in N} = (C_k)_{k\in N} \in \mathbf{R}^\infty. \quad (11.4.2)$$

Then the semidynamical system defined by (11.4.1) in $\mathbf{R}^\infty$ *is:*

(i) totally pressing in the sense of $(1,1,\cdots)$*-ordinary "Lebesgue measure" for every* $\gamma \in \mathbf{R}$.

(ii) totally pressing in the sense of $(1,1,\cdots)$*-standard "Lebesgue measure" for every* $\gamma \in \mathbf{R}$.

Chapter 12

On Uniformly Distributed Sequences on $[-\frac{1}{2}, \frac{1}{2}]$

12.1 Introduction

A first example of an infinite analog of Lebesgue measure, defined on the infinite-dimensional topological vector space $\mathbf{R}^\infty$ equipped with Tychonoff topology, has been constructed in [145][1]. The author has considered the product of Lebesgue measure and uniform probability measures on $\mathbf{R}^n \times [-\frac{1}{2}, \frac{1}{2}]^\infty$, and in the limit of $n \to \infty$ has obtained the requested measure μ on $\mathbf{R}^\infty$. Here (see Proposition 3.1, p. 701) has been shown that μ is $\mathbf{R}_0^\infty$-invariant, where $\mathbf{R}_0^\infty$ is a group of all eventually zero sequences. The author has shown that a special modification of μ gives an ℓ_2- invariant measure, and other modifications can give invariant measures with respect to a larger subspace of $\mathbf{R}^\infty$. Such measures have been adopted as infinite analogs of Lebesgue measures in $\mathbf{R}^\infty$.

Without having slightest information about existence of this excellent work of Y. Yamasaki [145], much of main results concerning this measure have been anew discovered by many authors in various manuscripts. For example, in [56], an essentially different technique (an inductive limit of measures) has been used to construct an example of such a σ-finite Borel measure in the Hilbert space ℓ_2 which is invariant under everywhere dense vector subspace of ℓ_2 and gets a numerical value one on the infinite-dimensional Hilbert cube $\prod_{k=1}^\infty [0, \frac{1}{k}]$. In [90](see, Theorem 1.4, p. 356) has been obtained a description of the group of all admissible translations(in the sense of invariance) for such measures, but this result was already 22 years old (see, [145], Proposition 5.1, p.704 and Formula (6.3), p. 707). In [92](see, Theorem 15.3.1, p. 208), Baker generator of shy sets in $\mathbf{R}^\infty$ has been described in terms of Kharazishvili's quasi-generators of shy sets in $\mathbf{R}^\infty$ which exactly coincides with Yamasaki's measure. Some another interesting applications of Kharazishvili's quasi-generators can be found in [96]. For example, Mankiewicz's generator of shy sets in $\mathbf{R}^\infty$ also has been described in terms of such measures (see, [96], Theorem 6.2., p. 253).

The purpose of the present chapter is to study a structure of uniformly distributed sequences on $[-\frac{1}{2}, \frac{1}{2}]$ in terms of the measure μ. More precisely, we are going to show that μ-almost every element of $\mathbf{R}^\infty$ is uniformly distributed on $[-\frac{1}{2}, \frac{1}{2}]$.

[1] Similar constructions of various invariant measures were considered by Moore [84].

The rest of the present chapter is the following.

Some auxiliary notions and facts are considered in Section 12.2. In Section 12.3 we give the proof of our main result.

12.2 Some Auxiliary Notions and Facts

Definition 12.2.1. Following [64], a sequence $s_1, s_2, s_3, \cdots$ of real numbers from an interval $[a,b]$ is said to be equidistributed, or uniformly distributed in that interval if the proportion of terms falling in a subinterval $[c,d]$ is proportional to the length of that interval.

We consider a certain extension of this notion as follows.

Definition 12.2.2. A sequence of real numbers $(x_k)_{k\in\mathbb{N}} \in \mathbf{R}^\infty$ is called uniformly distributed on $[a,b]$ if for each c,d with $a \le c < d \le b$ we have

$$\lim_{n\to\infty} \frac{\#(\{x_k : 1 \le k \le n\} \cap [c,d])}{n} = d - c, \tag{12.2.1}$$

where $\#(\cdot)$ denotes the counter measure of a set.

Lemma 12.2.1. *Let $\mathcal{B}[-\frac{1}{2},\frac{1}{2}]$ be a set of all bounded Borel measurable functions on $[-\frac{1}{2},\frac{1}{2}]$ and let $\lambda_\infty := \lambda^\infty$, where λ denotes a standard linear Lebesgue measure on $[-\frac{1}{2},\frac{1}{2}]$. Then for $f \in \mathcal{B}[-\frac{1}{2},\frac{1}{2}]$, we have*

$$\lambda_\infty(\{(x_k)_{k\in\mathbb{N}} : (x_k)_{k\in\mathbb{N}} \in [0,1]^\infty \ \& \ \lim_{N\to\infty} \frac{1}{N}\sum_{n=1}^{N} f(x_n) = \int_{-\frac{1}{2}}^{\frac{1}{2}} f(x)dx\}) = 1. \tag{12.2.2}$$

Proof. Note that it is sufficient to prove Lemma 12.2.1 for $f \in \mathcal{B}[-\frac{1}{2},\frac{1}{2}]$ with

$$\int_{-\frac{1}{2}}^{\frac{1}{2}} f(x)dx = 0. \tag{12.2.3}$$

Indeed, if $\int_{-\frac{1}{2}}^{\frac{1}{2}} f(x)dx = c \neq 0$, then we consider a function g defined by $g(x) = f(x) - c$ for $x \in [-\frac{1}{2},\frac{1}{2}]$. The validity of Lemma 12.2.1 for g implies

$$\lambda_\infty(\{(x_k)_{k\in\mathbb{N}} : (x_k)_{k\in\mathbb{N}} \in [0,1]^\infty \ \& \ \lim_{N\to\infty} \frac{1}{N}\sum_{n=1}^{N} (f(x_n) - c) = \int_{-\frac{1}{2}}^{\frac{1}{2}} (f(x) - c)dx\}) = 1, \tag{12.2.4}$$

which is equivalent to the condition

$$\lambda_\infty(\{(x_k)_{k\in\mathbb{N}} : (x_k)_{k\in\mathbb{N}} \in [0,1]^\infty \ \& \ \lim_{N\to\infty} \frac{1}{N}\sum_{n=1}^{N} f(x_n) = \int_{-\frac{1}{2}}^{\frac{1}{2}} f(x)dx\}) = 1. \tag{12.2.5}$$

For $N \ge 1$, we define F_N on $[-\frac{1}{2},\frac{1}{2}]^\infty$ as follows:

$$F_N(x_1, \cdots, x_n, \cdots) = \frac{1}{N}\sum_{n=1}^{N} f(x_n). \tag{12.2.6}$$

Then we get

$$\int_{[-\frac{1}{2},\frac{1}{2}]^\infty} F_N^2(x_1,\cdots)d\lambda_\infty(x_1,\cdots) = \frac{1}{N^2}\sum_{n=1}^{N}\int_{[-\frac{1}{2},\frac{1}{2}]^\infty} f^2(x_n)d\lambda_\infty(x_1,\cdots)+$$

$$\frac{2}{N^2}\sum_{1\le i<j\le N}\int_{[-\frac{1}{2},\frac{1}{2}]^\infty} f(x_i)f(x_j)d\lambda_\infty(x_1,\cdots) =$$

$$= \frac{1}{N^2}\sum_{n=1}^{N}\int_{-\frac{1}{2}}^{\frac{1}{2}} f^2(x)dx + \frac{2}{N}\sum_{1\le i<j\le N}\Big(\int_{[-\frac{1}{2},\frac{1}{2}]} f(x)d(x)\Big)^2 =$$

$$\frac{1}{N^2}\sum_{n=1}^{N}\int_{-\frac{1}{2}}^{\frac{1}{2}} f^2(x)dx. \tag{12.2.7}$$

Hence

$$\sum_{m=1}^{\infty}\int_{[-\frac{1}{2},\frac{1}{2}]^\infty} F_{m^2}^2(x_1,\cdots)d\lambda_\infty(x_1,\cdots) = \int_{-\frac{1}{2}}^{\frac{1}{2}} f^2(x)dx\sum_{m=1}^{\infty}\frac{1}{m^2} < \infty. \tag{12.2.8}$$

By Levi well known theorem, we deduce that the sequence $(F_{m^2})_{m\ge 1}$ tends to zero as well m tends to $+\infty$ λ_∞-almost everywhere on $[-\frac{1}{2}, \frac{1}{2}]$. For $N \ge 1$, we can choose $m \ge 1$ such that $m^2 \le N < (m+1)^2$. Then we get

$$|F_N| = \Big|\frac{1}{N}\big(f(x_1)+\cdots+f(x_{m^2})\big) + \frac{1}{N}\big(f(x_{m^2+1})+\cdots+f(x_N)\big)\Big| =$$

$$\Big|\frac{m^2}{N}\times\frac{1}{m^2}\big(f(x_1)+\cdots+f(x_{m^2})\big) + \frac{1}{N}\big(f(x_{m^2+1})+\cdots+f(x_N)\big)\Big| \le$$

$$\le |F_{m^2}| + \frac{2m}{N}||f|| = |F_{m^2}| + \frac{2m^2}{mN}||f|| \le |F_{m^2}| + \frac{2}{m}||f||, \tag{12.2.9}$$

where $||f|| = \sup_{x\in[-\frac{1}{2},\frac{1}{2}]} f(x)$.

Since the right side of (12.2.9) tends to zero when m tends to $+\infty$(equivalently, N tends to $+\infty$), we end the proof of Lemma 12,2.1.

□

Lemma 12.2.2. *Let S be a set of all elements of $[-\frac{1}{2}, \frac{1}{2}]^\infty$ which are uniformly distributed on $[-\frac{1}{2}, \frac{1}{2}]$. Then $\lambda_\infty(S) = 1$.*

Proof. Let $(f_k)_{k\in\mathbf{N}}$ be a countable subclass of $\mathcal{B}[-\frac{1}{2}, \frac{1}{2}]$ which defines a uniformly convergence on $[-\frac{1}{2}, \frac{1}{2}]$. [2] For $k \in N$, we set

$$B_k = \{(x_k)_{k\in\mathbb{N}} : (x_k)_{k\in\mathbb{N}} \in [0,1]^\infty \ \&\ \lim_{N\to\infty}\frac{1}{N}\sum_{n=1}^{N} f_k(x_n) = \int_{-\frac{1}{2}}^{\frac{1}{2}} f_k(x)dx\}. \tag{12.2.10}$$

[2] We say that a family $(f_k)_{k\in\mathbf{N}}$ of elements of $\mathcal{B}[-\frac{1}{2}, \frac{1}{2}]$ defines a uniformly convergence on $[-\frac{1}{2}, \frac{1}{2}]$, if for each $(x_n)_{n\in N} \in [-\frac{1}{2}, \frac{1}{2}]^\infty$ the validity of the condition $\lim_{N\to\infty}\frac{1}{N}\sum_{n=1}^{N} f_k(x_n) = \int_{-\frac{1}{2}}^{\frac{1}{2}} f_k(x)dx$ for $k \in N$ implies that $(x_n)_{n\in N}$ is uniformly distributed on $[-\frac{1}{2}, \frac{1}{2}]$. Indicator functions of closed subintervals of $[-\frac{1}{2}, \frac{1}{2}]$ with rational end-points is an example of such a family.

By Lemma 12.2.1, we have $\lambda_\infty(B_k) = 1$ for $k \in \mathbf{N}$, which implies $\lambda_\infty(\cap_{k\in\mathbb{N}} B_k) = 1$. Hence

$$\lambda_\infty(\{(x_k)_{k\in\mathbb{N}} : (x_k)_{k\in\mathbb{N}} \in [0,1]^\infty \ \& \ (\forall k)(k \in \mathbb{N} \to \lim_{N\to\infty} \frac{1}{N}\sum_{n=1}^{N} f_k(x_n) =$$

$$\int_{-\frac{1}{2}}^{\frac{1}{2}} f_k(x)dx\}) = 1. \tag{12.2.11}$$

The latter relation means that λ_∞-almost every elements of $[-\frac{1}{2}, \frac{1}{2}]^\infty$ is uniformly distributed on $[-\frac{1}{2}, \frac{1}{2}]$.

□

Lemma 12.2.3. *Let $(x_k)_{k\in\mathbb{N}}$ be uniformly distributed on $[-\frac{1}{2}, \frac{1}{2}]$. Then for each $n \in \mathbb{N}$ and each $(y_i)_{1\le i\le n} \in \mathbb{R}^n$, a sequence $(y_1, \cdots, y_n, x_{n+1}, x_{n+2}, \cdots)$ is also uniformly distributed on $[-\frac{1}{2}, \frac{1}{2}]$.*

Proof. Since a sequence of real numbers $(x_k)_{k\in\mathbb{N}}$ is uniformly distributed on $[-\frac{1}{2}, \frac{1}{2}]$, for each a, b with $-\frac{1}{2} \le a < b \le \frac{1}{2}$ we have

$$\lim_{N\to\infty} \frac{\#(\{x_k : 1 \le k \le N\} \cap [a,b])}{N} = b - a, \tag{12.2.12}$$

which implies

$$\lim_{N\to\infty} (\frac{\#(\{x_k : n+1 \le k \le N\} \cap [a,b])}{N} = b - a \tag{12.2.13}$$

for all fixed $n \in \mathbb{N}$.

The latter relation gives that

$$\lim_{N\to\infty} \frac{\#(\{y_1, \cdots, y_n\} \cup \{x_k : n+1 \le k \le N\} \cap [a,b])}{N} = b - a. \tag{12.2.14}$$

□

We use the notation $\mathbf{R}^{\mathbb{N}}$ for $\mathbf{R}^\infty$. We denote by $B(\mathbf{R}^{\mathbb{N}})$ the σ-algebra of all Borel subsets in $\mathbf{R}^{\mathbb{N}}$. Let $(a_i)_{i\in\mathbb{N}}$ and $(b_i)_{i\in\mathbb{N}}$ be sequences of real numbers such that

$$(\forall i)(i \in \mathbb{N} \to a_i < b_i). \tag{12.2.15}$$

We put

$$A_n = \mathbf{R}_1 \times \cdots \times \mathbf{R}_n \times (\prod_{i>n} \Delta_i),$$

for $n \in \mathbb{N}$, where

$$(\forall i)(i \in \mathbb{N} \to \mathbf{R}_i = \mathbf{R} \ \& \ \Delta_i = [a_i, b_i[). \tag{12.2.16}$$

We also put

$$\Delta = \prod_{i\in\mathbb{N}} \Delta_i.$$

For an arbitrary natural number $i \in \mathbb{N}$, consider the Lebesgue measure μ_i defined on the space $\mathbf{R}_i$ and satisfying the condition $\mu_i(\Delta_i) = 1$. Denote by λ_i the normalized linear Lebesgue measure on the interval Δ_i.

For an arbitrary $n \in \mathbb{N}$, denote by ν_n the measure defined by

$$\nu_n = \prod_{1 \le i \le n} \mu_i \times \prod_{i>n} \lambda_i, \tag{12.2.17}$$

and by $\overline{\nu}_n$ the Borel measure in the space $\mathbf{R}^{\mathbb{N}}$ defined by

$$(\forall X)(X \in B(\mathbf{R}^{\mathbb{N}}) \rightarrow \overline{\nu}_n(X) = \nu_n(X \cap A_n)). \tag{12.2.18}$$

Lemma 12.2.4. *([56]) For an arbitrary Borel set $X \subseteq \mathbf{R}^{\mathbb{N}}$ there exists the limit*

$$\nu_\Delta(X) = \lim_{n \to \infty} \overline{\nu}_n(X).$$

Moreover, the functional ν_Δ is a nontrivial σ-finite measure defined on the Borel σ-algebra $B(\mathbf{R}^{\mathbb{N}})$.

Remark 12.2.1. Setting $\Delta = [-\frac{1}{2},\frac{1}{2}[^\infty$, a measure $\mu := \nu_\Delta$ stands Yamasaki measure on $\mathbf{R}^\infty$ (cf. [145]). Note that for the measure μ we have

$$\mu(\mathbf{R}^\infty \setminus \cup_{n\in\mathbb{N}} A_n) = 0,$$

where

$$\cup_{n\in\mathbb{N}} A_n = \cup_{n\in\mathbb{N}} \cup_{z \in \mathbf{Z}^n \times (0,0,\cdots)} ([-\frac{1}{2},\frac{1}{2}[^\infty + z). \tag{12.2.19}$$

12.3 Main Result

Theorem 12.3.1. *Let μ be Yamasaki measure on $\mathbf{R}^\infty$ which gets a numerical value one on the set $[-\frac{1}{2},\frac{1}{2}]^\infty$. Then μ-almost every element of $\mathbf{R}^\infty$ is uniformly distributed on $[-\frac{1}{2},\frac{1}{2}]$.*

Proof. Let D be a set of all elements of $\mathbf{R}^\infty$ which are uniformly distributed on $[-\frac{1}{2},\frac{1}{2}]$. We have to show

$$\mu(\mathbf{R}^\infty \setminus D) = 0. \tag{12.3.1}$$

Let S be a set of all elements of $[-\frac{1}{2},\frac{1}{2}]^\infty$ which are uniformly distributed on $[-\frac{1}{2},\frac{1}{2}]$. By Lemma 12.2.3, a set

$$\cup_{n\in\mathbb{N}} \cup_{z \in \mathbf{Z}^n \times (0,0,\cdots)} (S + z) \tag{12.3.2}$$

is also a set of elements of $\mathbf{R}^\infty$ uniformly distributed on $[-\frac{1}{2},\frac{1}{2}]$. By Lemma 12.2.2, we have

$$\mu((\cup_{n\in\mathbb{N}} \cup_{z \in \mathbf{Z}^n \times (0,0,\cdots)} (S+z)) \Delta (\cup_{n\in\mathbb{N}} \cup_{z \in \mathbf{Z}^n \times (0,0,\cdots)} ([-\frac{1}{2},\frac{1}{2}[^\infty + z)) = 0. \tag{12.3.3}$$

Following Remark 12.2.1, we have

$$\mu(\mathbf{R}^\infty \setminus (\cup_{n\in\mathbb{N}} \cup_{z \in \mathbf{Z}^n \times (0,0,\cdots)} ([-\frac{1}{2},\frac{1}{2}[^\infty + z)) = 0, \tag{12.3.4}$$

which together with (12.3.3) implies

$$\mu(\mathbf{R}^\infty \setminus (\cup_{n\in\mathbb{N}} \cup_{z \in \mathbf{Z}^n \times (0,0,\cdots)} (S+z))) = 0. \tag{12.3.5}$$

Using the latter relation and the following simple inclusion

$$\cup_{n\in\mathbb{N}}\cup_{z\in\mathbf{Z}^n\times(0,0,\cdots)}(S+z)\subset D, \tag{12.3.6}$$

we deduce that

$$\mu(\mathbf{R}^\infty\setminus D)\leq\mu(\mathbf{R}^\infty\setminus(\cup_{n\in\mathbf{N}}\cup_{z\in\mathbf{Z}^n\times(0,0,\cdots)}(S+z)))=0. \tag{12.3.7}$$

□

Chapter 13

An Expansion into an Infinite-Dimensional Multiple Trigonometric Series of a Square Integrable Function in $\mathbf{R}^\infty$

13.1 Introduction

Let $\mathbf{R}^\infty$ be an infinite-dimensional topological vector space of all real-valued sequences equipped with Tychonoff topology. We denote by $\mathcal{B}(\mathbf{R}^\infty)$ a Borel σ-algebra of subsets of the $\mathbf{R}^\infty$ defined by the corresponding topology.

Let $\mathcal{R}$ be the class of all infinite dimensional rectangles $R \subset \mathbf{R}^\infty$ of the form

$$R = \prod_{i=1}^{\infty}[a_i, b_i), \quad -\infty < a_i < b_i < +\infty \tag{13.1.1}$$

with $0 < \prod_{i=1}^{\infty}(b_i - a_i) < +\infty$, where

$$\prod_{i=1}^{\infty}(b_i - a_i) := \lim_{n \to \infty} \prod_{i=1}^{n}(b_i - a_i). \tag{13.1.2}$$

In [2] has been constructed such an example of a shift-invariant non-σ-finite measure λ with $\mathcal{B}(\mathbf{R}^\infty) \subset dom(\lambda) = \mathcal{M}$ that for every $R = \prod_{i=1}^{\infty}[a_i, b_i) \in \mathcal{R}$ the following equality

$$\lambda(R) = \prod_{i=1}^{\infty}(b_i - a_i) \tag{13.1.3}$$

holds.

For every $R \in \mathcal{R}$ with $\lambda(R) > 0$, one can consider a Hilbert space $L_2(R, \mathcal{M}_R, \lambda_R)$ of all real-valued square integrable functions on R with respect to the measure λ_R, where $\mathcal{M}_R = \mathcal{M} \cap R$ and λ_R denotes a usual restriction of the λ to the $\mathcal{M}_R$. It is natural, that if $\lambda_R(R) = 0$, then $L_2(R, \mathcal{M}_R, \lambda_R)$ stands a zero-dimensional Hilbert space which consists only one zero function defined on R.

The purpose of the present chapter is an investigation of a question asking *whether one can construct a basis in $L_2(R,\mathcal{M}_R,\lambda_R)$ for $\lambda_R(R)>0$ which will be an infinite-dimensional analog of the basis in $L_2(\prod_{i=1}^n[a_i,b_i))$ consisting from the constant and multiple trigonometric functions.*

The present chapter is organized as follows.

In Section 13.2 we consider a certain auxiliary proposition from the functional analysis and establish the validity of some auxiliary facts. In Section 13.3 we present our main results.

13.2 Auxiliary Propositions from the Functional Analysis

Lemma 13.2.1 ([12], Theorem 14.25, p. 589). *Let $\{f_1,f_2,f_3,\dots\}$ be an orthonormal system in a separable Hilbert space H. Then the following assertions are equivalent:*

Maximality. *The orthonormal system $\{f_1,f_2,f_3,\dots\}$ is maximal;*

Denseness. *The set of finite linear combinations from the collection $\{f_1,f_2,f_3,\dots\}$ is dense in H;*

Parseval's identity *For every $f\in H$,*

$$\sum_{j=1}^{\infty}<f,f_j>_H^2=||f||_H^2; \tag{13.2.1}$$

Convergence of the Fourier series. *The series $\sum_{j=1}^{\infty}<f,f_j>_H f_j$ converges to f in the Hilbert space; that is,*

$$\lim_{n\longrightarrow\infty}||f-\sum_{j=1}^{n}<f,f_j>_H f_j||_H=0. \tag{13.2.2}$$

Lemma 13.2.2. *The family of functions of the form $\mathcal{X}_A$, where A is an elementary rectangle*[1] *in R and $\mathcal{X}_A$ denotes an indicator function of the A, has dense linear span in $L_2(R,\mathcal{M}_R,\lambda_R)$.*

Proof. Since functions of the form $\mathcal{X}_B$, $B\in M_B$, $\lambda_R(B)<+\infty$, have dense linear span in $L_2(R,M_R,\lambda_R)$, it suffices to show that any function of this form can be approximated arbitrarily close by linear combinations of functions of the form $\mathcal{X}_C$, where C is an elementary rectangle in R.

By Charatheodory construction, we claim that for every $\varepsilon>0$ there exists a sequence of elementary rectangles $(R_j)_{j\in IN}$ in R such that $B\subseteq A:=\cup_{j=1}^{+\infty}R_j$ and

$$\sum_{j=1}^{+\infty}\lambda_R(R_j)<\lambda_R(B)+[\frac{\varepsilon}{2}]^2. \tag{13.2.3}$$

Then we have

$$||\mathcal{X}_A-\mathcal{X}_B||^2_{L_2(R,\mathcal{M}_R,\lambda_R)}=\int_R(\mathcal{X}_A-\mathcal{X}_B)^2d\lambda_R=\int_R(\mathcal{X}^2_{(A\setminus B)})d\lambda_R$$

[1]A subset $C\subseteq R=\prod_{i=1}^{\infty}[a_i,b_i[$ is called elementary rectangle in R if there exists $n\in IN=N\setminus\{0\}$ and a Lebesque measurable set D in $\prod_{i=1}^{n}[a_i,b_i[$ (i.e., $D\in\mathcal{L}(\prod_{i=1}^{n}[a_i,b_i[))$ such that $C=D\times\prod_{i>n}[a_i,b_i[$.

$$= \int_R \mathcal{X}_{(A\setminus B)} d\lambda_R = \lambda_R(A\setminus B) = \sum_{j=1}^{+\infty} \lambda_R(R_j) - \lambda_R(B) \leq \left[\frac{\varepsilon}{2}\right]^2. \tag{13.2.4}$$

Hence, we have

$$||\mathcal{X}_A - \mathcal{X}_B||_{L_2(R,\mathcal{M}_R,\lambda_R)} \leq \sqrt{\left[\frac{\varepsilon}{2}\right]^2} = \frac{\varepsilon}{2}. \tag{13.2.5}$$

For $j \in IN$ we set $C_j = R_1^{'} \cap \cdots \cap R_{j-1}^{'} \cap R_j$, where $X^{'}$ is defined by $X^{'} = R \setminus X$.

Let us show that the following two conditions are satisfied:

(a) The $(C_j)_{j\in IN}$ is a family of pairwise disjoint elementary rectangles;

(b) $\cup_{j\in IN} C_j = A$.

If we consider two elements $C_{j_1} = R_1^{'} \cap \cdots \cap R_{j_1-1}^{'} \cap R_{j_1}$ and $C_{j_2} = R_1^{'} \cap \cdots \cap R_{j_2-1}^{'} \cap R_{j_2}$ for $j_1 < j_2$, then $C_{j_1} \subseteq R_{j_1}$ and $C_{j_2} \subseteq R_{j_1}^{'}$. This follows that the $(C_j)_{j\in IN}$ is a family of pairwise disjoint sets.

Now let $x \in A$. Then there exists at least one $j_0 \in IN$ such that $x \in R_{j_0}$. We set $\mathcal{J} := \{j : j \in IN \ \& \ x \in R_j\}$. We set $j_* = \min(\mathcal{J})$. Now it is obvious that $x \in C_{j_*}$. Hence, we get $A \subseteq \cup_{j\in IN} C_j$.

The converse relation $\cup_{j\in IN} C_j \subseteq A$ is obvious, because

$$C_j \subseteq R_j \subseteq \cup_{j\in IN} R_j = A. \tag{13.2.6}$$

Now, for each j, we have

$$\mathcal{X}_{C_j} = (1-\mathcal{X}_{R_1}) \cdots (1-\mathcal{X}_{R_{j-1}}) \cdot \mathcal{X}_{R_j}, \tag{13.2.7}$$

therefore, $\mathcal{X}_{C_j}$ is a linear combination of functions of the form $\mathcal{X}_{R_{j_1}} \cdots \mathcal{X}_{R_{j_k}} = \mathcal{X}_{R_{j_1} \cap \cdots \cap R_{j_k}}$, where $j_1, \ldots, j_k$ are positive integers.

But each set of the form $R_{j_1} \cap \cdots \cap R_{j_k}$ is again an elementary rectangle in R. Hence, each function $\mathcal{X}_{C_j}$ is a linear combination of indicator functions of elementary rectangles in R.

Select n so large that

$$\int_R \sum_{j=n+1}^{+\infty} \mathcal{X}_{C_j} d\lambda_R < \left[\frac{\varepsilon}{2}\right]^2. \tag{13.2.8}$$

Then we have

$$\left\|\mathcal{X}_B - \sum_{j=1}^{n} \mathcal{X}_{C_j}\right\|_{L_2(R,\mathcal{M}_R,\lambda_R)} = \left\|\sum_{j=n+1}^{+\infty} \mathcal{X}_{C_j}\right\|_{L_2(R,M_R,\lambda_R)} = \left[\int_R \Big(\sum_{j=n+1}^{+\infty} \mathcal{X}_{C_j}\Big)^2 d\lambda_R\right]^{\frac{1}{2}}$$

$$= \left[\int_R \Big(\sum_{j=n+1}^{+\infty} \mathcal{X}_{C_j}\Big) d\lambda_R\right]^{\frac{1}{2}} < \frac{\varepsilon}{2}. \tag{13.2.9}$$

Therefore, $f = \sum_{j=1}^{n} \mathcal{X}_{C_j}$ is a linear combination of elementary rectangles indicator functions and

$$\left\|\mathcal{X}_B - \sum_{j=1}^{n} \mathcal{X}_{C_j}\right\|_{L_2(R,\mathcal{M}_R,\lambda_R)} < \frac{\varepsilon}{2}. \tag{13.2.10}$$

□

Lemma 13.2.3. *Let $R=\prod_{i\in IN}[a_i,b_i[\in \mathcal{R}$ with $\lambda_R(R)>0$. Let $(f_i)_{i\in IN}$ be a sequence of functions such that for each i, $f_i\in L_1[a_i,b_i[$, and $0\leq f_i\leq \mathcal{X}_{[a_i,b_i[}$. Then $f=\prod_{i\in IN} f_i\in L_1(R,\mathcal{M}_R,\lambda_R)$ with*

$$\int_R f d\lambda_R=\prod_{i\in IN}\int_{[a_i,b_i[} f_i dx_i. \tag{13.2.11}$$

Proof. Since

$$\lim_{n\to+\infty}\prod_{i=1}^{n} f_i(x_i)\times \mathcal{X}_{\prod_{i=n+1}^{+\infty}[a_i,b_i)}(x_{n+1},\dots)=\Big(\prod_{i\in N} f_i\Big)(x_1,x_2,\dots) \tag{13.2.12}$$

for every $(x_1,x_2,\dots)\in \mathbf{R}^\infty$, by Fubini theorem and dominated convergence, we have

$$\int_R \lim_{n\to+\infty}\prod_{i=1}^{n} f_i(x_i)\times \mathcal{X}_{\prod_{i=n+1}^{+\infty}[a_i,b_i)}(x_{n+1},\dots)d\lambda_R$$

$$=\lim_{n\to+\infty}\int_R\prod_{i=1}^{n} f_i(x_i)\times \mathcal{X}_{\prod_{i=n+1}^{+\infty}[a_i,b_i[}(x_{n+1},\dots)d\lambda_R$$

$$=\lim_{n\to+\infty}\Big[\int_{\prod_{i=1}^{n}[a_i,b_i[}\Big(\prod_{i=1}^{n} f_i(x_i)\Big)dx_1\cdots dx_n\times\int_{\prod_{i=n+1}^{+\infty}[a_i,b_i[}\mathcal{X}_{\prod_{i=n+1}^{+\infty}[a_i,b_i[}(x_{n+1},\dots)d\lambda_{\prod_{i=n+1}^{+\infty}[a_i,b_i[}\Big]$$

$$=\lim_{n\to+\infty}\Big[\int_{\prod_{i=1}^{n}[a_i,b_i[}\Big(\prod_{i=1}^{n} f_i(x_i)\Big)dx_1\cdots dx_n\times\lambda_{\prod_{i=n+1}^{+\infty}[a_i,b_i[}\Big(\prod_{i=n+1}^{+\infty}[a_i,b_i[\Big)\Big]$$

$$=\lim_{n\to+\infty}\prod_{i=1}^{n}\int_{[a_i,b_i[} f_i(x_i)dx_i\times\lim_{n\to+\infty}\lambda_{\prod_{i=n+1}^{+\infty}[a_i,b_i[}\Big(\prod_{i=n+1}^{+\infty}[a_i,b_i[\Big)$$

$$=\prod_{i=1}^{+\infty}\int_{[a_i,b_i[} f_i(x_i)dx_i\times 1=\prod_{i=1}^{+\infty}\int_{[a_i,b_i[} f_i(x_i)dx_i. \tag{13.2.13}$$

□

For every $R=\prod_{j\in IN}[a_j,b_j[\in\mathcal{R}$ and $i\in IN$ we set

$$K_i=\Big\{\frac{1}{\sqrt{b_i-a_i}}\Big\}\cup\Big\{\sqrt{\frac{2}{b_i-a_i}}\sin\frac{n\pi x_i}{b_i-a_i},x_i\in\mathbf{R}_i:=\mathbf{R},n\in IN\Big\}$$

$$\cup\Big\{\sqrt{\frac{2}{b_i-a_i}}\cos\frac{n\pi x_i}{b_i-a_i},x_i\in\mathbf{R}_i:=\mathbf{R},n\in IN\Big\}. \tag{13.2.14}$$

We set

$$\Phi_R=\Big\{\phi=(\phi_i)_{i\in IN}\in\prod_{i=1}^{\infty}K_i\ \&\ (\exists i_\phi)(i_\phi\in IN)(\forall i)\Big(i\geq i_\phi\to\phi_i=\frac{1}{\sqrt{b_i-a_i}}\Big)\Big\}. \tag{13.2.15}$$

It is obvious that $card(\Phi_R)=\aleph_0$.

Lemma 13.2.4. *The family Φ_R is orthonormal system in $L_2(R,\mathcal{M}_R,\lambda_R)$.*

Proof. Let $\phi^{(1)}, \phi^{(2)} \in \Phi_R$ and $\phi^{(1)} \neq \phi^{(2)}$. Then there exists $i_0 \in IN$ such that $\phi^{(1)}_{i_0} \neq \phi^{(2)}_{i_0}$. We set

$$k_1 = \min\Big\{i : (\forall j)\Big(j \geq i \to \phi^{(1)}_j = \frac{1}{\sqrt{b_j - a_j}}\Big)\Big\} \tag{13.2.16}$$

and

$$k_2 = \min\Big\{i : (\forall j)\Big(j \geq i \to \phi^{(2)}_j = \frac{1}{\sqrt{b_j - a_j}}\Big)\Big\} \tag{13.2.17}$$

We put $m_0 = \max\{k_1, k_2\}$. It is obvious that $i_0 < m_0$. Take into account orthogonality of $\phi^{(1)}_{i_0}$ and $\phi^{(2)}_{i_0}$ in $L_2([a_{i_0}, b_{i_0}[)$ (i.e., $\int_{[a_{i_0}, b_{i_0}[} \phi^{(1)}_{i_0}(x) \cdot \phi^{(2)}_{i_0}(x)dx = 0$), by Fubini theorem we have

$$< \phi^{(1)}, \phi^{(2)} >_{L_2(R,\mathcal{M}_{\mathcal{R}},\lambda_R)} = \int_R \phi^{(1)} \cdot \phi^{(2)} d\overline{\lambda}_R = \int_{\prod_{i=1}^{i_0-1}[a_i,b_i[} \Big(\prod_{i=1}^{i_0-1} \phi^{(1)}_i(x_i) \cdot \phi^{(2)}_i(x_i)\Big) dx_1 \cdots x_{i_0-1}$$

$$\times \int_{[a_{i_0}, b_{i_0}[} \phi^{(1)}_{i_0}(x_{i_0}) \cdot \phi^{(2)}_{i_0}(x_{i_0}) dx_{i_0} \times \int_{\prod_{i=i_0+1}^{m_0-1}[a_i,b_i[} \Big(\prod_{i=i_0+1}^{m_0-1} \phi^{(1)}_i(x_i) \cdot \phi^{(2)}_i(x_i)\Big) dx_{i_0+1} \cdots dx_{m_0-1}$$

$$\times \int_{\prod_{i\geq m_0}[a_i,b_i[} \prod_{i\geq m_0} \phi^{(1)}_i(x_i) \cdot \phi^{(2)}_i(x_i) d\lambda_{\prod_{i\geq m_0}[a_i,b_i[}(x_{m_0}, \dots) = 0. \tag{13.2.18}$$

For $\phi \in \Phi_R$, we set

$$n_0 = \min\Big\{i : (\forall j)\Big(j \geq i \to \phi_j = \frac{1}{\sqrt{b_j - a_j}}\Big)\Big\}. \tag{13.2.19}$$

Take into account normality of the ϕ_i for $i \in IN$ (i.e., $\int_{[a_i,b_i[} \phi_i^2(x)dx = 1$), by Fubini theorem we have

$$||\phi||^2_{L_2(R,\mathcal{M}_{\mathcal{R}},\lambda_R)} = \int_R \phi^2 d\lambda_R = \int_{\prod_{i=1}^{n_0-1}[a_i,b_i[} \Big(\prod_{i=1}^{n_0-1} \phi_i^2(x_i)\Big) dx_1 \cdots dx_{n_0-1}$$

$$\times \int_{\prod_{i\geq n_0}[a_i,b_i[} \Big(\prod_{i\geq n_0} \phi_i^2(x_i)\Big) d\lambda_{\prod_{i\geq n_0}[a_i,b_i[}(x_{n_0}, \dots)$$

$$= \prod_{i=1}^{n_0-1} \Big(\int_{[a_i,b_i[} \phi_i^2(x_i)dx_i\Big) \times \prod_{i\geq n_0} \Big(\frac{1}{\sqrt{b_i - a_i}}\Big)^2 \cdot (b_i - a_i) = 1. \tag{13.2.20}$$

□

Lemma 13.2.5. *The set of finite linear combinations from the collection* $\{\phi : \phi \in \Phi_R\}$ *is dense in* $L_2(R, \mathcal{M}_{\mathcal{R}}, \lambda_R)$.

Proof. By Lemma 13.2.4., it suffices to prove that every function of the form $f = X_A$, where A is an elementary rectangle in R, can be approximated arbitrarily close by a finite linear combination from the collection $\{\phi : \phi \in \Phi_R\}$.

Let $A = B \times \prod_{n>m}[a_i, b_i[$, where $B \in L(\prod_{1\leq i\leq m}[a_i, b_i[)$. Then we have

$$X_A = X_B \times X_{\prod_{n>m}[a_i,b_i[}. \tag{13.2.21}$$

Since $\prod_{1\le i\le m} K_i$ is orthonormal basis in $L_2(\prod_{1\le i\le m}[a_i,b_i[,\ L\Big(\prod_{1\le i\le m}[a_i,b_i[\Big),\ \lambda_{\prod_{1\le i\le m}[a_i,b_i[})$ and

$$\Big(\prod_{i>m}\sqrt{b_i-a_i}\Big)\cdot X_B \in L_2\Big(\prod_{1\le i\le m}[a_i,b_i[,\ L(\prod_{1\le i\le m}[a_i,b_i[),\ \lambda_{\prod_{1\le i\le m}[a_i,b_i[}\Big), \tag{13.2.22}$$

wc claim that there exists a finite family of elements $(H_k)_{1\le k\le n}$ of the $\prod_{1\le i\le m} K_i$ and $(c_k)_{1\le k\le n}\in \mathbf{R}$ such that

$$\int_{\prod_{1\le i\le m}[a_i,b_i[}\Big(\Big(\prod_{i>m}\sqrt{b_i-a_i}\Big)\cdot X_B-\sum_{1\le k\le n}c_kH_k\Big)dx_1\cdots x_m<\varepsilon^2\cdot\prod_{i\ge m}(b_i-a_i). \tag{13.2.23}$$

Then we get

$$\Big\|X_A-\sum_{1\le k\le n}c_kH_k\times\prod_{i>m}\frac{1}{\sqrt{b_i-a_i}}X_{[a_i,b_i[}\Big\|^2_{L_2(R,\mathcal{M}_{\mathcal{R}},\lambda_R)}$$

$$=\int_{\prod_{i\in IN}[a_i,b_i[}\Big(X_A-\sum_{1\le k\le n}c_kH_k\times\prod_{i>m}\frac{1}{\sqrt{b_i-a_i}}X_{[a_i,b_i[}\Big)^2 d\lambda_{\prod_{i\in IN}[a_i,b_i[}$$

$$=\int_{\prod_{i\in IN}[a_i,b_i[}\Big(X_B\times\prod_{i>m}X_{[a_i,b_i[}-\prod_{i>m}\frac{1}{\sqrt{b_i-a_i}}\times\sum_{1\le k\le n}c_kH_k\times\prod_{n>m}X_{[a_i,b_i[}\Big)^2 d\lambda_{\prod_{i\in IN}[a_i,b_i[}$$

$$=\prod_{i>m}\frac{1}{b_i-a_i}\int_{\prod_{i\in IN}[a_i,b_i[}\Big(\Big(\prod_{i>m}\sqrt{b_i-a_i}\Big)\times X_B\times\prod_{n>m}X_{[a_i,b_i[}$$
$$-\sum_{1\le k\le n}c_kH_k\times\prod_{n>m}X_{[a_i,b_i[}\Big)^2 d\lambda_{\prod_{i\in IN}[a_i,b_i[}$$

$$=\Big(\prod_{i>m}\frac{1}{b_i-a_i}\Big)\int_{\prod_{i\in IN}[a_i,b_i[}\Big(\Big[\Big(\prod_{i>m}\sqrt{b_i-a_i}\Big)\times X_B-\sum_{1\le k\le n}c_kH_k\Big]\times\prod_{n>m}X_{[a_i,b_i[}\Big)^2 d\lambda_{\prod_{i\in IN}[a_i,b_i[}$$

$$=\Big(\prod_{i>m}\frac{1}{b_i-a_i}\Big)\int_{\prod_{1\le i\le m}[a_i,b_i[}\Big(\Big[\Big(\prod_{i>m}\sqrt{b_i-a_i}\Big)\times X_B-\sum_{1\le k\le n}c_kH_k\Big]\Big)^2 d\lambda_{\prod_{1\le i\le m}[a_i,b_i[}$$
$$\times\int_{\prod_{i>m}[a_i,b_i[}\Big(\prod_{i>m}X_{[a_i,b_i[}\Big)^2 d\lambda_{\prod_{i>m}[a_i,b_i[}$$

$$=\Big(\prod_{i>m}\frac{1}{b_i-a_i}\Big)\times\int_{\prod_{1\le i\le m}[a_i,b_i[}\Big[\Big(\prod_{i>m}\sqrt{b_i-a_i}\Big)\times X_B-\sum_{1\le k\le n}c_kH_k\Big]^2 d\lambda_{\prod_{1\le i\le m}[a_i,b_i[}$$
$$\times\prod_{i>m}(b_i-a_i)\le\prod_{i>m}\frac{1}{b_i-a_i}\times\prod_{i>m}(b_i-a_i)\times\varepsilon^2=\varepsilon^2. \tag{13.2.24}$$

□

13.3 Main Results

Combining results of Lemma 13.2.1 and Lemmas 13.2.4–13.2.5, we get the following assertion.

Theorem 13.3.1. *For every $R \in \mathcal{R}$, we have:*

I. *The family $\{\phi : \phi \in \Phi_R\}$ is orthonormal basis in $L_2(R, \mathcal{M}_{\mathcal{R}}, \lambda_R)$, i.e., the Φ_R is satisfying the conditions:*

1. *Orthogonality: Every two different elements of the Φ_R are orthogonal, i.e., for all ϕ_1, ϕ_2 in Φ_R with $\phi_1 \neq \phi_2$ we have*

$$< \phi_1, \phi_2 >_{L_2(R,\mathcal{M}_{\mathcal{R}},\lambda_R)} = 0; \tag{13.3.1}$$

2. *Normalization: Every element ϕ in Φ_R has norm 1, i.e.,*

$$||\phi||_{L_2(R,\mathcal{M}_{\mathcal{R}},\lambda_R)} = 0; \tag{13.3.2}$$

3. *Completeness: The linear span of the family Φ_R is dense in $L_2(R, \mathcal{M}_{\mathcal{R}}, \lambda_R)$.*

II. (**Maximality**) *The orthonormal system $\{\phi : \phi \in \Phi_R\}$ is maximal;*

III. (**Parseval's identity**) *For every $f \in L_2(R, \mathcal{M}_{\mathcal{R}}, \lambda_R)$,*

$$\sum_{\phi \in \Phi_R} < f, \phi >^2_{L_2(R,\mathcal{M}_{\mathcal{R}},\lambda_R)} = ||f||^2_{L_2(R,\mathcal{M}_{\mathcal{R}},\lambda_R)}. \tag{13.3.3}$$

IV. (**Convergence of the Fourier series**) *The series $\sum_{\phi \in \Phi_R} < f, \phi >_{L_2(R,\mathcal{M}_{\mathcal{R}},\lambda_R)} \phi$ converges to f in $L_2(R, \mathcal{M}_{\mathcal{R}}, \lambda_R)$; that is, for $\varepsilon > 0$ there exists a finite subset $\Phi_\varepsilon \subset \Phi_R$ such that for every finite subset Φ_1 with $\Phi_\varepsilon \subseteq \Phi_1$ we have*

$$\left\| f - \sum_{\phi \in \Phi_1} < f, \phi >_{L_2(R,\mathcal{M}_{\mathcal{R}},\lambda_R)} \phi \right\|_{L_2(R,\mathcal{M}_{\mathcal{R}},\lambda_R)} < \varepsilon. \tag{13.3.4}$$

Recall, that if H_i is a family of Hilbert spaces indexed by $i \in I$, then the direct sum of the H_i, denoted

$$\bigoplus_{i \in I} H_i \tag{13.3.5}$$

consists of the set of all sequences

$$x = (x_i)_{i \in I} \in \prod_{i \in I} H_i \tag{13.3.6}$$

such that

$$\sum_{i \in I} ||x_i||^2_{H_i} < +\infty. \tag{13.3.7}$$

For $x = (x_i)_{i \in I}, y = (y_i)_{i \in I} \in \prod_{i \in I} H_i$, the inner product is defined by

$$< x, y >_{\bigoplus_{i \in I} H_i} = \sum_{i \in I} < x_i, y_i >_{H_i} . \tag{13.3.8}$$

Each of the H_j is included as a closed subspace in the direct sum $\bigoplus_{i \in I} H_i$. Moreover, the family $(H_i)_{i \in I}$ are pairwise orthogonal. Conversely, if there is a system of closed subspaces $(V_i)_{i \in I}$ in a Hilbert space H which are pairwise orthogonal and whose union is dense in H, then H is canonically isomorphic to the direct sum $\bigoplus_{i \in I} V_i$. In this case, H is called the internal direct sum of the family $(V_i)_{i \in I}$ and is denoted by $\bigoplus^*{}_{i \in I} V_i$.

Theorem 13.3.2. *The following representation*

$$L_2(\mathbf{R}^\infty, \mathcal{M}, \lambda) = \bigcup_{(R_i)_{i\in N} \in \mathcal{R}^{(*,\infty)}} \overset{*}{\bigoplus}_{R\in\{R_i:i\in N\}} L_2(R, \mathcal{M}_R, \lambda_R) \tag{13.3.9}$$

is valid, where $\mathcal{R}^{(*,\infty)}$ *denotes a set of sequences* $(R_i)_{i\in N}$ *such that* $(R_i)_{i\in IN}$ *is a set of pairwise disjoint sequence of elements of the* $\mathcal{R}$, $R_0 = \mathbf{R}^\infty \setminus \cup_{i\in IN} R_i$, *and*

$$L_2(R_0, \mathcal{M}_{R_0}, \lambda_{R_0}) = \Big\{f : f \in \mathbf{R}^{R_0} \ \& \ \int_{R_0} f^2(x)d\lambda = 0\Big\}. \tag{13.3.10}$$

Proof. Let us show that

$$L_2(\mathbf{R}^\infty, \mathcal{M}, \lambda) \subseteq \cup_{(R_i)_{i\in N}\in\mathcal{R}^{(*,\infty)}} \overset{*}{\bigoplus_{R\in\{R_i:i\in N\}}} L_2(R, \mathcal{M}_R, \lambda_R). \tag{13.3.11}$$

Let $f \in L_2(\mathbf{R}^\infty, \mathcal{M}, \lambda)$. This means that there exists a countable family $\{F_j : j \in IN\}$ of elements of the $\mathcal{R}$ that

$$\int_{\mathbf{R}^\infty \setminus \cup_{j\in IN} F_j} f^2(x)d\lambda(x) = 0 \tag{13.3.12}$$

and

$$0 \leq \int_{\cup_{j\in IN} F_j} f^2(x)d\lambda(x) < +\infty. \tag{13.3.13}$$

For $j \in IN$ we set $R_j = F_1' \cap \cdots \cap F_{j-1}' \cap F_j$, where X' is defined by $X' = \mathbf{R}^\infty \setminus X$.
It is obvious that the following two conditions are satisfied:
(a) The $(R_j)_{j\in IN}$ is a family of pairwise disjoint elements of the $\mathcal{R}$;
(b) $\cup_{j\in IN} R_j = \cup_{j\in IN} F_j$.
It is obvious that the following expansion

$$f(x) = \sum_{i\in IN} \mathcal{X}_{R_i} f(x) + \mathcal{X}_{\mathbf{R}^\infty \setminus \cup_{i\in IN} R_i}(x) f(x) \tag{13.3.14}$$

is valid for all $x \in \mathbf{R}^\infty$.

On the other hand,

$$f^2(x) = \sum_{i\in IN} \mathcal{X}_{R_i} f^2(x) + \mathcal{X}_{\mathbf{R}^\infty \setminus \cup_{i\in IN} R_i}(x) f^2(x). \tag{13.3.15}$$

Since $0 \leq \int_{\mathbf{R}^\infty} f^2(x)d\lambda(x) < +\infty$, we have

$$+\infty > ||f||_{L_2(\mathbf{R}^\infty, \mathcal{M}, \lambda)} = \int_{\mathbf{R}^\infty} f^2(x)d\lambda(x) = \int_{\mathbf{R}^\infty} \sum_{i\in N} \mathcal{X}_{R_i} f^2(x)d\lambda(x)$$

$$= \sum_{i\in N} \int_{\mathbf{R}^\infty} \mathcal{X}_{R_i} f^2(x)d\lambda(x) = \sum_{i\in N} ||\mathcal{X}_{R_i} f^2(x)||^2_{L_2(R_i, \mathcal{M}_{R_i}, \lambda_{R_i})}. \tag{13.3.16}$$

The validity of the converse relation is obvious. □

Let $(R_i)_{i\in N} \in \mathcal{R}^{(*,\infty)}$. We set $\Phi_{R_0} = \{O_{R_0}\}$, where O_{R_0} denotes the zero of the $\mathbf{R}^{R_0}$. For every $i \in N$ and $\phi \in \Phi_{R_i}$, we set $\overline{\phi}(x) = \phi(x)$ if $x \in R_i$, and $= 0$, otherwise. We put

$$\overline{\Phi}_{R_i} = \{\overline{\phi} \,:\, \phi \in \Phi_{R_i}\}. \tag{13.3.17}$$

The main result is formulated as follows.

Theorem 13.3.3. *For $f \in L_2(\mathbf{R}^\infty, \mathcal{M}, \lambda)$, there exists $(R_i)_{i\in N} \in \mathcal{R}^{(*,\infty)}$ such that*

$$||f||^2_{L_2(\mathbf{R}^\infty,\mathcal{M},\lambda)} = \sum_{i\in N} \sum_{\overline{\phi}\in\overline{\Phi}_{R_i}} < f, \overline{\phi} >^2_{L_2(\mathbf{R}^\infty,\mathcal{M},\lambda)} . \tag{13.3.18}$$

Proof. By Theorem 4.3.2, for $f \in L_2(\mathbf{R}^\infty, \mathcal{M}, \lambda)$, there exists $(R_i)_{i\in N} \in \mathcal{R}^{(*,\infty)}$ such that

$$f \in \overset{*}{\bigoplus_{R\in\{R_i : i\in N\}}} L_2(R, \mathcal{M}_R, \lambda_R). \tag{13.3.19}$$

We have

$$||f||^2_{L_2(\mathbf{R}^\infty,\mathcal{M},\lambda)} = \sum_{i\in N} \int_{R_i} f^2(x) X_{R_i} d\lambda = \sum_{i\in N} \int_{R_i} f^2(x) X_{R_i} d\lambda_{R_i}$$

$$= \sum_{i\in N} ||f X_{R_i}||^2_{L_2(R_i,\mathcal{M}_{R_i},\lambda_{R_i})} = \sum_{i\in N} \sum_{\phi\in\Phi_{R_i}} < f X_{R_i}, \phi >^2_{L_2(R_i,\mathcal{M}_{R_i},\lambda_{R_i})}$$

$$= \sum_{i\in N} \sum_{\overline{\phi}\in\overline{\Phi}_{R_i}} < f, \overline{\phi} >^2_{L_2(\mathbf{R}^\infty,\mathcal{M},\lambda)} . \tag{13.3.20}$$

□

Chapter 14

On Questions of U. Darji and D. Fremlin

14.1 On a Question of U. Darji

Let G be a complete separable metric group, by which we mean a group with a complete metric for which the transformation (from $G \times G$ onto G) which sends (x,y) into $x^{-1}y$ is continuous. Let $B(G)$ denotes the σ-algebra of Borel subsets of G.

A Borel subset $X \subseteq G$ will be called Haar null set if there exists a Borel probability measure m over G such that $m(gXh) = 0$ for all $g, h \in G$. Such an m will be called left and right transversal to X. Each subset of X is also called Haar null set in G.

Following [83], the class of all Haar null sets in G constitute a σ-ideal. A two-valued probability measure $\mu_{(H,G)}$ defined by that σ-ideal is called a two-valued Haar measure in G.

We say that a set $X \subseteq G$ is nowhere dense (in G) if $\text{int}(\text{cl}(X)) = \emptyset$.

We say that a set $Y \subseteq G$ is a meagre subset(also, meager set or a set of the first category) in G if Y can be represented in the form $Y = \cup_{n\in\omega} Y_n$, where ω denotes a set of all natural numbers and all $Y_n (n \in \omega)$ are nowhere dense subsets of G. Note that each subset of a meagre set is also a meagre subset.

We say that a set $Z \subseteq G$ is a second category subset of G if Z is not a meagre subset in G.

It is well known that the class of all meagre subsets in G constitute a σ-ideal. A two-valued probability measure $\mu_{(C,G)}$ defined by that σ-ideal is called a two-valued category measure in G.

A subset $X \subseteq G$ is called small in the sense of measure if it is Haar null, i.e., $\mu_{(H,G)}(X) = 0$. Analogously, a subset $Y \subseteq G$ is called small in the sense of category if it is a meagre subset in G, i.e., $\mu_{(C,G)}(X) = 0$. Further, let P be a sentence formulated only by using the notions of Haar null sets, meagre sets and of purely set-theoretical notions. We say that the duality principle between the measure and category is valid with respect to the sentence P if the sentence P is equivalent to the sentence P^* obtained from the sentence P by interchanging in it the notions of the above small sets.

Theorem 14.1.1. *Let P be a sentence defined by :*

"For every two Polish groups G_1 and G_2, and for every Haar null set $Y \subset G_1$, we have $(\forall X)(X \subseteq G_2 \rightarrow Y \times X$ is Haar null in $G_1 \times G_2)$."

Then the duality principle between the measure and category is valid with respect to the sentence P.

Proof. We present the proof of Theorem 14.1.1 in two steps.

Step 1. Let us show that the sentence P is valid.

Without loss of generality, we can assume that X is Borel. Let us show that $Y \times X$ is Haar null in $G_1 \times G_2$. Since $Y \subset G_1$ is Haar null in G_1 we have $\mu_{(H,G_1)}(Y) = 0$. On the other hand, there is a Borel probability measure m over G_1 such that $m(gYh) = 0$ for all $g,h \in G$. Let consider a Borel set $Y \times G_2$. For a Borel measure $\mu \times \lambda$, where λ is an arbitrary Borel probability in G_2 (for example, a Dirac measure on G_2 concentrated on any singleton of G_2) we have

$$\forall (g_1, f_1), (g_2, f_2) \in G_1 \times G_2 \rightarrow \mu \times \lambda((g_1, f_1)(Y \times G_2)(g_2, f_2)) =$$

$$\mu \times \lambda((g_1 Y \times G_2)(g_2, f_2)) = \mu \times \lambda(g_1 Y g_2 \times G_2) = \mu(g_1 Y g_2) \times \lambda(G_2) = 0.$$

The latter relation means that $Y \times G_2$ is Haar null in $G_1 \times G_2$ as well its all subset. Hence $\mu_{(H,G_1 \times G_2)}(Y \times X) = 0$ for every $X \subseteq G_2$.

Step 2. We have to show that the sentence P^* obtained from the sentence P by interchanging in it the notion of the Haar null subset by the notion of the meagre subset holds also; in other words we have to show that the sentence P^* defined by

"For every two Polish groups G_1 and G_2, and for every meagre set $Y \subset G_1$ we have $(\forall X)(X \subseteq G_2 \rightarrow Y \times X$ is a meagre set in $G_1 \times G_2)$"

is valid.

If $Y \subseteq G_1$ is a meagre subset in G_1 then Y can be represented in the form $Y = \cup_{n \in \omega} Y_n$, where all $Y_n (n \in \omega)$ are nowhere dense subsets of G_1.

Let us show that $Y_n \times G_2$ is the nowhere dense subset of $G_1 \times G_2$ for each $n \in \omega$.

Since $Y_n \subseteq G_1$ is nowhere dense (in G_1) we have $\mathrm{int}(\mathrm{cl}(Y_n)) = \emptyset$. It is obvious to show that $\mathrm{int}(\mathrm{cl}(Y_n \times G_2)) = \emptyset$. If we assume the contrary then there will be non-empty open subsets $U_1 \subseteq G_1$ and $U_2 \subseteq G_2$ such that $U_1 \times U_2 \subseteq \mathrm{int}(\mathrm{cl}(Y_n \times G_2))$. So $\mathrm{int}(\mathrm{cl}(Y_n \times G_2)) = \mathrm{int}(\mathrm{cl}(Y_n)) \times G_2$ we claim that $U_1 \subseteq \mathrm{int}(\mathrm{cl}(Y_n))$ which contradicts to the condition that Y_n is nowhere dense subset of G_1.

Finally we get

$$Y \times G_2 = (\cup_{n \in \omega} Y_n) \times G_2 = \cup_{n \in \omega}(Y_n \times G_2),$$

which means that $Y \times G_2$ is a meagre subset in $G_1 \times G_2$ as well its all subsets. Hence $Y \times X$ is a meagre set in $G_1 \times G_2$ which implies that $\mu_{(C,G_1 \times G_2)}(Y \times X) = 0$.

This ends the proof of Theorem 14.1.1. □

Recall that a locally compact Hausdorff topological group G whose left Haar measure coincides with its right Haar measure is called a unimodular group. Clearly, every abelian locally compact σ-compact Hausdorff topological groups are unimodular. There exist an examples of unimodular groups that are not abelian (see, for example [71]). The definition of a unimodular group follows that its every Haar measure is two sided invariant. Indeed, if λ is a left Haar measure, then it must be its right Haar measure. Correspondingly, for every two elements $f,g \in G$ and for every Borel subset E in G we get $\mu(fEg) = \mu(fE) = \mu(E)$.

Darji [25] posed a question asking *whether a Polish group G can be written as the union of two sets, one meagre and the other Haar null.* If we denote by μ_H and μ_C the restrictions of measures $\mu_{(H,G)}$ and $\mu_{(C,G)}$ to the σ-algebra $\mathrm{dom}(\mu_{(H,G)}) \cap \mathrm{dom}(\mu_{(C,G)})$, respectively, then Darji's question will be equivalent to the question asking *whether the probability measures μ_H and μ_C are mutually singular.* By using the technique of openly Haar null sets elaborated by Solecki [131], Cohen [22] showed that every infinite product of locally compact non-compact Polish groups decomposes into the disjoint union of a Haar null set and a meagre set, which gives a partial positive answer to a question of Darji. We give a different proof Cohen's above mentioned result for the family of unimodular Polish groups that are not compact.

By the scheme presented in [89](see the proof of Theorem 16.5) one can get the validity of the following assertion.

Lemma 14.1.1. *Every unimodular Polish group may be written as the disjoint union of a Haar null set and a meagre set.*

By Theorem 14.1.1 and Lemma 14.1.1 we get the validity of the following corollary.

Corollary 14.1.1. *For $i \in \mathbb{N} \setminus \{1\}$, let G_i be a Polish group and G_1 be a unimodular Polish group. Then $G = \prod_{i \in \mathbb{N}} G_i$ decomposes into the disjoint union of a Haar null set and a meagre set.*

Proof. By Lemma 14.1.1 we know that G_1 may be represented as the disjoint union of a Haar null set X_1 (in G_1) and a meagre set X_2 (in G_2). We consider the sets $Y_1 := X_1 \times \prod_{k>1} G_k$ and $Y_2 := X_2 \times \prod_{k>1} G_k$. By Theorem 14.1.1 we know that Y_1 is Haar null set in G and Y_2 is the meagre set in G. So $G = Y_1 \cup Y_2$ and $Y_1 \cap Y_2 = \emptyset$, we end the proof of Corollary 14.1.1.
□

We present a certain application of the Corollary 14.1.1 which allows us to study the topological structure of the carrier [1] of the Baker measure [3] defined on the product of unimodular Polish groups that are not compact.

Let $(X_i, M_i, \mu_i)_{i \in \mathbb{N}}$ be a sequence of measure spaces and for each i ($i \in \mathbb{N}$), let ρ_i be a metric on the set X_i. Assume that the following conditions are satisfied.

(i) Each (X_i, ρ_i) is a locally compact metric space.

(ii) Each M_i contains the family, $\mathcal{B}(X_i)$, of Borel subsets of X_i and μ_i is a regular Borel measure on X_i (cf. [125], Def. 2.15).

(iii) For all i, and for every $\delta > 0$, there exists a sequence (A_j) of Borel subsets of X_i such that $d_i(A_j) < \delta$ and $X_i = \cup_{j=1}^{\infty} A_j$, where $d_i(A_j)$ is the diameter of A_j in X_i.

[1] Let μ be a Borel measure defined in G. A Borel subset $A \subseteq G$ is called carrier of μ if $\mu(G \setminus A) = 0$.

Define $X = \prod_{i=1}^{\infty} X_i$ and let equip X with the product topology. Let denote by $\mathcal{R}$ the family of all rectangles $R \subseteq X$ of the form $R = \prod_{i=1}^{\infty} R_i$, $R_i \in \mathcal{B}(X_i)$, and

$$0 \leq \prod_{i=1}^{\infty} \mu_i(R_i) := lim_{n\to\infty} \prod_{i=1}^{n} \mu_i(R_i) < +\infty.$$

For $R \in \mathcal{R}$, we define $\tau(R) = \prod_{i=1}^{\infty} \mu_i(R_i)$.

Let τ^* be the set function on the powerset $\mathcal{P}(X)$ defined by

$$\tau^*(E) = \inf\{\sum \tau(R_j) : R_j \in \mathcal{R} \,\&\, E \subseteq \cup R_j\}$$

for $E \subseteq X$.

Also, let use the convention that $0 \cdot +\infty = +\infty \cdot 0 = 0$, $+\infty \cdot +\infty = +\infty$, and that the infimum taken over the empty set has the value $+\infty$.

Lemma 14.1.2. *([3], Theorem I, p. 2579) The set function τ^* is an outer measure in X. Let M be the σ-algebra of subsets of X which are measurable with respect to τ^*, and let μ be the measure on X obtained by restricting τ^* to $\mathcal{B}(X)$. Then, for all $R = \prod_{i=1}^{\infty} R_i \in \mathcal{R}$, we have $\mu(R) = \prod_{i=1}^{\infty} \mu_i(R_i)$. If each space X_i contains disjoint subsets A_i, B_i such that $\mu_i(A_i) = \mu_i(B_i) = 1$, then the measure μ is not σ-finite. Finally, assume that each (X_i, ρ_i) is an M_i-measurable group. If each μ_i is left-invariant measure on M_i, then μ is left-invariant Borel measure in X. Similarly, if each μ_i is right-invariant measure on M_i, then μ is right-invariant Borel measure in X.*

The following lemma is an immediate consequence of Lemma 14.1.2.

Lemma 14.1.3. *Under conditions of Lemma 14.1.2, if each X_i is a Polish group and μ_i is a two sided invariant Borel measure in X_i $(i \in \mathbb{N})$, then μ is a two sided invariant Borel measure in X.*

Proof. By the definition of the measure μ, for $g = (g_i)_{i\in\mathbb{N}}, f = (f_i)_{i\in\mathbb{N}} \in X$ and for $E \in \mathcal{B}(X)$, we get

$$\mu(gEf) = \inf\{\sum \tau(R_j) : R_j \in \mathcal{R} \,\&\, gEf \subseteq \cup R_j\} =$$
$$\inf\{\sum \tau(R_j) : R_j \in \mathcal{R} \,\&\, E \subseteq \cup g^{-1}R_jf^{-1}\} =$$
$$\inf\{\sum \tau(g^{-1}R_jf^{-1}) : g^{-1}R_jf^{-1} \in \mathcal{R} \,\&\, E \subseteq \cup g^{-1}R_jf^{-1}\} =$$
$$\inf\{\sum \tau(R_j) : R_j \in \mathcal{R} \,\&\, E \subseteq \cup R_j\} = \mu(E).$$

□

Lemma 14.1.4. *([111], Theorem 13.3.1, p. 160) For $i \in \mathbb{N}$, let G_i be a unimodular Polish group that is not compact and let μ_i be a Haar measure in G_i. Then the Borel measure μ defined by Lemma* 14.1.2 *is a two sided invariant generator of shy (Haar null) sets in the product-group $G = \prod_{i\in\mathbb{N}} G_i$.*

Let G be a complete separable metric group and μ be a Borel measure in G. We say that μ is generator of shy (Haar null) sets in G if the condition $\overline{\mu}(X) = 0$ implies that X is shy in G, where $\overline{\mu}$ denotes a usual completion of μ. It is obvious that if μ is a quasi-finite [2] two-sided invariant Borel measure in G then μ is generator of shy (Haar null) sets in G.

Lemma 14.1. *([111], Theorem 13.4.1) For $k \in \mathbb{N}$, let $(G_k)_{k\in\mathbb{N}}$ be a family of unimodular Polish groups that are not compact. Let λ_k be a Haar measure in G_k. Let $\mathcal{H}$ be a class of Borel measurable rectangles of the form $\prod_{k\in\mathbb{N}} Y_k$ such that*

$$card(\{k : 0 \leq \lambda_k(Y_k) < +\infty\} = \omega.$$

If a subset $X \subseteq \prod_{i\in\mathbb{N}} G_i$ is covered by the union of countable family of elements of $\mathcal{H}$ then X is shy in the product-group $G = \prod_{i\in\mathbb{N}} G_i$.

Lemma 14.1.5. *([111], Corollary 13.3.1, p.161) For $k \in \mathbb{N}$, let G_k be a unimodular Polish group that is not compact. Then any compact subset (and hence any K_σ subset) of the product-group $\prod_{k\in\mathbb{N}} G_k$ is shy (Haar null) generated by any two sided invariant generator of shy (Haar null) sets on the entire group.*

Lemma 14.1.6. *For $i \in \mathbb{N}$, let G_i be a unimodular Polish group that is not compact and let μ_i be a Haar measure on G_i. Then the Borel measure μ defined by Lemma 14.1.4 is the generator of shy (Haar null) sets in the product-group $G = \prod_{i\in\mathbb{N}} G_i$.*

Theorem 14.1.2. *For $i \in \mathbb{N}$, let G_i be a unimodular Polish group that is not compact and let μ_i be a Haar measure on G_i. Then the Borel measure μ defined by Lemma 14.1.4 is concentrated on a meagre set which is not covered by a countable family of compact sets in $G = \prod_{i\in\mathbb{N}} G_i$.*

Proof. By Lemma 14.1.1 we know that G_1 may be written as the disjoint union of a Haar null set X_1 and a meagre set X_2. Since $\mu_1(X_1) = 0$ and $\mu(X_1 \times \prod_{k>1} G_k) = 0$, we claim that $A := X_1 \times \prod_{k>1} G_k$ is shy (Haar null) in G. Hence μ is concentrated on a set $B := X_2 \times \prod_{k>1} G_k$, which by Theorem 14.1.1 is meagre in G. Let us show that B is not covered by a countable family of compact sets in G. Indeed, assume the contrary and $(F_k)_{k\in\mathbb{N}}$ be a countable family of compact sets in G which covers B. By Lemma 14.1.6 we know that B is shy (Haar null) and we obtain that G as the union of two shy sets A and B is again shy (Haar null) which is the contradiction. □

14.2 On Questions of D. Fremlin

Let E be a Polish space and $\mathcal{B}(E)$ be a Borel σ-algebra of subsets of E equipped with a σ-finite atomless (diffused) Borel measure μ. Let $\mathcal{P}(E)$ be a power set of E. For $A \in \mathcal{P}(E)$ we set

$$\mu^*(A) - \inf\{\mu(B) : A \subseteq B \ \&B \in \mathcal{B}(E)\}$$

[2]A Borel measure μ in G is called quasi-finite if there is a compact set $K \subseteq G$ such that $0 < \mu(K) < +\infty$.

and

$$\mu_*(A) = \sup\{\mu(B) : B \subseteq A \,\&B \in \mathcal{B}(E)\}.$$

μ^* and μ_* are called outer and inner measures generated by the measure μ, respectively.

Following [14], a compact, perfect, totally disconnected, measure zero subset of the real line $\mathbb{R}$ is called a Cantor set.

D.Fremlin posed the following problems.

Problem 14.2.1 ([34], Problem 1.8) Let $X \subseteq \mathbb{R}$, and let λ^* denotes Lebesgue outer measure.

$$\lambda^*(X) = 0 \Leftrightarrow \exists \text{ a Borel probability measure } \mu \text{ such that } (\forall t)(t \in \mathbb{R} \to \mu^*(X+t) = 0)?$$

Problem 14.2.2 ([34], Problem 1.9) Let $X \subseteq \mathbb{R}$.

$$\forall t \in \mathbb{R} \;\; \mu^*_{\text{Cantor}}(X+t) = 0 \to \lambda^*(X) = 0?$$

The validity of the following result has been underlined by D.Fremlin.

Lemma 14.2.1. *([34], Claim 1.10) (ZFC+CH) The answers to Problem 14.2.1 and Problem 14.2.2 are in the negative.*

Therefore the real questions were whether one can find counterexamples in ZFC, that is, without resorting to extra set-theoretic assumptions.

M. Elekes and J. Steprāns extended Fremlin's above mentioned results as follows.

Lemma 14.2.2. *([34], Theorem 1.12 (First Main Theorem)) (ZFC) Problem 14.2.1 has a negative answer.*

Definition 14.2.1. ([34], Definition 1.13) Let $X \subseteq \mathbb{R}$ with $\lambda^*(X) > 0$ and μ be a Borel probability measure. We say that μ reflects the positive measure of X if there exists $t \in \mathbb{R}$ such that $\mu^*(X+t) > 0$. Hence, taking Lemma 14.2.1 into account, one can reformulate Fremlin's second problem as

Problem 14.2.3 ([34], Problem 1.14) Is it consistent that μ_{Cantor} reflects the positive measure of every X?

Lemma 14.2.2 states that there exist a Borel probability measure μ and a subset $X \subseteq \mathbb{R}$ such that μ does not reflect the positive measure of X. Problem 14.2.3, which was still open, asks if a fixed measure μ_{Cantor} consistently reflects. So it is natural to ask the same question about other (fixed) measures (which was also open).

Problem 14.2.4 ([34], Problem 1.15) Is it consistent that there exist an atomless singular Borel probability measure μ such that for every $X \subseteq \mathbb{R}$ with $\lambda^*(X) > 0$ there exists $t \in \mathbb{R}$ such that $\mu^*(X+t) > 0$?

M. Elekes and J. Steprāns considered the following category analogue.

Problem 14.2.5 ([34], Problem 1.16) Is it consistent that there exist a Cantor set K such that for every non-meagre set $X \subseteq \mathbb{R}$ there exists a $t \in \mathbb{R}$ such that $K \cap (X+t)$ is relatively non-meagre in K ?

Their result (see, [34], Theorem 3.35) is result motivates the following Baire category analogue. It is consistent with ZFC that there exist an abelian Polish group G and a Cantor set $C \subset G$ such that for every non-meagre set $X \subseteq G$ there exists a $t \in G$ such that $C \cap (X+t)$ is relatively non-meagre in C. This essentially generalizes results of Bartoszynski [7] and Burke-Miller [13].

We say that the strict duality principle between the measure and category is valid if the duality between the measure and the category is valid for all P sentences formulated only by using the notions of Haar null sets, meagre sets and of purely set-theoretical notions.

The following result is known as the Erdös-Sierpiński duality principle.

Lemma 14.2.3. *([89], (duality principle)) If the continuum hypothesis is true, then the strict duality between a linear Lebesgue measure and the Baire category of the real axis R is valid.*

Remark 14.2.1. Let X be a subset of $\mathbb{R}$. X is called a Sierpiński set if X is uncountable and for every Lebesgue measure zero subset Y of $\mathbb{R}$ the intersection $X \cap Y$ is at most countable. The family of all Sierpiński sets is a $\sigma-$ideal invariant under all transformations of the real line which preserve the $\sigma-$ideal of all Lebesgue measurable subsets of $\mathbb{R}$ (in particular, this family is invariant under all translations of $\mathbb{R}$); Note that the assumption $(MA)\&(2^{\omega} > \omega_1)$ implies that there are no Sierpiński sets. Sierpiński proved that under the Continuum Hypothesis there exists a Sierpiński subset of the real line $\mathbb{R}$ (see, for example, [20], Theorem 8.10, p. 157). Note that every Sierpiński set is a meagre subset of the real line $\mathbb{R}$ and no uncountable subset of a Sierpiński set is Lebesgue measurable (see, [20], Theorem 8.11, p. 157) which implies that $\lambda^*(X) > 0$. It is obvious that $\forall t \in \mathbb{R}$ $\mu^*_{\text{Cantor}}(X+t) = 0$. Indeed, since μ_{Cantor} is concentrated on a set C which is of measure zero and $X+t$ is Sierpiński set for each $t \in \mathbb{R}$, we deduce that

$$\mu^*_{\text{Cantor}}(X+t) = \mu^*_{\text{Cantor}}(X+t \cap C) = 0.$$

An analogous result is valid for every atomless singular Borel probability measure μ in $\mathbb{R}$. Hence every Sierpiński set X is a counterexample in $ZFC+CH$ to Problems 14.2.1-14.2.4.

Let denote by P a sentence defined by :

$$(\exists X)(X \subseteq \mathbb{R}\ \&\ \lambda^*(X) > 0 \rightarrow (\forall Y)(\forall t)(Y \text{is a meagre subset in } \mathbb{R}\ \&\ t \in \mathbb{R}) \rightarrow \text{card}(X+t \cap Y) \leq \omega).$$

Then P^* will be defined by :

$$(\exists X)(X \subseteq \mathbb{R}\ \&\ X \text{ is a non-meagre subset in } \mathbb{R} \rightarrow (\forall Y)(\forall t)(Y \text{is Lebesgue null subset in } \mathbb{R}\ \&\ t \in \mathbb{R}) \rightarrow \text{card}(X+t \cap Y) \leq \omega).$$

Since the validity of the sentence P in the theory $ZFC+CH$ is realized by a Sierpiński set, by the Erdös-Sierpiński duality principle (see Lemma 14.2.3) we deduce that the sentence P^* is also valid in $ZFC+CH$. Since each cantor set K in $\mathbb{R}$ is Lebesgue null and every

A function $\sigma : \omega^{(\omega)} \to \omega$ is called a winning strategy for the player I if $(a_0, a_1, a_2, \ldots) \in A$ for an arbitrary strategy τ for the player II.

Analogously, a function $\tau : \omega^{(\omega)} \to \omega$ is called a winning strategy for the player II if $(a_0, a_1, a_2, \ldots) \notin A$ for an arbitrary strategy σ for the player I.

An infinite game of type G_A is called determined if there exists a winning strategy for at least one of two players in this game.

The Axiom of Determinateness (denoted by AD) is the assertion that *every infinite game of type G_A (for arbitrary $A \subseteq \omega^\omega$) is determined.*

The classical result of Mycielski and Swierczkowski [82][3] is formulated as follows.

Lemma 14.3.1. *(ZF+DC+AD) Every subset of the real axis $\mathbb{R}$ is Lebesgue measurable.*

Remark 14.3.1. The proof of Lemma 14.3.1 is given in Appendix. An example of the another model of set theory in which every set of reals is Lebesgue measurable, can be found in [132].

Lemma 14.3.2. *([50], Fact 6) (ZF+DC) A set $S \subseteq \mathbb{R}$ is Haar null if and only if it has Lebesgue measure zero.*

Proof. Necessity. Let $S \subseteq \mathbb{R}$ be a Haar null set and we have to show that S has Lebesgue measure zero. We need only consider Borel sets, because every Lebesgue measure zero set is contained in a Borel set with measure zero. On the other hand, if a Borel set S is Haar null, there is a Borel probability measure μ on $\mathbb{R}$ such that $\mu(x+S+y) = 0$ for all $x, y \in \mathbb{R}$. Since $\mathbb{R}$ is abelian Polish group, the latter relation is equivalent to the condition that $\mu(S-h) = 0$ for all $h \in \mathbb{R}$.

Let λ be Lebesgue measure on $\mathbb{R}$. Though λ is not finite, it is σ-finite, so by the Tonelli theorem [32] we have that

$$0 = \int_{\mathbb{R}} \mu(S-y)d\lambda(y) = \int_{\mathbb{R}} \lambda(S-x)\mu d(x) = \mu(\mathbb{R})\lambda(S).$$

In other words, S has Lebesgue measure zero.

Sufficiency. Let S has Lebesgue measure zero. We define a Borel probability measure μ in $\mathbb{R}$ as follows:

$$(\forall X)(X \in \mathcal{B}(\mathbb{R}) \to \mu(X) = \lambda(X \cap [0,1])).$$

Let us show that $\mu(S-h) = 0$ for all $h \in \mathbb{R}$. Indeed, we have

$$(\forall h)(h \in \mathbb{R} \to \mu(S-h) = \lambda(S-h \cap [0,1]) \leq \lambda(S-h) = \lambda(S) = 0).$$

□

Let (E_1, S_1, μ_1) and (E_2, S_2, μ_2) be two measure spaces. The measures μ_1 and μ_2 are called isomorphic if there exists a measurable isomorphism φ from E_1 onto E_2 such that

$$(\forall X)(X \in S_1 \to \mu_1(X) = \mu_2(\varphi(X))).$$

[3]The proof of Mycielski and Swierczkowski above mentioned classical result can be found in the Appendix of the present monograph.

Lemma 14.3.3. *([20]) Let E_1 and E_2 be any two Polish spaces. Let μ_1 be a probability diffused Borel measure in E_1 and let μ_2 be a probability diffused Borel measure in E_2. Then there exists a Borel isomorphism $\varphi : (E_1, B(E_1)) \to (E_2, B(E_2))$ such that*

$$\mu_1(X) = \mu_2(\varphi(X))$$

for every $X \in B(E_1)$.

Lemma 14.3.4. *(ZF+DC+AD) Let E be a Polish space and let μ be a probability diffused Borel measure on E. Then the completion $\overline{\mu}$ of μ is defined on the power set of E.*

Proof. Let b_1 be the standard Lebesgue measure in $[0,1]$. According to Lemma 14.3.3, the measure b_1 is Borel isomorphic to the probability measure μ. Denote this isomorphism by φ and consider an arbitrary set $W \subset E$. We have to show that $W \in \mathrm{dom}(\overline{\mu})$. By Lemma 14.3.1 we know that $\varphi^{-1}(W) = X \cup Y$, where $X \in B([0,1])$ and $(\exists Z)(Y \subset Z \in B([0,1])\ \&\ b_1(Z) = 0)$. Let us take the set $X \cup Z$. The Borel isomorphism between the measures b_1 and μ implies that both $\varphi(X)$ and $\varphi(Z)$ are Borel subsets in E such that $\mu(\varphi(Z)) = 0$. On the one hand, we can write $W = \varphi(X) \cup \varphi(Y)$, where $\varphi(Y) \subset \varphi(Z)$. Clearly, $\overline{\mu}(\varphi(Y)) = 0$ since $\mu(\varphi(Z)) = 0$. So $\varphi(X), \varphi(Y) \in \mathrm{dom}(\overline{\mu})$ we claim that $W = \varphi(X) \cup \varphi(Y) \in \mathrm{dom}(\overline{\mu})$. □

Lemma 14.3.5. *(ZF+DC+AD) Let E be a Polish space and let μ be a sigma-finite Borel measure in E. Then*

$$(\forall X)(X \subseteq E \to \mu^*(X) = \mu_*(X) = \overline{\mu}(X)),$$

where $\overline{\mu}$ denotes a usual completion of μ.

Lemma 14.3.6. *(ZF+DC+AD) Problem 14.2.1 has an affirmative answer, i.e., for a set $X \subseteq \mathbb{R}$, the condition $\lambda^*(X) = 0$ holds if and only if there is a Borel probability measure μ such that $\mu^*(X+t) = 0$ for all $t \in \mathbb{R}$.*

Proof. Necessity. Let $X \subseteq \mathbb{R}$ be a set with $\lambda^*(X) = 0$. Since $\lambda^*(X) = \lambda(X) = 0$, there is a Borel set Y which contains X and for which $\lambda(Y) = 0$. By Lemma 14.3.2, there is a Borel probability measure μ such that $\mu(Y-h) = 0$ for each $h \in \mathbb{R}$. Since $X - h \subseteq Y - h$ and $Y - h$ is Borel for $h \in \mathbb{R}$, we deduce that $\mu^*(X-h) \leq \mu(Y-h) = 0$ for every $h \in \mathbb{R}$.

Sufficiency. Suppose that there is a Borel probability measure μ such that $\mu^*(X+t) = 0$ for all $t \in \mathbb{R}$. We have to show that $\lambda^*(X) = 0$. Assume the contrary and let $\lambda^*(X) > 0$. Since $\lambda^*(X) = \lambda_*(X)$, we claim that there is a Borel set $Y \subseteq X$ such that $\lambda(Y) > 0$. By Lemma 14.3.2 we know that for a Borel probability measure μ there is a shift t_0 such that $\mu(Y+t_0) > 0$. Using Lemma 14.3.5, we have

$$\mu^*(X+t_0) = \mu_*(X+t_0) \geq \mu(Y+t_0) > 0$$

which contradict to the condition that $\mu^*(X+t) = 0$ for all $t \in \mathbb{R}$. □

Corollary 14.3.1. *(ZF+DC+AD) Problem 14.2.4 has an affirmative answer. Moreover, an arbitrary Borel probability measure reflects the positive measure of every X, i.e., for every Borel probability measure μ in $\mathbb{R}$ and for every $X \subseteq \mathbb{R}$ with $\lambda^*(X) > 0$ there exists $t \in \mathbb{R}$ such that $\mu^*(X+t) > 0$.*

Proof. Assume the contrary. Then there will be a Borel probability measure μ in $\mathbb{R}$ and a subset $X \subseteq \mathbb{R}$ with $\lambda^*(X) > 0$ such that for every $t \in \mathbb{R}$ we have $\mu^*(X+t) = 0$.

By Lemma 14.3.6 we claim that $\lambda^*(X) = 0$ which is a contradiction. □

We have the following corollaries of Corollary 14.3.1.

Corollary 14.3.2. *(ZF+DC+AD) Problem 14.2.2 has an affirmative answer, i.e., if* $X \subseteq \mathbb{R}$ *and* $\mu^*_{Cantor}(X+t) = 0$ *for every* $t \in \mathbb{R}$, *then* $\lambda^*(X) = 0$.

Corollary 14.3.3. *(ZF+DC+AD) Problem 14.2.3 has an affirmative answer, i.e.,* μ_{Cantor} *reflects the positive measure of every* X.

14.4 Main Results

The purpose of the present section is to show that Problems 14.2.1-14.2.4 are not solvable within $ZF+DC$ theory. In other words, Problems 14.2.1-14.2.4 are independent from $ZF+DC$ theory.

Theorem 14.4.1. *Problem 14.2.1 is independent of* $ZF+DC$ *theory.*

Proof. On the one hand, by Remark 2.1 we know that the answer to Problem 14.2.1 is negative in the theory ZFC. On the other hand, by Lemma 14.3.6 we know that Problem 14.2.1 has an affirmative answer in the theory $ZF+DC+AD$. Hence Problem 14.2.1 is independent of $ZF+DC$ theory. □

Theorem 14.4.2. *Problem 14.2.2 is independent of* $ZF+DC$ *theory.*

Proof. On the one hand, by Remark 14.2.1 we know that the answer to Problem 14.2.2 is negative in the theory ZFC. On the other hand, by Corollary 14.3.2 we know that Problem 14.2.2 has an affirmative answer in the theory $ZF+DC+AD$. Hence Problem 14.2.2 is independent of $ZF+DC$ theory. □

Theorem 14.4.3. *Problem 14.2.3 is independent of* $ZF+DC$ *theory.*

Proof. On the one hand, by Remark 14.2.1 we know that the answer to Problem 14.2.3 is negative in the theory ZFC. On the other hand, by Corollary 14.3.3 we know that Problem 14.2.3 has an affirmative answer in the theory $ZF+DC+AD$. Hence Problem 14.2.3 is independent of $ZF+DC$ theory. □

Theorem 14.4.4 *Problem 14.2.4 is independent of* $ZF+DC$ *theory.*

Proof. On the one hand, by Remark 14.2.1 we know that the answer to Problem 14.2.4 is negative in the theory ZFC. On the other hand, by Corollary 14.3.1 we know that Problem 14.2.4 has an affirmative answer in the theory $ZF+DC+AD$. Hence Problem 14.2.4 is independent of $ZF+DC$ theory. □

Chapter 15

On a Certain Modification of P. Erdős Problem for Translation-Invariant Quasi-Finite Diffused Borel Measures in Polish Groups that are not Locally Compact

15.1 Introduction

Following [80], many years ago Erdős noted that if E is a Lebesgue measurable subset of the plane $\mathbb{R}^2$ with infinite measure, then for every $c > 0$, E contains the vertices of a triangle of area c. As several people have noted, this remains true if E has positive measure and is unbounded. In [36] and again in [37], [38], [39], Erdős poses what he said was an interesting and perhaps difficult problem.

Problem 15.1.1. Is there a finite constant c such that if a Lebesgue measurable set E has measure greater than c, then E contains the vertices of a triangle of area 1? Moreover, is it true that the best constant is $c_0 = \frac{4\pi}{3\sqrt{3}}$, the area of the disk such that the area of the inscribed equilateral triangle is 1?

Remark 15.2.1. It is well known in folklore that Erdős problem is equivalent to the following problem.

Problem 15.1.1$'$. Let m^2 denotes a two-dimensional Lebesgue measure in $\mathbb{R}^2$. Is there a finite constant c such that if a Lebesgue measurable set E has positive Lebesgue measure, then for each s with $0 < s < \frac{m^2(E)}{c}$ the set E contains the vertices of a triangle of area s? Moreover, is it true that the best constant is $c_0 = \frac{4\pi}{3\sqrt{3}}$?

On the one hand, it is obvious to show that if c is a solution of the Problem 15.1.1 then it will be also a solution of the Problem 15.1.1$'$. Indeed, assume the contrary and let c be not a solution of the Problem 15.1.1$'$. Then there will be a Lebesgue measurable set E_0 with positive Lebesgue measure and a positive number s_0 with $0 < s_0 < \frac{m^2(E_0)}{c}$ such that the set

E_0 contains no the vertices of a triangle of area s_0. Let consider a mapping $T : \mathbb{R}^2 \to \mathbb{R}^2$ defined by

$$T(x,y) = (\frac{x}{\sqrt{s_0}}, \frac{y}{\sqrt{s_0}}) \tag{15.1.1}$$

for $(x,y) \in \mathbb{R}^2$. Let consider a set $T(E_0)$. It is obvious that

$$m^2(T(E_0)) = \frac{m^2(E_0)}{s_0} > m^2(E_0) \times \frac{c}{m^2(E_0)} = c. \tag{15.1.2}$$

By our assumption $T(E_0)$ must contain the vertices M_1, M_2, M_3 of a triangle of area 1. It is obvious that the points $T^{-1}(M_1), T^{-1}(M_2), T^{-1}(M_3)$ belong to the set E_0 and they are the vertices of a triangle of area s_0 which is a contradiction.

On the other hand, if c is a solution of the Problem 15.1.1$'$ then for each s with $0 < s < \frac{m^2(E)}{c}$ the set E contains the vertices of a triangle of area s. Since $\frac{m^2(E)}{c}$ is bigger then 1, we deduce that for $s_1 = 1$, the set E contains the vertices of a triangle of area s_1 and we obtain that c is a solution of the Problem 15.1.1.

Freiling and Mauldin [80] have studied Erdös above mentioned problem. Using some standard approximations in measure theory, they proved that the Problem 15.1.1 is equivalent to the following problem.

Problem 15.1.2. Is there a finite constant c such that for every $n \in \mathbb{N}$ if E is the union of the interiors of no more than n compact convex sets and E has measure greater than c, then E contains the vertices of a triangle of area 1. Moveover, is c_0 the best possible constant?

Freiling and Mauldin [79] showed that the constant c_0 is the best possible if n is 1. More lately, they proved in [80] that the constant c_0 remains the best possible if n is 2 or 3.

Note that the Problem 15.1.1 still remains open.

In [100] has been considered more delicate question for an extension λ of the n-dimensional Lebesgue measure ℓ_n defined in $\mathbb{R}^n$ $(n > 1)$ asking *whether there is a positive constant* c_λ *such that every* λ*-measurable set in* $\mathbb{R}^n$ *with the* λ*-measure greater than* c_λ *contains the vertices of a triangle of area one.* That question has been answered negatively in [100]. More precisely, it has been proved an existence of an extension λ of the Lebesgue measure ℓ_n such that for every positive constant c there exists a λ-measurable set E_c for which the λ-measure of E_c is greater than c and the set E_c contains no the vertices of a triangle of area one.

If we consider Erdös above-mentioned question for a quasi-finite [1] translation-invariant diffused [2] Borel measure in an infinite-dimensional separable Banach space then we observe that the answer to that question is no. Indeed, following [50] (see,p. 218), in an infinite-dimensional separable Banach space, every non-trivial translation-invariant Borel measure has the property that all open sets have infinite measure. Now it is obvious that for every quasi-finite translation-invariant diffused Borel measure μ in an infinite-dimensional separable Banach space X there is no a positive constant c such that every measurable set in X with μ measure greater than c contains the vertices of a triangle of area onc. In order

[1] A Borel measure μ is called quasi-finite if there is a Borel subset Y such that $0 < \mu(Y) < \infty$.

[2] A Borel measure is called diffused if it vanishes on all singletons.

to see it we can consider any open ball B of radius $1/2$. Then $\mu(B) = +\infty$ and every triangle with the vertices in B has an area less than 1. Note that the similar result fails in finite-dimensional Euclidean space for Lebesgue measure because, as we have mentioned above, every set of infinite Lebesgue measure contains the vertices of a triangle of area one.

Let $\mathcal{B}(X)$ denotes the σ-algebra of Borel subsets of the infinite-dimensional separable Banach space X. Following [50], a Borel subset $Y \subseteq X$ is called shy if there exists a Borel probability measure μ in X such that $\mu(Y+h) = 0$ for all $h \in X$. A Borel subset $W \subseteq X$ is called *prevalent* iff the complement of W is shy.

It well known that if a planar set A has a positive measure and A is unbounded, then A contains the vertices of a triangle of area one(see, for example, [79], p.1). We have the following infinite analogue of that result: each everywhere dense non-shy Borel subset E (for example, prevalence) in an infinite-dimensional separable Banach space X contains the vertices of a triangle of area one. Indeed, let L be such one dimensional vector space for which exists a sequence $(z_k)_{k\in\mathbb{N}}$ of elements of E that a distance between z_k and L tends to $+\infty$ as well k tends to $+\infty$. Let m be a linear Lebesgue measure on L and $L^{\perp}$ be its linear complement in X. We set

$$(\forall Y)(Y \in \mathcal{B}(X) \to \mu(Y) = \sum_{x\in L^{\perp}} m(Y + x \cap L)). \tag{15.1.3}$$

Obviously, μ is a translation-invariant quasi-finite diffused Borel measure in X. Following [96](see, Corollary 2.1, p. 241), the measure μ is generator of shy sets provided that the validity of the condition $\mu(Y) = 0$ implies that Y is a shy set in X. Note that $\mu(E) > 0$. Indeed, assume the contrary and let $\mu(E) = 0$. Since μ is generator of shy sets, we deduce that E is shy and we get a required contradiction. So there is $x_0 \in L^{\perp}$ such that $m(E + x_0 \cap L) > 0$. Obviously $E \cap (L - x_0) \subseteq E$ and by well known fact (see, [58], Exercise 124, p. 189) a set $(E \cap (L - x_0)) - (E \cap (L - x_0))$ contains any open interval (with center at zero) in L. Let $r > 0$ be such a real number that for a closed ball $B[\mathbf{0}, r]$ the condition $B[\mathbf{0}, r] \cap L \subseteq (E \cap (L - x_0)) - (E \cap (L - x_0))$ holds. Since distant between z_k and L tends to $+\infty$ as well k tends to $+\infty$ we deduce that there is a point $x_1 \in \{z_k : k \in \mathbb{N}\}$ such that the distance between x_1 and $L - x_0$ is greater than $1/r$. Now we can choose two points $x_2, x_3 \in E \cap (L - x_0)$ such that $||x_3 - x_2|| = 2/h$. Obviously points x_1, x_2, x_3 are in E and

$$S = \frac{1}{2}||x_3 - x_2||\rho(x_1, L - x_0) = \frac{1}{2} \times \frac{2}{h} \times h = 1, \tag{15.1.4}$$

where S denotes area of the triangle with vertices x_1, x_2, x_3.

By the scheme presented above one can get the validity of the following assertion.

Theorem 15.1.1. *Each everywhere dense Borel non-shy set E in an infinite-dimensional separable Banach space X, for $n > 2$ contains the vertices of an $n-1$-dimensional polyhedron of $n-1$-dimensional Lebesgue measure one.*

Let $\mathbf{G}$ be a Polish group, by which we mean a separable group with a complete metric ρ for which the transformation (from $\mathbf{G} \times \mathbf{G}$ onto $\mathbf{G}$) which sends $(\mathbf{x}, \mathbf{y})$ into $\mathbf{x}^{-1}\mathbf{y}$ is continuous. Let $\mathcal{B}(\mathbf{G})$ denotes the σ-algebra of Borel subsets of $\mathbf{G}$.

Definition 15.1.1. A Borel measure μ in G is called left(or right) invariant if

$$(\forall X)(\forall g)(X \in B(G) \& g \in G \to \mu(gX) = \mu(X)(\text{ or } \mu(Xg) = \mu(X))). \tag{15.1.5}$$

Definition 15.1.2. A Borel measure μ in G is called two sided invariant if

$$(\forall X)(\forall g, f)(X \in B(G) \& g, f \in G \to \mu(gXf) = \mu(X)). \tag{15.1.6}$$

Definition 15.1.3. Let K be the class of measures in G. We say that a measure $\mu \in K$ has the property of uniqueness in the class K if μ and λ are equivalent for every $\lambda \in K$.

Definition 15.1.4. Let A_1, A_2, A_3 be three points in G. Under area defined by points A_1, A_2, A_3 we understand area of planar triangle in $\mathbb{R}^2$ with vertices B_1, B_2, B_3 such that $\rho(A_i, A_j) = ||B_i - B_j||_2$ for $1 \leq i \leq j \leq 3$, where $||\cdot||_2$ denotes a usual norm in the two-dimensional Euclidean topological vector space $\mathbb{R}^2$.

In Section 6 has been presented a method of the construction of a reach class of non-σ-finite quasi-finite translation-invariant diffused Borel measures in infinite-dimensional separable Banach spaces.

The following lemmas due to Oxtoby contain more interesting information about existence of invariant Borel measures in such Polish groups which strictly contains a class of all infinite-dimensional separable Banach spaces.

Lemma 15.1.1. *([88], Theorem 3) In any uncountable complete separable metric group which is dense in itself there exists a left(or right)-invariant quasi-finite diffused Borel measure and no any left(or right)-invariant quasi-finite diffused Borel measure possesses a uniqueness property.*

The main purpose of the present section is to consider the following generalizations and modifications of the Problem 15.1.1.

Question 15.1.1 Let G be an uncountable complete separable metric group which is dense in itself. Let μ be a left(or right)-invariant quasi-finite diffused Borel measure. Does there exist a positive constant c such that each Borel set E with the μ-measure greater than c contains three points such that area defined by that points is equal to one?
Question 15.1.2 Let G be an uncountable complete separable metric group with two-sided invariant metric which is dense in itself. Let μ be a two-sided-invariant quasi-finite diffused Borel measure in G. Does there exist a positive constant c such that each Borel set E with the μ-measure greater than c contains three points such that area defined by that points is equal to one?

Example 15.1.1. It is obvious that if a diameter of the group G is less than one the answer to Questions 15.1.1-15.1.2 is no. For example, if we consider a topological vector space of all real valued sequences $\mathbb{R}^\infty$ equipped with Tychonoff metric ρ defined by

$$\rho((x_k)_{k\in\mathbb{N}}, (y_k)_{k\in\mathbb{N}}) = \sum_{k\in\mathbb{N}} \frac{|x_k - y_k|}{2^k(1 + |x_k - y_k|)} \tag{1.7}$$

for $(x_k)_{k\in\mathbb{N}}, (y_k)_{k\in\mathbb{N}} \in \mathbb{R}^\infty$, then we observe that the distance between any two points is less than 1. Hence an answer to Questions 15.1.1-15.1.2 is no because area defined by each three points in $\mathbb{R}^\infty$ is always less than 1.

Example 15.1.2. Concerning with Questions 15.1.1-15.1.2, it can be noted that there exists a two-sided-invariant quasi-finite diffused Borel measure in the two-dimensional topological vector space $\mathbb{R}^2$ (with a usual Euclidian metric) for which the answer to Questions 15.1.1-15.1.2 is no. In this direction it is sufficient under such a measure to take a measure μ defined by the formula (15.1.3) for $X = \mathbb{R}^2$.

The chapter is organized as follows.

In Section 15.2 we consider some auxiliary facts from measure theory and functional analysis. In Section 15.3 we present a negative answers to the Questions 15.1.1 for left(or right)-invariant quasi-finite diffused Borel measures in uncountable complete separable metric groups which are dense in itself and are not locally compact. In Section 15.4 we present another approach for a solution of the Question 15.1.1 for the infinite-dimensional separable Hilbert space ℓ_2 and consider narrow related topics.

15.2 Some Auxiliary Facts from Measure Theory and Functional Analysis

Let (X_i, M_i, μ_i) be a sequence of measure spaces and for each i let ρ_i be a metric on the set X_i. Assume that the following conditions are satisfied:

(i) Each (X_i, ρ_i) is a locally compact metric space;

(ii) Each M_i contains the family, $B(X_i)$, of Borel subsets of X_i and μ_i is a regular Borel measure on X_i;

(iii) For all i, and for every $\delta > 0$, there exists a sequence (A_j) of Borel subsets of X_i such that $d_i(A_j) < \delta$ and $X_i = \cup_{j=1}^{\infty} A_j$, where $d_i(A_j)$ is the diameter of A_j in X_i.

Define $X = \prod_{i=1}^{\infty} X_i$ and let equip X with the product topology. Let denote by $\mathcal{R}$ the family of all rectangles $R \subseteq X$ of the form $R = \prod_{i=1}^{\infty} R_i$, $R_i \in \mathcal{B}(X_i)$, and

$$0 \le \prod_{i=1}^{\infty} \mu_i(R_i) := \lim_{n \to \infty} \prod_{i=1}^{n} \mu_i(R_i) < +\infty. \tag{15.2.1}$$

For $R \in \mathcal{R}$, we define

$$\tau(R) = \prod_{i=1}^{\infty} \mu_i(R_i). \tag{15.2.2}$$

Let τ^* be the set function on the powerset $\mathcal{P}(X)$ defined by

$$\tau^*(E) = \inf\left\{ \sum \tau(R_j) : R_j \in \mathbb{R} \,\&\, E \subseteq \cup R_j \right\} \tag{15.2.3}$$

for $E \subseteq X$.

Also, let use the convention that $0 \cdot +\infty = +\infty \cdot 0 = 0$, $+\infty \cdot +\infty = +\infty$, and that the infimum taken over the empty set has the value $+\infty$.

Lemma 15.2.1. *([3], Theorem I, p. 2579) The set function τ^* is an outer measure on X. Let M be the σ-algebra of subsets of X which are measurable with respect to τ^*, and let μ be the measure on M obtained by restricting τ^* to M. Then $B(X) \subseteq M$, and for all $R = \prod_{i=1}^{\infty} R_i \in \mathcal{R}$, we have $\mu(R) = \prod_{i=1}^{\infty} \mu_i(R_i)$. If each space X_i contains disjoint subsets*

A_i, B_i such that $\mu_i(A_i) = \mu_i(B_i) = 1$, then the measure μ is not σ-finite. Finally, assume that each (X_i, ρ_i) is an M_i-measurable group. If each μ_i is left-invariant measure on M_i, then μ is a left-invariant measure on M. Similarly, if each μ_i is right-invariant measure on M_i, then μ is right-invariant measure on M.

Following R. Baker [3], a measure λ being the completion of a translation invariant Borel measure in $\mathbb{R}^\infty$ is called a "Lebesgue measure" if for any measurable rectangle $\prod_{i=1}^{\infty} R_i$, $R_i \in \mathcal{B}(\mathbb{R})$ with $0 \leq \prod_{i=1}^{\infty} m(R_i) < \infty$, the following equality

$$\lambda\Big(\prod_{i=1}^{\infty} R_i\Big) = \prod_{i=1}^{\infty} m(R_i) \tag{15.2.4}$$

holds, where m denotes a linear Lebesgue measure on $\mathbb{R}$.

Setting $(X_i, M_i, \mu_i) := (\mathbb{R}, \mathcal{B}(\mathbb{R}), m)$ for $i \in \mathbb{N}$, where $\mathcal{B}(\mathbb{R})$ denotes a Borel σ-algebra of subsets of the real axis $\mathbb{R}$ and m denotes a linear Lebesgue measure on $\mathbb{R}$, then the measure defined by Lemma 15.2.1 will be "Lebesgue measure" in $\mathbb{R}^\infty$. We denote that measure by λ.

Lemma 15.2.2. *[35, Theorem 2.3.6] Let $(Y_i)_{i\in I}$ be a family of topological spaces and $Y = \prod_{i\in I} Y_i$ be their cartesian product endowed with product topology. Let $p_j : Y \to Y_j$ be the j-projection for $j \in I$, defined by*

$$p_j((y_i)_{i\in I}) = y_j \quad (y_i)_{i\in I} \in Y. \tag{15.2.5}$$

If X is a topological space, then a mapping $f : X \to Y$ is continuous if and only if each composition $p_j \circ f$, $j \in I$, is continuous.

Lemma 15.2.3. *(Lusin-Souslin [55, Theorem 15.1]) Let X, Y be Polish spaces and $f : X \to Y$ be continuous. If $A \subseteq X$ is Borel and $f|_A$ is injective then $f(A)$ is Borel.*

Let ℓ_2 be an infinite-dimensional separable Hilbert space and $(e_k)_{k\in\mathbb{N}}$ be the standard basis.

If we define an operator $T : \ell_2 \to \mathbb{R}^{\mathbb{N}}$ by

$$T(\sum_{k\in\mathbb{N}} \alpha_k e_k) = (\alpha_1, \alpha_2, 2\alpha_3, \cdots, n\alpha_{n+1}, \cdots) \tag{15.2.6}$$

then T will be clearly linear, injective and continuous (by Lemma 2.2).

Lemma 15.2.4. *([44, Lemma 2.5]) The subspace $S := T(\ell_2) \in \mathcal{B}(\mathbb{R}^{\mathbb{N}})$ and the operator $T : \ell_2 \to S$ is a Borel isomorphism.*

Lemma 15.2.5. *If we set*

$$(\forall X)(X \in \mathcal{B}(\ell_2) \to m_\infty(X) = \lambda(T(X)), \tag{15.2.7}$$

then m_∞ will be a translation-invariant inner regular non-σ-finite Borel measure in ℓ_2 which gets the numerical value one on the infinite-dimensional modified Hilbert cube P defined by

$$P = \{\sum_{k\in\mathbb{N}} \alpha_k e_k : 0 \leq \alpha_1 \leq 1, 0 \leq \alpha_2 \leq 1, 0 \leq \alpha_3 \leq 1/2, \cdots, \alpha_{n+1} \leq 1/n, \cdots\}. \tag{15.2.8}$$

In the sequel we preserve a notation m_∞ for its usual completion.

For every different points $x, y \in \ell_2$ and a positive number r, we put

$$C_{(x,y,r)} = \{z : z \in \ell_2 \ \& \ \rho(z, L(x,y)) = r\}, \tag{15.2.9}$$

where ρ denotes a standard metric in the separable Hilbert space ℓ_2 and $L(x,y)$ denotes a line drawing on points x and y. A set $C_{(x,y,r)}$ is called a cylinder (of a diameter $2r$) and a line $L(x,y)$ is called an axis of its symmetry.

We now state a result that will be of use later.

Lemma 15.2.6. *Let $C_{(x,y,r)}$ be a cylinder in ℓ_2 and let $L(x,y)$ be an axis of its symmetry. Let l be a line in ℓ_2 which is not parallel to the line $L(x,y)$. Then $card(C_{(x,y,r)} \cap l) \leq 2$.*

Proof. Without loss of generality we can assume that $x = (0,1,0,\cdots)$ and $y = (0,0,0,\cdots)$. Suppose that a line l is defined by the following equation

$$\frac{z_1 - b_1}{c_1} = \cdots = \frac{z_n - b_n}{c_n} = \cdots, \tag{15.2.10}$$

where $b = (b_1, \cdots, b_n, \cdots) \in \ell_2$ and vectors $c = (c_1, \cdots, c_n, \cdots) \in \ell_2$ and $e_2 = (0,1,0,0,\cdots)$ are not collinear.

Then for the cylinder $C_{(x,y,r)}$ we have the following representation

$$C_{(x,y,r)} = \{(z_1, z_2, \cdots) : z_1^2 + \sum_{i=3}^{\infty} z_i^2 = r^2\}. \tag{15.2.11}$$

Let us consider a parametric equation of the line l. It has the following form

$$z_k = b_k + c_k t \ (k \leq 1, t \in \mathbb{R}). \tag{15.2.12}$$

To find coordinates of points belonging to the intersection $C_{(x,y,r)} \cap l$ we must consider the following equation

$$z_1^2 + \sum_{i=3}^{\infty} z_i^2 = r^2 \tag{15.2.13}$$

with respect to a parameter t. It turns out that the set $C_{(x,y,r)} \cap l$ contains no more than two points. □

Lemma 15.2.7. *Let F be a Borel measurable set in ℓ_2 with $0 < m_\infty(F) < \infty$. Then F is not covered by the family of cylinders $(C_i)_{i \in I}$ such that $card(I) < c$, where c denotes the cardinality of the continuum.*

Proof. Without loss of generality we can assume that $T(F) \subseteq R = \prod_{i \in \mathbb{N}} R_i \in \mathcal{R}$. Assume the contrary and let F can be covered by the family of cylinders $(C_i)_{i \in I}$ such that $card(I) < c$. Let L_i be an axis of the symmetry of the cylinder C_i for $i \in I$. Let consider projection $(L_i^*)_{i \in I}$ of lines $(L_i)_{i \in I}$ to the $\mathbb{R}^2$.

For $x = (x_3, x_4, \cdots) \in \mathbb{R}^{\mathbb{N} \setminus \{1,2\}}$ we denote by $T(F)_x$ x-section of $T(F)$ defined by

$$T(F)_x = \{(x_1, x_2) : (x_1, x_2, x_3, \cdots) \in T(F)\}. \tag{15.2.14}$$

By Fubini theorem we have that

$$\lambda(T(F)) = \int_{\prod_{i>2} R_i} m^2(T(F)_x)d\lambda^{(2)}(x) > 0, \tag{15.2.15}$$

where $\lambda^{(2)}$ is the Baker measure in $\prod_{i>2} R_i$. Hence there is $(x_1,x_2,x_3,\cdots) \in T(F)$ such that a set $T(F)_x = \{(y_1,y_2) : (y_1,y_2,x_3,\cdots) \in T(F)\}$ has m^2 positive measure. Hence there exist real numbers a,b,d such that a vector (a,b) is not perpendicular to lines $(L_i^*)_{i\in I}$ and a set

$$D_0 = \{(y_1,y_2) : (y_1,y_2,x_3,\cdots) \in T(F) \ \& \ ay_1 + by_2 + d = 0\} \tag{15.2.16}$$

has a positive linear measure.

Let consider a set $F_0 = D_0 \times \{(x_3,x_4,\cdots)\} = D_0 \times \{(0,0,\cdots)\} + (0,0,x_3,x_4,\cdots)$. Obviously

$$F \supset T^{-1}(F_0) = T^{-1}(D_0 \times \{(0,0,\cdots)\} + (0,0,x_3,x_4,\cdots)) =$$
$$T^{-1}(D_0 \times \{(0,0,\cdots)\}) + T^{-1}(0,0,x_3,x_4,\cdots)). \tag{15.2.17}$$

Hence, $card(F \cap l)) = c$, where the line l is defined by

$$l = T^{-1}(D_0 \times \{(0,0,\cdots)\}) + T^{-1}(0,0,x_3,x_4,\cdots). \tag{15.2.18}$$

On the other hand, by Lemma 15.2.6 we deduce that

$$card((F \cap l) \cap (\cup_{i\in I} C_i)) \leq card(l \cap (\cup_{i\in I} C_i)) \leq 2card(I) + \aleph_0 < c. \tag{15.2.19}$$

The latter relation means that the set F is not covered by the set $\cup_{i\in I} C_i$. □

Lemma 15.2.8. *There exists a set $D \subset \ell_2$ such that the following two conditions are fulfilled:*

a) D intersects every Borel measurable set of positive m_∞ measure;

b) D does not contain the vertices of a triangle of area one.

Proof. Let $(F_\xi)_{\xi<\phi}$ be a sequence of Borel measurable sets of finite positive measure m_∞, where ϕ denotes the first ordinal number of a cardinality of the continuum.

Let τ be an operator of a global choice. We set $x_0 = \tau(F_0)$, $x_1 = \tau(F_1 \setminus \{x_0\})$ and $x_2 = \tau(F_2 \setminus (\{x_0\} \cup \{x_1\} \cup C_{(x_0,x_1,r(x_0,x_1))}))$, where $r(x_0,x_1) = \frac{2}{\rho(x_0,x_1)}$.

Assume that for $\xi < \phi$ a partial sequence of different points $(x_\theta)_{\theta<\xi}$ is constructed such that the following two conditions are fulfilled:

(i) For $\theta < \xi$ we have that $x_\theta \in F_\theta$;

(ii) A set $\{x_\theta : \theta < \xi\}$ does not contain the vertices of a triangle of area one.

We put

$$A_\xi = F_\xi \setminus (\cup_{(x,y)\in\{x_\theta:\theta<\xi\}^2} C_{(x,y,r(x,y))} \cup \cup_{\theta<\xi}\{x_\theta\}). \tag{15.2.20}$$

By Lemma 15.2.7, this difference is not empty set. Now x_ξ can be defined by $x_\xi = \tau(A_\xi)$. Thus, a transfinite ϕ-sequence of different points $(x_\xi)_{\xi<\phi}$ is constructed such that:

(i*) For $\xi < \phi$ we have that $x_\xi \in F_\xi$;

(ii*) A set $\{x_\xi : \theta < \xi\}$ does not contain the vertices of a triangle of area one.

We set $D = \{x_\xi : \xi < \phi\}$.

An inner regularity of the measure μ_∞(cf. Lemma 15.2.5) implies the validity of the condition a).

Let us check the validity of the condition b). Assume the contrary and let $x_{\xi_1}, x_{\xi_2}, x_{\xi_3}$ be three points of the set D such that $\xi_1 < \xi_2 < \xi_3$ and an area of the triangle with vertices $x_{\xi_1}, x_{\xi_2}, x_{\xi_3}$ is equal to one. Following our construction, x_{ξ_3} does not belong to the set $C_{(x_{\xi_1}, x_{\xi_2}, r(x_{\xi_1}, x_{\xi_2}))}$, and hence, we get a required contradiction. □

In the sequel we need the following lemmas.

Lemma 15.2.9. *([88], Theorem 2, p. 217) Let G be any complete separable metric group which is not locally compact, and let μ be any left-invariant Borel measure in G. Then every neighborhood contains non-denumerably many disjoint mutually congruent compact sets of equal finite positive measure.*

Lemma 15.2.10. *([20], Theorem 4.12, p. 81) Let E_1 and E_2 be any two Polish topological spaces. Let μ_1 be a probability diffused Borel measure on E_1 and let μ_2 be a probability diffused Borel measure on E_2. Then there exists a Borel isomorphism $\varphi : (E_1, B(E_1)) \to (E_2, B(E_2))$ such that*

$$\mu_1(X) = \mu_2(\varphi(X)) \tag{15.2.21}$$

for every $X \in B(E_1)$.

15.3 On Erdös Problem for Translation-Invariant Quasi-Finite Diffused Borel Measures in Uncountable Non-Locally-Compact Polish Groups

Theorem 15.3.1. *Let G be an uncountable non-locally-compact complete separable metric group which is dense in itself. Let μ be a left (or right)-invariant quasi-finite diffused Borel measure. Then there exists an increasing family of Borel subsets $(E_c)_{c>0}$ such that:*

a) $\mu(E_c) = c$ for $c > 0$;

b) $\cup_{c>0} E_c$ does not contain three points such that area defined by that points is equal to one.

Proof. Let us consider any open ball B of radius $1/2$. On the one hand, an area defined by arbitrary three points in B is less than one. On the other hand, following Lemma 15.2.9, B contains non-denumerably many disjoint mutually congruent compact sets of equal finite positive μ measure. Hence there is a compact set F and a countable family of elements $(h_k)_{k \in \mathbb{N}}$ of G such that

(i) $0 < \mu(F) < +\infty$;

(ii) $F + h_k \cap F + h_n = \emptyset$ for each different natural numbers $k, n \in \mathbb{N}$;

(iii) $\cup_{k \in \mathbb{N}} F + h_k \subseteq B$.

Let consider a Borel probability measure μ_1 defined by $\mu_1(Y) = \frac{\mu(Y \cap F)}{\mu(F)}$ for $Y \in \mathcal{B}(F)$. Let μ_2 be a linear Lebesgue measure defined on the Borel subsets of $[0.1]$. By Lemma 15.2.10 there exists a Borel isomorphism $\varphi : (F, \mathcal{B}(F)) \to ([0,1], \mathcal{B}([0,1]))$ such that

$$\mu_1(X) = \mu_2(\varphi(X)) \tag{15.3.1}$$

for every $X \in \mathcal{B}(F)$.

For each $0 \le A \le 1$ we set $F_A = \varphi^{-1}([0,A])$. Then we get

$$\mu_1(F_A) = \mu_2(\varphi(F_A)) = \mu_2(\varphi(\varphi^{-1}([0,A]))) = \mu_2([0,A]) = A. \tag{15.3.2}$$

The latter relation means that for each $0 < C < \mu(F)$ the set $F_{\frac{C}{\mu(F)}}$ has μ measure C.

Now let $c > 0$. Then there exist a natural number n and a non-negative number C with $0 < C < \mu(F)$ such that $n\mu(F) + C = c$. We set $E_c = \cup_{0 \le k \le n}(F + h_k) \cup (F_{\frac{C}{\mu(F)}} + h_{n+1})$. Thus a sequence $(E_c)_{c>0}$ is constructed such that conditions a)-b) of Theorem 15.3.1 are fulfilled. □

We have the following corollaries of Theorem 15.3.1.

Corollary 15.3.1. *Let G be an uncountable non-locally-compact complete separable metric group which is dense in itself. Let μ be a left(or right)-invariant quasi-finite diffused Borel measure. Then there exists no a positive constant c such that each Borel set E with the μ-measure greater than c contains three points such that an area defined by that points is equal to one.*

Corollary 15.3.2. *Let μ be a quasi-finite translation-invariant diffused Borel measure in infinite-dimensional separable Banach space $(E, ||\cdot||)$. Then there exists an increasing family of Borel subsets $(E_c)_{c>0}$ such that:*

a) $\mu(E_c) = c$ for $c > 0$;

b) $\cup_{c>0}E_c$ does not contain the vertices of a triangle of area one.

15.4 On Erdös Problem for an Extension of the Measure m_∞ in Infinite-Dimensional Separable Hilbert Space ℓ_2

Theorem 15.4.1. *There exist an extension ψ of the measure m_∞ such that:*

a) D is carrier of ψ, where D becomes from Lemma 15.2.8;

b) There exists an increasing family of ψ-measurable sets $(E_c)_{c>0}$ such that:

i) $\psi(E_c) = c$ for $c > 0$;

ii) $\cup_{c>0}E_c$ is everywhere dense in ℓ_2;

iii) $\cup_{c>0}E_c$ does not contain the vertices of a triangle of area one.

Proof. Let $\mathcal{S}$ be a σ-algebra of subsets of ℓ_2 defined by

$$\mathcal{S} = \{(A \cap D) \cup (B \cap (\ell_2 \setminus D)) : A, B \in \mathcal{B}(\ell_2)\}. \tag{15.4.1}$$

We set

$$\psi((A \cap D) \cup (B \cap (\ell_2 \setminus D)) = m_\infty(A). \tag{15.4.2}$$

Then it is not hard to show that ψ is an extension of the measure m_∞ and D is carrier of ψ.

For $c > 0$, we set $E_c = T^{-1}([0,c] \times [0,1]^{N \setminus \{1\}}) \cap D$. It is obvious that $\psi(E_c) = c$ and following Lemma 15.2.8, E_c as a subset of D does not contain the vertices of a triangle of area one. □

Remark 15.4.1. It is well known that if F is a Borel subset of the $\mathbb{R}^n$ $(n > 1)$ of positive Lebesgue measure m^n, then a set $F - F$ defined by

$$F - F = \{x - y : x, y \in F\} \tag{15.4.3}$$

contains an open ball with a center at the zero of the $\mathbb{R}^n$(see, [58], Exercise 124, p. 189). In [100] has been demonstrates that a similar result is not valid for every extensions of the Lebesgue measure ℓ_n.

Theorem 15.3.1 demonstrates that ψ stands such an example of the extension of the m_∞ that $m_\infty(E_c) > 0$ and the $E_c - E_c$ does not contain any open ball with center at the zero of the ℓ_2 for every $c > 0$.

We have the following consequence of Corollary 15.3.2.

Corollary 15.4.1. *There exists an increasing family of Borel subsets $(E_c)_{c>0}$ in ℓ_2 such that:*

a) $m_\infty(E_c) = c$ for $c > 0$;

b) $\cup_{c>0} E_c$ does not contain the vertices of a triangle of area one.

Theorem 15.4.2. *If F is of finite positive m_∞-measure and is everywhere dense in ℓ_2, then it contains the vertices of a triangle of area one.*

Proof. Without loss of generality we can assume that $T(F) \subseteq R = \prod_{i \in \mathbb{N}} R_i \in \mathcal{R}$. For $x = (x_2, x_3, \cdots) \in \mathbb{R}^{\mathbb{N} \setminus \{1\}}$ we denote by $T(F)_x$ x-section of $T(F)$ defined by

$$T(F)_x = \{x_1 : (x_1, x_2, x_3, \cdots) \in T(E)\}. \tag{15.4.4}$$

By Fubini theorem we have that

$$\lambda(T(F)) = \int_{\prod_{i>1} R_i} m(T(F)_x) d\lambda^{(1)}(x) > 0, \tag{15.4.5}$$

where $\lambda^{(1)}$ is the Baker measure in $\prod_{i>1} R_i$. Hence there is $(x_1, x_2, x_3, \cdots) \in T(F)$ such that a set $T(F)_x = \{y_1 : (y_1, x_2, x_3, \cdots) \in T(F)\}$ is of positive m measure. It is well known that a set $T(F)_x - T(F)_x$ defined by

$$T(F)_x - T(F)_x = \{x - y : x, y \in T(F)_x\} \tag{15.4.6}$$

contains an open interval with a center at the zero of the $\mathbb{R}$. Since

$$T(F)_x \times \{(x_2, x_3, \cdots)\} \subseteq T(F), \tag{15.4.7}$$

we deduce

$$T^{-1}(T(F)_x \times \{(x_2, x_3, \cdots)\}) \subseteq F. \tag{15.4.8}$$

Since $T(F)_x - T(F)_x$ contains an open interval with a center at the zero of the real axis $\mathbb{R}$, we get that a set

$$T^{-1}(T(F)_x \times \{(x_2, x_3, \cdots)\}) - T^{-1}(T(F)_x \times \{(x_2, x_3, \cdots)\}) =$$

$$T^{-1}((T(F)_x - T(F)_x) \times \{(0,0,\cdots)\}) \tag{15.4.9}$$

also contains a set $(-a,a) \times \{(0,0,\cdots)\} \subseteq F - F$ for some positive real number a. Since F is everywhere dense in ℓ_2, we can choose a point M_3 such that distance A between M_3 and the line $L = \{\alpha(1,0,0,\cdots) : \alpha \in \mathbb{R}\}$ will be greater than $1/a$. If we choose two points M_1 and M_2 in F such that $||\overrightarrow{M_1M_2}|| = 2/A$, then M_1, M_2, M_3 will be vertices of a triangle with area one. □

Remark 15.4.2. We can not extend Corollary 15.4.1 such that $E := \cup_{c>0} E_c$ will be everywhere dense in ℓ_2. Indeed, assume the contrary and let Y be a countable subset of the set E which is everywhere dense in ℓ_2. It is obvious that the set E as well its any subset does not contain the vertices of a triangle of area one. Hence a Borel subset $E_1 \cup Y$ does not contain the vertices of a triangle of area one. On the other hand, the Borel set $E_1 \cup Y$ has a finite positive m_∞ measure and is everywhere dense in ℓ_2. Following Theorem 15.4.2, $E_1 \cup Y$ contains the vertices of a triangle of area one and we get a required contradiction.

Chapter 16

On a Certain Version of the Erdös Problem

16.1 Introduction

In the present chapter we consider some modifications of the Problem 15.1.1.

Problem 16.1.1. Is there a finite constant c such that if a plane set $E \subseteq \mathbb{R}^2$ has an outer Lebesgue measure greater than c, then E contains the vertices of a triangle of area 1?

It is well known that if Lebesgue measure of a plane set E is equal to $+\infty$, then for each positive number s, E contains the vertices of a triangle of area s. We are going to show that an analogous result is not valid for every plane set whose outer Lebesgue measure is equal to $+\infty$.

Let X be an infinite-dimensional Banach space and $\mathcal{B}(X)$ be the σ-algebra of Borel subsets of X. Following Christensen [19], a Borel subset $Y \subseteq X$ is called shy set if there exists a Borel probability measure μ in X such that $\mu(Y+h) = 0$ for all $h \in X$. A Borel measure μ is called a testing measure for Y. A complement of a Borel shy set is called a prevalence. If X is an infinite-dimensional separable Banach space then the class of Borel shy sets constitute a σ-ideal. Let consider a σ-algebra $L(X)$ of subsets of X generated by that σ-ideal and let us define a measure μ on elements of $L(X)$ as follows: $\mu(Y) = 0$ if Y is shy and $\mu(Y) = 1$ otherwise. μ is a probability measure defined on $L(X)$ and is called shy-measure in X.
We consider Erdös problem for the shy-measure μ in an infinite-dimensional separable Banach space X as follows.

Problem 16.1.2. Is there a finite constant c such that if shy-measure of a set E is greater than c, then E contains the vertices of a triangle of area 1? What is the best constant in that cases?

Each constant which solves Problem 16.1.2 is called Erdös constant for the shy-measure in X.

Let ℓ_2 denotes the infinite-dimensional separable Hilbert space of all real-valued square summing sequences.

Definition 16.1.1. A Borel subset Y of ℓ_2 is called non-shy if Y is not shy.

Definition 16.1.2. A subset Z of ℓ_2 is called thick (in the sense of shyness) if an intersection of Z and any Borel non-shy subset of ℓ_2 is non-empty.

Problem 16.1.3. Does there exist a thick (in the sense of shyness) subset A of ℓ_2 which does not contain vertices of a triangle of area 1.

The rest of the chapter is the following.

In Section 16.2 we consider some auxiliary facts of the measure theory on the plane. In Section 16.3 we present our main result concerning with Problem 16.1.1. In Section 16.4 we consider Erdös problem for shy measures in infinite-dimensional separable Banach spaces. An example of a thick (in the sense of shyness) subset A of the infinite-dimensional separable Hilbert space ℓ_2 is constructed in Section 16.5 such that the set A does not contain vertices of a triangle of area 1.

16.2 Auxiliary Propositions from the Measure Theory on the Plane

Let $L(x,y)$ be a line drawing on points x and y. For $r>0$ we set

$$C_{(x,y,r)}=\{z\in R^2:\rho(z,L(x,y))=r\},$$

where ρ denotes a usual metric in R^2. $C_{(x,y,r)}$ is called a cylinder in R^2 and $L(x,y)$ is called an axis of its symmetry.

Lemma 16.2.1. *Let $C_{(x,y,r)}$ be a cylinder in R^2 and let $L(x,y)$ be an axis of its symmetry. Let l be a line in R^2 which is not parallel to the line $L(x,y)$. Then $card(C_{(x,y,r)}\cap l)\leq 2$.*

Proof. Without loss of generality we can assume that $x=(0,1)$ and $y=(0,0)$. Suppose that a line l is defined by the following equation

$$\frac{z_1-b_1}{c_1}=\frac{z_2-b_2}{c_2} \tag{16.2.1}$$

where $b=(b_1,b_2)\in R^2$ and vectors $c=(c_1,c_2)(\in R^2)$ and $e_2=(0,1)$ are not collinear.

Then for the cylinder $C_{(x,y,r)}$ we have the following representation

$$C_{(x,y,r)}=\{(z_1,z_2)\in R^2:z_1^2=r^2\}. \tag{16.2.2}$$

Let us consider a parametric equation of the line l. It has the following form

$$z_k=b_k+c_kt\ (k=1,2;\ t\in\mathbb{R}). \tag{16.2.3}$$

To find coordinates of points belonging to the intersection $C_{(x,y,r)}\cap l$ we must consider the following equation

$$z_1^2=r^2 \tag{16.2.4}$$

with respect to a parameter t. It turns out that the set $C_{(x,y,r)}\cap l$ contains no more than two points. □

Lemma 16.2.2. *Let F be a Borel measurable set in R^2 with $0 < m_2(F) < \infty$, where m_2 denotes a two-dimensional Lebesgue measure in R^2. Then F is not covered by the family of cylinders $(C_i)_{i \in I}$ such that $card(I) < c$, where c denotes the cardinality of the continuum.*

Proof. Assume the contrary and let F can be covered by the family of cylinders $(C_i)_{i \in I}$ such that $card(I) < c$. Let L_i be an axis of the symmetry of the cylinder C_i for $i \in I$. Let L be a line which is not parallel to lines $(L_i)_{i \in I}$. By Fubini theorem, there is a shift $L+a$ of the line L such that $m_a(L+a \cap F) > 0$, where m_a denotes a linear Lebesgue measure supported on $L+a$. Hence, cardinality of the intersection $L+a \cap F$ is equal to c, where c denotes the cardinality of the continuum. By Lemma 16.2.1 we know that the cardinality of the intersection $L+a \cap C_i$ is not bigger than 2. Hence $\text{card}((L+a \cap F) \setminus \cup_{i \in I}(L+a \cap C_i)) = c$. The latter relation means that $\text{card}((L+a \cap (F \setminus \cup_{i \in I} C_i)) = c$, which means that F is not covered by the family of cylinders $(C_i)_{i \in I}$. □

Lemma 16.2.3. *There exists a set $D \subset R^2$ such that the following two conditions are fulfilled:*

a) D intersects every Borel measurable plane set of positive m_2- measure;

b) D does not contain the vertices of a triangle of area one.

Proof. Let $(F_\xi)_{\xi < \phi}$ be a sequence of Borel measurable plane sets of finite positive measure m_2, where ϕ denotes the first ordinal number of a cardinality of the continuum.

Let τ be an operator of a global choice. We set $x_0 = \tau(F_0)$, $x_1 = \tau(F_1 \setminus \{x_0\})$ and $x_2 = \tau(F_2 \setminus (\{x_0\} \cup \{x_1\} \cup C_{(x_0, x_1, r(x_0, x_1))}))$, where $r(x_0, x_1) = \frac{2}{\rho(x_0, x_1)}$ and $\rho(x_0, x_1)$ denotes a usual distance between points x_0 and x_1 in R^2.

Assume that for $\xi < \phi$ a partial sequence of different points $(x_\theta)_{\theta < \xi}$ is constructed such that the following two conditions are fulfilled:

(i) For $\theta < \xi$ we have that $x_\theta \in F_\theta$;

(ii) A set $\{x_\theta : \theta < \xi\}$ does not contain the vertices of a triangle of area one.

We put

$$A_\xi = F_\xi \setminus (\cup_{(x,y) \in \{x_\theta : \theta < \xi\}^2} C_{(x,y,r(x,y))} \cup \cup_{\theta < \xi} \{x_\theta\}). \tag{16.2.5}$$

By Lemma 2.2, this difference is not empty set. Now x_ξ can be defined by $x_\xi = \tau(A_\xi)$. Thus, a transfinite ϕ-sequence of different points $(x_\xi)_{\xi < \phi}$ is constructed such that:

(i*) For $\xi < \phi$ we have that $x_\xi \in F_\xi$;

(ii*) A set $\{x_\xi : \theta < \xi\}$ does not contain the vertices of a triangle of area one.

We set $D = \{x_\xi : \xi < \phi\}$.

The validity of the condition a) is obvious.

Let us check the validity of the condition b). Assume the contrary and let $x_{\xi_1}, x_{\xi_2}, x_{\xi_3}$ be three points of the set D such that $\xi_1 < \xi_2 < \xi_3$ and an area of the triangle with vertices $x_{\xi_1}, x_{\xi_2}, x_{\xi_3}$ is equal to one. Following our construction, x_{ξ_3} does not belong to the set $C_{(x_{\xi_1}, x_{\xi_2}, r(x_{\xi_1}, x_{\xi_2}))}$, and hence, we get a required contradiction. □

16.3 Solution of the Problem 16.1.1

Theorem 16.3.1. *There does not exist a finite constant c such that each set E whose outer Lebesgue measure is greater than c contains the vertices of a triangle of area 1.*

Proof. Let D_a be a disc with center at the zero of R^2 such that $m_2(D_a) = a$ for each $a \in R^+$(It is sufficient to take under radius r_a a number $\sqrt{\frac{a}{\pi}}$). We set $E_a = D \cap D_a$, where D becomes from Lemma 16.2.3. It is obvious that $m_2^*(E_a) = a$ for each $a \in R^+$.

Suppose that there is a finite constant c such that each set E whose outer Lebesgue measure is greater than c contains the vertices of a triangle of area 1. Now let consider a set E_{2c}. It is clear that $m_2^*(E_{2c}) = 2c > c$ but E_{2c} as a subset of D(see Lemma 16.2.3) does not contain the vertices of a triangle of area 1. This is a required contradiction and Theorem 3.1 is proved. □

Theorem 16.3.2. *There exist a plane set E with Lebesgue outer measure $+\infty$ which does not contain the vertices of a triangle of area 1.*

Proof. It is sufficient to take under E the set D which becomes from Lemma 16.2.3. □

We recall the reader that Solovay model (cf. [132]) is the following system of axioms:

$$(ZF)\&(DC)\&(\text{every subset of } \mathbf{R} \text{ is measurable in the Lebesgue sense}),$$

where (ZF) denotes the Zermelo-Fraenkel set theory and (DC) denotes the axiom of Dependent Choices (cf. [20]).

Remark that the result of Theorem 16.3.2 fails in Solovay model. More precisely, in Solovay model each plane set E with Lebesgue outer measure $+\infty$ contains the vertices of a triangle of area 1. Hence, we get the validity of the following assertion.

Theorem 16.3.3. *A sentence "each plane set E with Lebesgue outer measure $+\infty$ contains the vertices of a triangle of area 1" is independent from the theory $(ZF)\&(DC)$.*

Here naturally arises a question asking whether the result of Theorem 16.3.1 is valid in Solovay model. As we know that question still remains open as well the Problem 16.1.1.

16.4 Solution of the Problem 16.1.2

Definition 16.4.1. (see, [50]) We say that an 1-dimensional subspace $L \subset X$ is an 1-dimensional probe for a Borel set $S \subset X$ if

$$(\forall h)(h \in X \to \mu_L(S + h \cap L) = 0),$$

where μ_L denotes an 1-dimensional Lebesgue measure supported on L.

Lemma 16.4.1. *If a Borel set $S \subset X$ has an 1-dimensional probe then S is shy set.*

Proof. Let consider an 1-dimensional Gaussian measure Γ_L defined on the subspace $L \subset X$. We put $\Gamma(Y) = \Gamma_L(Y \cap L)$ for $Y \in \mathcal{B}(X)$. Let us show that Γ is a testing measure for a Borel set $S \subset X$. Since $\mu_L(S + h \cap L) = 0$ for $h \in X$ and μ_L and Γ_L are equivalent measures, we deduce that

$$(\forall h)(h \in X \to \Gamma(S + h) = \Gamma_L(S + h \cap L) = \mu_L(S + h \cap L) = 0).$$

This ends the prove of Lemma 16.4.1. □

Theorem 16.4.1. *Each number $c \in [0,1[$ is Erdös constant for the shy-measure μ in an infinite-dimensional separable Banach space X.*

Proof. It is sufficient to show that the zero 0 is Erdös constant for the shy-measure μ in an infinite-dimensional separable Banach space X. Let E be such a Borel subset of X for which $\mu(E) > 0$. It is obvious that then E will be prevalence. Now let L be such one dimensional vector space for which exists a sequence $(z_k)_{k\in\mathbb{N}}$ of elements of E that a distance between z_k and L tends to $+\infty$ as well k tends to $+\infty$. Such one dimensional vector space L always exists because E is prevalence. By Lemma 16.4.1, there is $x_0 \in X$ such that $\mu_L(E+x_0\cap L) > 0$(In other cases, by Lemma 16.4.1 we will get that E is shy which contradicts to the condition that E is prevalence).

Obviously $E\cap(L-x_0) \subseteq L-x_0$ and by well known fact (see, [58], Exercise 124, p. 189) a set $(E\cap(L-x_0))-(E\cap(L-x_0))$ contains any open interval (with center at zero) in L. Let $r > 0$ be such a real number that for a closed ball $B[\mathbf{0},r]$ the condition $B[\mathbf{0},r]\cap L \subseteq (E\cap(L-x_0))-(E\cap(L-x_0))$ holds. Since distant between z_k and L tends to $+\infty$ as well k tends to $+\infty$ we deduce that there is a point $x_1 \in \{z_k : k \in \mathbb{N}\}$ such that the distance h between x_1 and $L-x_0$ is greater than $2/r$. We will get

$$\frac{1}{2}\times r\times\rho(x_1,L-x_0) = \frac{1}{2}\times r\times h > \frac{1}{2}\times r\times\frac{2}{r} = 1.$$

Now we can choose two points $x_2,x_3 \in E\cap(L-x_0)$ such that $||x_3-x_2|| = 2/h$. Obviously points x_1,x_2,x_3 are in E and

$$S = \frac{1}{2}||x_3-x_2||\rho(x_1,L-x_0) = \frac{1}{2}\times\frac{2}{h}\times h = 1, \tag{4.1}$$

where S denotes area of the triangle with vertices x_1,x_2,x_3. □

16.5 Solution of the Problem 16.1.3

Lemma 16.5.1. *Let F be a Borel non-shy set in ℓ_2. Then F is not covered by the family of cylinders $(C_i)_{i\in I}$ such that card$(I) < c$, where c denotes the cardinality of the continuum.*

Proof. Let L_i be an axis of the symmetry of the cylinder C_i for $i \in I$. Let L be a line which is not parallel to lines $(L_i)_{i\in I}$.

Since F is non-shy, by Lemma 16.4.1 there is $a \in \ell_2$ such that $\mu_a(L+a\cap F) > 0$, where μ_a denotes one-dimensional Lebesgue measure supported on $L+a$.

Hence, cardinality of the intersection $L+a\cap F$ is equal to c, where c denotes the cardinality of the continuum. By Lemma 15.2.6 we know that the cardinality of the intersection $L+a\cap C_i$ is not bigger than 2. Hence card$((L+a\cap F)\setminus\cup_{i\in I}(L+a\cap C_i)) = c$. The latter relation means that card$((L+a\cap(F\setminus\cup_{i\in I}C_i)) = c$, which means that F is not covered by the family of cylinders $(C_i)_{i\in I}$. □

Theorem 16.5.1. *There exists a thick (in the sense of shyness) subset A of ℓ_2 which does not contain vertices of a triangle of area 1.*

Proof. Let $(F_\xi)_{\xi<\phi}$ be a sequence of Borel non-shy sets, where ϕ denotes the first ordinal number of a cardinality of the continuum.

Let τ be an operator of a global choice. We set $x_0 = \tau(F_0)$, $x_1 = \tau(F_1 \setminus \{x_0\})$ and $x_2 = \tau(F_2 \setminus (\{x_0\} \cup \{x_1\} \cup C_{(x_0,x_1,r(x_0,x_1))}))$, where $r(x_0,x_1) = \frac{2}{\rho(x_0,x_1)}$.

Assume that for $\xi < \phi$ a partial sequence of different points $(x_\theta)_{\theta<\xi}$ is constructed such that the following two conditions are fulfilled:

(i) For $\theta < \xi$ we have that $x_\theta \in F_\theta$;

(ii) A set $\{x_\theta : \theta < \xi\}$ does not contain the vertices of a triangle of area one.

We put

$$A_\xi = F_\xi \setminus (\cup_{(x,y)\in\{x_\theta:\theta<\xi\}^2} C_{(x,y,r(x,y))} \cup \cup_{\theta<\xi}\{x_\theta\}).$$

By Lemma 16.5.1, this difference is not empty set. Now x_ξ can be defined by $x_\xi = \tau(A_\xi)$. Thus, a transfinite ϕ-sequence of different points $(x_\xi)_{\xi<\phi}$ is constructed such that:

(i*) For $\xi < \phi$ we have that $x_\xi \in F_\xi$;

(ii*) A set $\{x_\xi : \theta < \xi\}$ does not contain the vertices of a triangle of area one.

We set $A = \{x_\xi : \xi < \phi\}$.

By the construction of the set A we deduce that it is thick.

We have to show that A does not contain vertices of a triangle of area 1. Assume the contrary and let $x_{\xi_1}, x_{\xi_2}, x_{\xi_3}$ be three points of the set A such that $\xi_1 < \xi_2 < \xi_3$ and an area of the triangle with vertices $x_{\xi_1}, x_{\xi_2}, x_{\xi_3}$ is equal to one. Following our construction, x_{ξ_3} does not belong to the set $C_{(x_{\xi_1}, x_{\xi_2}, r(x_{\xi_1}, x_{\xi_2}))}$, and hence, we get a required contradiction. □

Chapter 17

Appendix

17.1 Proof of the Mycielski-Swierczkowski Theorem

We will consider some important results and facts from measure theory which have various interesting applications to the theory of invariant and quasiinvariant measures in infinite-dimensional vector spaces.

The following notion of a specific kind of partial ordering frequently turns out useful in studying various questions of measure theory and general topology.

Definition 17.1.1. A partial ordering $(T, \preceq)$ is called a **tree** if T has the least element and, for each $y \in T$ the set $\{x \in T : x \preceq y\}$ is well-ordered by $\preceq$. The least element of T is called the **root** of T. For any ordinal number α the α-th level of T is the set

$$T_\alpha = \{y : \{x \in T : x \prec y\} \text{ has order type } \alpha\}.$$

The height of a tree is a least ordinal α such that the α-th level of T is empty.

Let A be a non-empty set and let α be an ordinal. A complete A-ary tree of height α, which consists of all functions from $\cup_{\beta \prec \alpha} A^\beta$ and is ordered by inclusion, is denoted by $A^{\prec \alpha}$. If $A = \{0; 1\}$, then complete A-ary trees are called **binary** trees.

Any linearly ordered subset of the tree $(T, \preceq)$ is called a **branch** in T. A subset P of T is called a **path** through the tree T if P is a branch and contains exactly one element from each nonempty level of T.

The following fundamental statement was established by Kőnig.

Theorem 17.1.1. *Suppose that $(T, \preceq)$ is a tree of height ω such that all levels of T are finite. Then there exists a path through T.*

Proof. Let x_0 be the root of T. For each $n \in \omega \setminus \{0\}$, we can recursively pick an element $x_n \in T_n$ such that $x_n \succ x_{n-1}$ and the set

$$\{y \in T : x_n \prec y\}$$

is finite. This is possible since every level T_n of T is infinite. Then $(x_n)_{n \in N}$ is a path through T. □

We continue our discussion with the classical theorem of Mycielski and Swierczkowski. This theorem concerns the Lebesgue measurability of all linear subsets.

Theorem 17.1.2. *(ZF & DC & AD) Every subset of the real axis R is Lebesgue measurable.*

Proof. To prove Theorem 17.1.2 it is sufficient to show that every subset X of the real axis R has measure zero or contains a measurable subset with a positive Lebesgue measure.

It can be assumed, without loss of generality, that $X \subseteq [0;1]$.

The collection S of all subsets of $[0;1]$ represented by a finite union of intervals, half-intervals or segments with rational end points is countable. Let W_β be an element from the class S with a numerical number β.

Let us fix a number $\varepsilon > 0$. We say that an infinite sequence of natural numbers $\alpha = (a_0, a_1, \cdots)$ is ε-correct if the following two conditions hold:

1) every number a_{2n} is equal to 0 or 1;

2) The Lebesgue measure of the subset $W_{a_{2m+1}}$ is less than $\varepsilon \times 2^{-2m}$ for all $m \in \omega$.

To every such ε-correct α-sequence there correspond a number $F(\alpha)$ from the interval $[0;1]$ with a binary representation $0, a_0 a_2 a_4 a_6 \cdots$ and a set

$$V_\alpha = W_{a_1} \cup W_{a_3} \cup W_{a_5} \cup \cdots$$

Let $A(\varepsilon)$ be the set of sequences $\alpha = (a_0, a_1, \cdots)$ satisfying at least one of the following two conditions:

a) a sequence α is ε-correct and a point $F(\alpha)$ belongs to the difference $X \setminus V_\alpha$,

b) a sequence α is not ε-correct and the smallest m, for which at least one of conditions 1) or 2) is not valid, does not satisfy condition 2), but satisfies condition 1).

According to the AD, two cases are possible.

Case I. There exists a winning strategy τ_ε for player II in a game of type $G_{A(\varepsilon)}$, where ε is a fixed positive number. In that case the set X has measure zero. Indeed, for the fixed number ε, we can construct a covering of the set X by a countable union of elements of S, whose Lebesgue measure is not greater than ε.

Let us denote by C_m the collection of all sequences with length $m+1$ consisting of 0 or 1 for all natural number m. Let $U = (a_0, a_2, a_4, \cdots, a_{2m}) \in C_m$. Denote the conjugate sequence of numbers $a_1, a_3, \cdots, a_{2m+1}$ by

$$a_{2i+1} = \tau_\varepsilon(a_0, a_1, a_2, \cdots, a_{2i-1}, a_{2i}),$$

where $i \in \omega$ (the definition of such a sequence is possible by using mathematical recursion). Since every number a_{2i} with $i \leq m$ is equal to 0 or 1, and the function τ_ε is a winning strategy for player II in a game of type $G_{A(\varepsilon)}$, the Lebesgue measure of the set $W_{a_{2i+1}}$ must be less than or equal to $\varepsilon \times 2^{-2i}$ for arbitrary $i \in \omega$. In particular, the Lebesgue measure of the set $W(u) = W_{a_{2m+1}}$ is not greater than $\frac{\varepsilon}{2^{2m}}$. Hence the union

$$Z = \cup_{m \in \omega} \cup_{u \in C_m} W(u)$$

has a Lebesgue measure no greater than

$$\sum_{m=0}^{\infty} \sum_{u \in C_m} \varepsilon \times 2^{-2m} = \varepsilon \sum_{m=0}^{\infty} 2^{m+1} 2^{-2m} = 4\varepsilon.$$

Finally, we need verify that $X \subseteq Z$. Let $x \in X$. Assuming that $X \subseteq [0;1]$, the number xX can be represented by

$$x = 0, a_0 a_2 a_4 \cdots,$$

where every symbol a_{2i} is equal to 0 or 1.

Using mathematical recursion, let us construct a sequence $a_1, a_3, a_5, \cdots$ of numbers of type a_{2i+1} by

$$a_{2i+1} = \tau_\varepsilon(a_0, a_1, \cdots, a_{2i}),$$

where $i \in \omega$.

Using the property of the winning strategy τ_ε of player II in a game of type $G_{A(\varepsilon)}$, we conclude that $\alpha \notin A(\varepsilon)$.

Note that all numbers a_{2i}, being the leads of player I, are equal to 0 or 1. On the one hand, using the definition of the set $A(\varepsilon)$, we conclude that the sequence α is ε-correct and $F(\alpha) \notin X \setminus V_\alpha$. On the other hand, the point $F(\alpha)$ coincides with the point $x \in X$ and $x \in V_\alpha$. By the definition of the set Z, we have $V_\alpha \subseteq Z$, and, correspondingly, $x \in Z$. As x was an arbitrary point of the set X, the inclusion $X \subseteq Z$ is proved.

Case 2. For a fixed positive number ε, there exists a winning strategy σ_ε for player I in a game of type $G_{A(\varepsilon)}$. Let $m \in \omega$. Denote by ω_m the collection of all indexes $a \in \omega$, which correspond to the set W_a whose measure is not greater than $\frac{\varepsilon}{2^{2m}}$. Let us fix the numeration

$$\omega_m = \{a[m,l] : l \in \omega\}$$

by natural numbers l for the whole set ω_m.

To every infinite sequence $\gamma = (l_0, l_1, l_2, \cdots)$ of natural numbers l_m let there correspond the following subjects:

a) The sequence $a_1, a_3, a_5, \cdots$ of numbers defined by $a_{2i+1} = a[i, l_i]$;

b) The sequence of numbers $a_0, a_2, a_4, \cdots$ defined by induction (with respect to i) using the mathematical recursion formula

$$a_{2i} = \sigma_\varepsilon(a_0, a_1, a_2 \cdots, a_{2i-2}, a_{2i-1}),$$

where $i \in \omega$ is arbitrary (in particular, $a_0 = \sigma_\varepsilon(\emptyset)$);

c) The resulting sequence

$$\alpha(\gamma) = (a_0, a_1, \cdots);$$

d) The point $H(\gamma) = F(\alpha(\gamma))$ (i.e., the point $0, a_0 a_2 a_4 \cdots$ in binary representation).

The sequence of numbers $a_1, a_3, a_5, \cdots$ arising in this construction obviously satisfies condition 2) in the definition of an ε-correct sequence for any $m \in \omega$. This means that by choosing σ_ε as a winning strategy, the resulting sequence $\alpha(\gamma)$ (equal to $\sigma_\varepsilon^*(a_1, a_3, a_5, \cdots)$) must be ε-correct, while the point $H(\gamma)$ must be an element of the difference $X \setminus V_{\alpha(\gamma)}$. Hence the function H maps the Baire space N of all infinite sequences of natural numbers into the set X so that

$$H(\gamma) \in X \setminus V_{\alpha(\gamma)}$$

for any $\gamma \in N$.

We are going to prove that the set $P = H(N)$ contains a measurable subset of positive Lebesgue measure. Let us begin by verifying that the upper measure of the set P is greater than ε. Assume the contrary. Then the set P can covered by the countable union $W = \cup_{m\in\omega} W_m$ of pairwise nonintersecting elements W_m of the family S and the total measure of these elements is not greater than ε. Using the validity of

$$\sum_{m=0}^{\infty} \varepsilon \times 2^{-2m} = \frac{4}{3}\varepsilon > \varepsilon,$$

it can be assumed, without loss of generality, that the measure of each set W_m is not greater than $\varepsilon \times 2^{-2m}$. Under this assumption, there exists a sequence $\gamma = (l_0, l_1, l_2, \cdots)$ satisfying the equality $W_m = W_{a[m,l_m]}$ for all m. Now, on virtue of the above arguments, we have $H(\gamma) \in P \setminus V_{\alpha(\gamma)}$. Therefore, $H(\gamma) \in P \setminus W$. But this contradicts the choice of W and the inclusion $P \subseteq W$ is proved.

Thus, the upper measure P is actually greater than ε. Note that the set P coincides with the union of the countable family of sets

$$P(k) = \{H(\gamma) : \gamma = (l_0, l_1, l_2, \cdots) \in N \text{ and } l_0 \leq k\},$$

which generates an $\subseteq$-increasing sequence. So there exists a number k_0 such that the set $P_0 = P(k_0)$ has a Lebesgue measure greater than ε. Analogously, there exists a natural number k_1 such that the set

$$P_1 = \{H(\gamma) : \gamma = (l_0, l_1, l_2, \cdots) \in P_0 \text{ and } l_1 \leq k_1\}$$

has a measure greater than ε. If we continue this process, then we can construct a sequence of natural numbers $k_0, k_1, k_2, \cdots$ such that the set

$$P_m = \{H(\gamma) : \gamma \in N \ \& \ \gamma(i) \leq k_i \text{ for all } i \leq m\}$$

has a Lebesgue measure greater than ε for any $m \in \omega$. Assuming the extreme left term to be equal to zero, we denote by $\gamma(i)$ the i-th term l_i of the infinite sequence $\gamma = (l_0, l_1, \cdots, l_i, \cdots)$.

We introduce the family of sets:

$$Y_m = \{y \in [0;1] : \text{there exists } \mathbf{x} \in \mathbf{P_m} \text{ such that the first } \mathbf{m} \text{ symbols of the}$$
$$\text{binary representations of } \mathbf{x} \text{ and } \mathbf{y} \text{ coincide with deficiency } \}.$$

Every set of Y_m with its point y also contains some half-interval having the form $[p, q[$ containing the point y and having a nonzero length for $y \neq 1$ (this half-interval is generated by the points of the unit interval whose first m symbols of the binary representation are fixed.) Hence all Y_m are Borel sets and ,therefore, measurable. The intersection $Y = \cap_{m\in\omega} Y_m$ of all these sets is thus also measurable.

Let us estimate the measure of the set Y. Since the set Y_m is measurable and consists of the set P_m (it is clear from the definition of the set Y_m), the Lebesgue measure of Y_m is greater than ε. By the construction of the sets P_m and P_{m+1}, we obtain

$$P_{m+1} \subseteq P_m$$

for all $m \in \omega$, and therefore $Y_{m+1} \subseteq Y_m$. So Y has a Lebesgue measure greater than ε.

It remains to verify $Y \subseteq P$. Let $y \in Y$. To each natural number m there corresponds a sequence $\gamma = (l_{m0}, l_{m1}, l_{m2}, \cdots) \in N$ such that the inequality $l_{mi} \leq k_i$ is fulfilled for $i \leq m$ and the first m binary symbols of numbers x and $H(\gamma_m)$ coincide. The family T of all finite sequences

$$\gamma_m \lceil i = (l_{m0}, l_{m1}, \cdots, l_{m,i-1})$$

where $m \in \omega$ and $i \leq m$, forms an infinite tree of height ω, such that all branches of T are finite. (The number of branches at each vertex of the tree is limited by the corresponding value k_m). Hence, by using the König theorem (Theorem 17.1.1), the tree T has an infinite path; in other words, there exists an infinite sequence $\gamma = (l_0, \cdots)$ of natural numbers l_i such that the finite sequence $\gamma \lceil i = (l_0, \cdots, l_{i-1})$ belongs to the tree T for all $i \in \omega$. By the definition of T, we can choose an index $m(i) \geq i$ for arbitrary i such that the equality

$$\gamma_{m(i)} \lceil i = \gamma \lceil i$$

will be fulfilled.

By the definition of the function H, the latter equality implies the coincidence of the first i binary symbols of the real numbers $H(\gamma)$ and $H(\gamma_{m(i)})$. At the same time, by virtue of the inequality $m(i) \geq i$ and the choice of the sequence γ_m, the first i binary symbols of the numbers $H(\gamma_{m(i)})$ and y coincide. Tending i to infinity, we obtain $y = H(\gamma)$, i.e., $y \in P$.

Thus, the set P (respectively, the set X since $P \subseteq X$) contains a measurable subset Y of positive Lebesgue measure in Case 2.

This completes the proof of the Mycielski and Swierczkowski theorem.

□

The following result is a simple consequence of Theorem 17.1.2.

Theorem 17.1.3. *The Axiom of Determinacy contradicts the Axiom of Choice.*

Proof. On the one hand, by the Axiom of Choice (in particular, by the Vitali theorem based on an uncountable form of the Axiom of Choice (AC), we can easily construct such a subset X of the real axis R that is not measurable in the Lebesgue sense. On the other hand, using the Mycielski-Swierczkowski theorem (Theorem 17.1.2), we have that every subset of real axis (including X) must be measurable in the Lebesgue sense. This is a contradiction, and Theorem 17.1.3 is proved. □

17.2 Proof of the Martin-Solovay Theorem

To formulate Martin's Axiom, we need some notations and definitions.

Let P be an arbitrary nonempty set.

A binary relation of $G \subseteq P \times P$ is called an equivalence relation on P if the following three conditions hold:

1) $(p,p) \in G$ for all elements $p \in P$;
2) $(p,q) \in G$ and $(q,r) \in G$ imply $(p,r) \in G$;
3) $(p,q) \in G$ implies $(q,p) \in G$.

If G is an equivalence relation on P, then the pair (P,G) is called a set equipped with an equivalence relation.

Obviously, if G is an equivalence relation on P, then we have a partition of P canonically associated with G. This partition consists of the sets

$$G(p) \quad (p \in P),$$

where $G(p)$ denotes the section of G corresponding to an element $p \in P$; in other words,

$$G(p) = \{q : (p,q) \in G\}.$$

Conversely, every partition of the set P canonically defines an equivalence relation on P.

Let P be an arbitrary set and let G be a binary relation on P. We say that G is a partial order on P if the following three conditions hold:

1) $(p,p) \in G$ for each element p of P;
2) $(p,q) \in G$ and $(q,r) \in G$ imply $(p,r) \in G$;
3) $(p,q) \in G$ and $(q,p) \in G$ imply $p = q$.

Suppose that G is a partial order on a set P. As usual, we write

$$p \preceq q \quad \text{iff } (p,q) \in G.$$

The pair $(P, \preceq)$ is called a partially ordered set.

Let $(P, \preceq)$ be a partially ordered set. We say that a set $D \subseteq P$ is dense in P if for each $p \in P$ there is $q \in D$ with $p \preceq q$. (This is actually the density in the topological sense for a suitable topology on P, viz., that having as a base all sets of the form $\{p : p \succeq q\}$, where $q \in P$).

A set $D \subseteq P$ is called a subnet if for arbitrary two elements $p_1 \in P$ and $p_2 \in P$ there exists an element $q \in D$ such that $p_1 \preceq q$ and $p_2 \preceq q$. Two elements p and q are called compatible if there is an element $r \in P$ such that $p \preceq r$ and $q \preceq r$. Finally, we say that $(P, \preceq)$ satisfies the countable chain condition (or, simply, c.c.c.) if every uncountable subset of P contains at least two compatible elements.

Martin's Axiom usually denoted by MA is the following statement:

If $(P, \preceq)$ is a partially ordered set satisfying c.c.c. and $(D_i)_{i \in I}$ is a family of dense subsets in P with $card(I) < 2^{\omega}$, then there is a subnet $D \subseteq P$ which intersects each D_i.

The Continuum Hypothesis readily implies Martin's Axiom. Indeed, let us assume CH. Let $(P, \preceq)$ be any partially ordered set and let $(D_n)_{n \in \omega}$ be any sequence of dense subsets of P. Then we can recursively construct an increasing sequence $(p_n)_{n \in \omega}$ of elements of P such that $p_n \in D_n$ for each $n \in \omega$. Now, put

$$D = \{p \in P : (\exists n \in \omega)(p_n \preceq p)\}.$$

Evidently, D is a subnet in P which intersects every D_n.

Remark 17.2.1. Gödel showed that theory **ZFC** & **CH** (equivalently, theory **ZFC** & **MA** & **CH** because **CH** implies Martin's Axiom **MA**) is consistent, since it is valid in the Constructible Universe (cf. [20], p 21). Martin and Solovay [78] proved that models **ZFC** & **MA** and **ZFC** & **MA** & $\neg$**CH** are consistent.

Theorem 17.2.1 (Martin & Solovay [78]). *Let ℓ_n be an n-dimensional Lebesgue measure defined in R^n. If $(E_i)_{i \in I}$ is a family of ℓ_n-measure zero subsets of R^n, such that $card(I) < c$, then in the system of axioms* **ZFC** & **MA** *the set $E = \bigcup_{i \in I} E_i$ is Lebesgue null in $\mathbb{R}^n$.*

Proof. Let $\varepsilon > 0$. We must find an open set including E which has ℓ_n-measure $\leq \varepsilon$. Let P be the collection of all open sets of ℓ_n measure $< \varepsilon$; and for $p, q \in P$ let $p \preceq q$ mean $p \subseteq q$. We first show that the set P satisfies the *c.c.c.* Let Q be a pairwise incompatible subset of P. Let

$$Q_n = \left\{ p \in Q : \ell_n(p) \leq \left(1 - \frac{1}{2^n}\right)\varepsilon \right\}.$$

It is enough to show that Ω_n is countable.

Let $E' \subset R^n$ be a countable set everywhere dense in R^n.

For each $p \in Q_n$, choose $\overline{p} \subseteq p$ so that $\overline{p}$ is a finite union of open balls with center at E' and rational radius such that

$$\ell_n(p - \overline{p}) < \frac{\varepsilon}{2^n}.$$

Since all such finite unions constitute only a countable family, it suffices to show that if p and q are distinct members of Q_n, then $\overline{p} \neq \overline{q}$. Suppose $\overline{p} = \overline{q}$. Then

$$p \cup q \subseteq (p \setminus \overline{p}) \cup q,$$

so

$$\ell_n(p \cup q) < \frac{\varepsilon}{2^n} + \left(1 - \frac{1}{2^n}\right)\varepsilon = \varepsilon.$$

This implies that p and q are compatible, which contradicts $p, q \in Q$.

For $i \in I$, let $D_i = \{p \in P : E_i \subseteq p\}$. Using the relation $\ell_n(E_i) = 0$, we easily conclude that D_i is dense in P. By Martin's axiom it follows that there is a subnet D of P which meets every D_i.

Let G be the union of the members of D. Then G is open. From $D_i \cap D \neq \emptyset$ we find that $E_i \subseteq G$, so $E \subseteq G$.

It remains to show that $\ell_n(G) > \varepsilon$ leads to a contradiction. By Lindelöf's theorem, G is a countable union of sets in D. It follows that there is a finite union G_1 of sets in D such that $\ell_n(G_1) > \varepsilon$. But since D is a subnet, some member of D includes G_1 and hence has ℓ_n measure $\geq \varepsilon$. This is a contradiction, and Theorem 17.2.1 is proved. □

17.3 Proof of the Liouville Theorem

First, we need formulate a precise definition of differentiable functions on the Euclidean space R^n.

Definition 17.3.1. A real-valued function $f : R^n \to R$ is called differentiable at a point $x = (x_1, \cdots, x_n)$ if there exist a family of real numbers $A_1, \cdots, A_n$ and a real-valued function h in R^n such that

$$f(x_1 + \Delta x_1, \cdots, x_n + \Delta x_n) - f(x_1, \cdots, x_n) = \\ = \sum_{i=1}^{n} A_i \Delta x_i + h(\Delta x_1, \cdots, \Delta x_n), \tag{17.3.1}$$

where

$$\lim_{\rho \to 0} \frac{h(\Delta x_1, \cdots, \Delta x_n)}{\rho} = 0, \ \rho = \sqrt{\sum_{i=1}^{n} (\Delta x_i)^2}. \tag{17.3.2}$$

Note that if the function f is differentiable, then its coefficients $A_i(1 \leq i \leq n)$ have an interesting interpretation; in particular, simple equality $A_i = \frac{\partial}{\partial x_i} f(x_1, \cdots, x_n)$ holds for $1 \leq q \leq n$, where $\frac{\partial}{\partial x_i} f$ denotes the partial derivative with respect to the variable x_i. Hence

$$\Delta f(x_1, \cdots, x_n) = \sum_{i=1}^{n} \frac{\partial}{\partial x_i} f(x_1, \cdots, x_n) \Delta x_i + h(\Delta x_1, \cdots, \Delta x_n). \qquad (17.3.3)$$

Definition 17.3.2. A function $f : R^n \to R$ is called differentiable if it is differentiable at an arbitrary point in R^n.

Definition 17.3.3. A transformation $U = (U_1, \cdots, U_n) : R^n \to R^n$ is called differentiable if its all components are differentiable.

Definition 17.3.4. Let $U = (U_1, \cdot, U_n)$ be a differentiable transformation. The determinant of the matrix $(\frac{\partial U_i}{\partial x_j})_{1 \leq i,j \leq n}$ is called the Jacobian of the transformation U and is denoted by $I(U)$.

The following geometrical proof of the formula of substituting variables in the Lebesgue multiple integral is essentially due to Ostrogradsky.

Theorem 17.3.1. *Let $U : R^n \to R^n$ be a differentiable transformation. $f : R^n \to R$ be a real-valued Lebesgue integrable function. Then the formula*

$$(\forall D)(D \in B(R^n) \to \int_{U(D)} f dl_n = \int_D I(U) f(U) dl_n) \qquad (17.3.4)$$

is valid.

Proof. Let us consider the parallelepiped

$$D_X = \prod_{i=1}^{n} [x_i; x_i + \Delta x_i] \qquad (17.3.5)$$

generated by the family of vectors $(\mathbf{P_i})_{1 \leq i \leq n}$ at the point $x = (x_1, \cdot, x_n)$, where

$$P_i = A_i B_i, \qquad (17.3.6)$$

$$A_i = (x_1, \cdots, x_n), \qquad (17.3.7)$$

$$B_i = (x_1, \cdots, x_{i-1}, x_i + \Delta x_i, x_{i+1}, \cdots, x_n). \qquad (17.3.8)$$

Under the transformation U the vector P_i transforms into the vector $U(P_i) = U(A_i)U(B_i)$. If we ignore the numbers of first order with respect to dx_i $(1 \leq i \leq n)$, then the Lebesgue measure of $U(D_X)$ is equal to the Lebesgue measure of the parallelepiped constructed on the family of vectors $(U(A_i)U(b_i))_{1 \leq i \leq n}$.

In our assumptions,

$$U(A_i)U(B_i) = (\frac{\partial U_i}{\partial x_j} \Delta x_j)_{1 \leq j \leq n} \ (1 \leq i \leq n). \qquad (17.3.9)$$

Hence

$$\det((\frac{\partial U_i}{\partial x_j}\Delta x_j)_{1\le i,j\le n}) = \det((\frac{\partial U_i}{\partial x_j})_{1\le i,j\le n} \times (\Delta_{ij})_{1\le i,j\le n}) =$$

$$= \det((\frac{\partial U_i}{\partial x_j})_{1\le i,j\le n}) \cdot \det((\Delta_{ij})_{1\le i,j\le n}) = I(U) \cdot l_n(\Delta), \quad (17.3.10)$$

where

$$\Delta_{ij} = \begin{cases} \Delta x_i, & \text{if } i = j, \\ 1, & \text{if } i \neq j. \end{cases} \quad (17.3.11)$$

For $D = \Delta_X$, we obtain the validity of the formula (17.3.4) when the function f is constant and equal to 1.

Note that this relation can be easily extended to the class of all Lebesgue measurable subsets $D \subseteq R^n$ and Lebesgue integrable functions $f : R^n \to R$.

□

Corollary 17.3.1. *Let $U : R^n \to R^n$ be a differentiable transformation. Then the formula*

$$(\forall D)(D \in B(R^n) \to l_n(U(D)) = \int_D I(U) dl_n) \quad (17.3.12)$$

is valid.

Now, we will easily get the proof of Liouville's theorem.

Let us consider the system of first order linear differential equations

$$\frac{dy(x)}{dx} = A(x)y(x), \quad (17.3.13)$$

i.e.,

$$\frac{dy_i(x)}{dx} = \sum_{i=1}^{n} a_{ij}(x)y_j(x), \ \ i = 1, 2, \cdots, n. \quad (17.3.14)$$

Let $((y_i^{(j)}(x))_{1\le i\le n})_{1\le j\le n}$ be a fundamental system of solutions of the system (17.3.13). Then the function $W(x)$ defined by

$$W(x) = \det((y_i^{(j)}(x))_{1\le i,j\le n}) \quad (17.3.15)$$

is called the Wronskian determinant of the system (17.3.13).

Theorem 17.3.2. *(Liouville) The Wronskian determinant of the system* (17.3.13) *satisfies the condition*

$$W(x) = W(x_0)e^{\int_{x_0}^{x} Tr(A(\tau))d\tau}, \quad (17.3.16)$$

where $Tr(A(\tau)) = \sum_{i=1}^{n} a_{ii}(\tau)$, $\tau, x_0, x \in R$.

Proof. Using the well-known rule of the differentiability of determinants, we obtain

$$W'(x) = \begin{vmatrix} \frac{dy_1^{(1)}}{dx} & \frac{dy_1^{(2)}}{dx} & \cdots & \frac{dy_1^{(n)}}{dx} \\ y_2^{(1)} & y_2^{(2)} & \cdots & y_2^{(n)} \\ \cdots & \cdots & \cdots & \cdots \\ y_n^{(1)} & y_n^{(2)} & \cdots & y_n^{(n)} \end{vmatrix} +$$

$$\begin{vmatrix} y_1^{(1)} & y_1^{(2)} & \cdots & y_1^{(n)} \\ \frac{dy_2^{(1)}}{dx} & \frac{dy_2^{(2)}}{dx} & \cdots & \frac{y_2^{(n)}}{dx} \\ \cdots & \cdots & \cdots & \cdots \\ y_n^{(1)} & y_n^{(2)} & \cdots y_n^{(n)} & \end{vmatrix} + \cdots + \begin{vmatrix} y_1^{(1)} & y_1^{(2)} & \cdots & y_1^{(n)} \\ y_2^{(1)} & y_2^{(2)} & \cdots & y_2^{(n)} \\ \cdots & \cdots & \cdots & \cdots \\ \frac{y_n^{(1)}}{dx} & \frac{dy_n^{(2)}}{dx} & \cdots & \frac{dy_n^{(n)}}{dx} \end{vmatrix} \quad (17.3.17)$$

If we replace the values $\frac{dy_i^{(k)}}{dx}$ standing in the right side of the above equality by the values defined by (17.3.13) for $1 \le i, j \le n$, and use the property of a determinant asserted that the determinant does not change when we add to the elements of some row (column) the linear combination of elements of other rows (columns), then we obtain

$$W'(x) = \begin{vmatrix} a_{11}y_1^{(1)} & a_{11}y_1^{(2)} & \cdots & a_{11}y_1^{(n)} \\ y_2^{(1)} & y_2^{(2)} & \cdots & y_2^{(n)} \\ \cdots & \cdots & \cdots & \cdots \\ y_n^{(1)} & y_n^{(2)} & \cdots & y_n^{(n)} \end{vmatrix} + \cdots + \begin{vmatrix} y_1^{(1)} & y_1^{(2)} & \cdots & y_1^{(n)} \\ y_2^{(1)} & y_2^{(2)} & \cdots & y_2^{(n)} \\ \cdots & \cdots & \cdots & \cdots \\ a_{nn}y_n^{(1)}y_n^{(1)} & a_{nn}y_n^{(2)} & \cdots & a_{nn}y_n^{(n)} \end{vmatrix}, \quad (17.3.18)$$

i.e.,

$$W'(x) = \sum_{i=1}^{n} a_{ii}(x)W(x). \quad (17.3.19)$$

The integration of the above equality completes the proof of Liuoville theorem.

□

References

[1] AHMED N. U., DING X., On invariant measures of nonlinear Markov processes. *J. Appl. Math. Stochastic Anal.* 6 (1993), no. 4, 385–406.

[2] BAKER R.,"Lebesgue measure" on $\mathbf{R}^\infty$, *Proc. Amer. Math. Soc.,* **113(4)** (1991), 1023–1029.

[3] BAKER R., "Lebesgue measure" on $\mathbf{R}^\infty$, *Proc. Amer. Math. Soc.,* **132(9)**, (2004), 2577–2591.

[4] HA HUY BANG. A remark on differential operators of infinite order. *Acta Math. Vietnam.* 21 (1996), no. 2, 289-294.

[5] BARNABY N., KAMRAN N. Dynamics with infinitely many derivatives: variable coefficient equations. J. High Energy Phys. 2008, no. 12, 022, 27 pp.

[6] BARNABY, N., KAMRAN N. Dynamics with infinitely many derivatives: the initial value problem. *J. High Energy Phys.* 2008, no. 2, 2008, 39 pp.

[7] BARTOSZYŃSKI T., On perfectly meager sets, *Proc. Amer. Math. Soc.* 130 (2002), no. 4, 11891195.

[8] BECKER H., KECHRIS A.S., *The Descriptive Set Theory of Polish Group Actions,* London Mathematical Society Lecture Note Series, **232**, Cambridge University Press, 1996.

[9] BECKER H., Polish group actions: Dichotomies and generalized elementary embeddings, *J. Amer. Math. Soc.* **11(2)** (1998), 397–449.

[10] BIERKENS J., VAN GAANS ONNO, LUNEL S.V., Existence of an invariant measure for stochastic evolutions driven by an eventually compact semigroup.*J. Evol. Equ.* 9 (2009), no. 4, 771–786.

[11] BIRKHOFF G., ROTA G. , *Ordinary Differential Equations,* New York: John Wiley and Sons, Inc., 1978.

[12] BRUCKNER A. M., BRUCKNER J. B. AND THOMSON B. S. , *Real Analysis,* Prentice-Hall, Englewood Cliffs, NJ, 1996.

[13] BURKE M. AND MILLER A. W., Models in which every nonmeager set is nonmeager in a nowhere dense Cantor set, *Canad. J. Math.* 57 (2005), 11391154.

[14] CABRELLI C. A., HARE K. E. , MOLTER U. M., Classifying Cantor sets by their fractal dimensions, *Proc. Amer. Math. Soc.,* **138(11)** (2010), 3965–3974.

[15] CANNARSA P., DA PRATO G., FRANKOWSKA H., Invariant measures associated to degenerate elliptic operators, *Indiana Univ. Math. J.,* 59 (2010), no. 1, 53–78.

[16] CARATHÉODORY C.,*Vorlesungen iuber reellen Funktionen,* Leipzig and Berlin, 1918.

[17] CARMICHAEL R. D., Linear differential equations of infinite order. *Bull. Amer. Math. Soc.* **42** (1936), no. 4, 193–218

[18] CHRISTENSEN J. P. R. , On sets of Haar measure zero in abelian Polish groups, *Israel J. Math.* **13** (1972), 255-260.

[19] CHRISTENSEN J.R., Measure theoretic zero sets in infinite dimensional spaces and applications to differentiability of Lipschitz mappings. Actes du Deuxieme Colloque d'Analyse Fonctionnelle de Bordeaux (Univ. Bordeaux, 1973), I, pp. 29–39. *Publ. Dep. Math.* (Lyon), **10(2)** (1973), 29–39.

[20] CICHON J., KHARAZISHVILI A., WEGLORZ B., *Subsets of the real line,* Wydawnictwo Uniwersytetu Lodzkiego, Lodz (1995).

[21] COHEN P. J., The independence of the continuum hypothesis, *Proc. Nat. Acad. Sci.,* **50**, (1963), 1143–1148; **51** (1964), 105–110.

[22] COHEN M. P., NOTES ON HAAR NULL SETS, (2012), 1-18. Retrieved February 7, 2013 from

http://www.math.unt.edu/ mpc0061/preprints/HaarNullNotes.pdf

[23] CRAVEN T., CSORDAS G. Differential operators of infinite order and the distribution of zeros of entire functions. *J. Math. Anal. Appl.* 186 (1994), no. 3, 799-820.

[24] CSÖRNYEI M., Aronszajn null and Gaussian null sets coincide, *Israel J. Math.*, **111** (1999), 191–201.

[25] DARJI U., Haar negligible comeagre sets, in Problems presented at the conference Integration, Vector Measures, and Related Topics IV, ed. D. H. Fremlin (2011). Retrieved Oct. 26, 2012 from

http://www.um.es/beca/Murcia2011/openproblems/openproblems.pdf

[26] DAVIS H. T., Differential Equations of Infinite Order with Constant Coefficients. *Amer. J. Math.* **52** (1930), no. 1, 97–108.

[27] DAWIDOWICZ A. L., POSKROBKO A., On asymptotic behaviour of the dynamical systems generated by von Foerster-Lasota equations, *Control Cybernet.* **35(4)** (2006), 803–813

[28] DAWIDOWICZ A. L., POSKROBKO A., Age-dependent population dynamics with the delayed argument, *Progr. Nonlinear Differential Equations Appl.,* **75** (2007), 165174.

[29] DAWIDOWICZ A. L., POSKROBKO, A., On chaotic and stable behaviour of the von Foerster-Lasota equation in some Orlicz spaces, *Proc. Est. Acad. Sci.,* **57(2)** (2008), 61–69.

[30] DICKSON D. G., Infinite order differential equations. *Proc. Amer. Math. Soc.* 15 1964 638–641.

[31] DOUGHERTY, R., Examples of non-shy sets, *Fund. Math.*, **144** (1994), 73–88.

[32] DUNFORD N., SCHWARTZ J. T., Linear operators, Part 1, Interscience, New York, 1958.

[33] TRÛN DÚC VÂN; DINH NHO HÀO. *Differential operators of infinite order with real arguments and their applications.* World Scientific Publishing Co., Inc., River Edge, NJ, 1994. viii+240 pp. ISBN: 981-02-1611-4.

[34] ELEKES M., STEPRāNS J., Haar null sets and the consistent reflection of non-meagreness (to appear in *Canad. J. Math.*)

[35] ENGELKING R., *Outline of general topology*, PWN, Warsaw; North-Holland, Amsterdam, 1968.

[36] ERDöS P., Set-theoretic, measure-theoretic, combinatorial, and number-theoretic problems concerning point sets in Euclidean space, Real Anal. Exchange, 4(2), (1978/79), 113–138.

[37] ERDöS P., *My Scottish Book Problems, in: The Scottish Book, Mathematics from the Scottish Café.* Edited by R. Daniel Mauldin, Birkhäuser, Boston, Mass., 1981.

[38] ERDöS P., *Some combinatorial, geometric and set theoretic problems in Measure Theory,* Oberwolfach 1983, Lecture Notes in Mathematics 1089, Springer-Verlag (1984).

[39] ERDöS P., Problems and results in combinatorial geometry. Discrete geometry and convexity , New York (1982), 1–11, Ann. New York Acad. Sci., 440, New York Acad. Sci., New York (1985).

[40] GANTMACHER F. R., *Theorie des matrices.* Tome 1: Theorie generale. (French) Traduit du Russe par Ch. Sarthou. Collection Universitaire de Mathematiques, No. 18 Dunod, Paris 1966 xiii+370 pp

[41] GHERMANESCO M., *Bulletin de l'AcadÉmie Royale de Belgique,* vol. 19 (1933), pp. 387-394;

[42] GILL T.L., ZACHARY W.W., Lebesgue measure on a version of $\mathbb{R}^\infty$, *Real Analysis Exchange*, Summer Symposium (2009), 42–49.

[43] GILL TEPPER L., PANTSULAIA G.R. AND ZACHARY W.W.,*Constructive Analysis In Infinitely Many Variables,* Communications in Mathematical Analysis, Volume 13, Number 1 (2012), 107-141. 19389787

[44] GILL T., KIRTADZE A., PANTSULAIA G., PLICHKO A., The existence and uniqueness of translation invariant measure in separable Banach spaces, *Functiones et Approximatio, Commentarii Mathematici*(in preparation)

[45] GURTIN M. E., MACCAMY R. C., Non-linear age-dependent population dynamics, *Arch. Rat. Mech. Anal.,* **54** (1974), 281300.

[46] HALMOS P. R. , *Measure theory*, Princeton, Van Nostrand, 1950.

[47] HARDY G. H., LITTLEWOOD J. E., *ÍSome problems of diophantine approximation Í*, Acta Math. **41** no. 1 (1916), 196.

[48] HARDY G. H., LITTLEWOOD J. E., *Some problems of diophantine approximation.* Acta Math. **37 (1)** (1914), 193–239.

[49] HAUSDORFF F, *Mengenlehre*, 3rd ed., Berlin, 1935.

[50] HUNT B.R. , SAUER T. , YORKE J.A. , Prevalence: a translation-invariant "almost every" on infinite-dimensional spaces, *Bull. Amer. Math. Soc.* (N.S.), **27** (2), (1992), 217–238.

[51] JECH I., *Lectures in set theory*, Berlin, Springer (1973).

[52] JOSKOT K. , Turbulent solutions of first order partial differential equation, *J. Differential Equations,* **58** (1985) 114.

[53] KAKUTANI S. , Ober die Metrisation der topologischen Gruppen, *Proc. Imp. Acad. Tokyo,***12** (1936), 82–84.

[54] KAKUTANI S., On equivalence of infinite product measures. *Ann. Math.*, **49** (1948), 214–224.

[55] KECHRIS A.S., *Classical descriptive set theory,* Graduate Texts in Mathematics, 156. Springer-Verlag, New York, 1995. xviii+402 pp.

[56] KHARAZISHVILI A.B., On invariant measures in the Hilbert space, *Bull. Acad. Sci. Georgian SSR.* **114(1)** (1984), 41–48 (in Russian).

[57] KHARAZISHVILI A.B., *Topological aspects of measure theory*, Naukova Dumka, Kiev, 1984.(Russian).

[58] KIRILLOV A. A., GVISHIANI A. D., *Theorems and exercises in functional analysis,* Nauka, Moscow (1979) (in Russian).

[59] KOROBEĬNIK YU. F. Boundary properties of analytic solutions of infinite-order differential equations. (Russian) *Mat. Sb. (N.S.)* 115(157) (1981), no. 3, 364–390, 495.

[60] KOROBEĬNIK YU. F., KORSHIKOVA T. I.,Analytic solutions of an infinite-order differential equation which are smooth on the boundary. (Russian) *Izv. Severo-Kavkaz. Nauchn. Tsentra Vyssh. Shkoly Estestv. Nauk.* 1984, no. 4, 32–36, 102.

[61] KOWALEWSKI A. Time-optimal control of infinite order hyperbolic systems with time delays.*Int. J. Appl. Math. Comput. Sci.* 19 (2009), no. 4, 597–608.

[62] VON KOCH H., *Arkiv för Matematik, Astronomi och Fysik,* vol. 15, No. 26 (1921), pp. 1-16;

[63] KURATOWSKI K., MOSTOWSKI A., *Set theory*, Nauka, Moscow, 1980. (Russian).

[64] KUIPERS L., NIEDERREITER H., *Uniform distribution of sequences,* Pure and Applied Mathematics. Wiley-Interscience [John Wiley & Sons], New York-London-Sydney, 1974.

[65] LALESCO T., *Journal de Mathématiques,* (6), vol. 4 (1908), pp. 125-202;

[66] LANG G. B., *Illinois dissertation,* 1935 (unpublished).

[67] LASOTA A., PIANIGIANI G., Invariant measures on topological spaces, *Boll. Un. Mat. Ital.*, **15-B** (1977), 592603.

[68] LASOTA A., MACKEY M. C., WAZEWSKA-CZYZEWSKA M., Minimizing therapeutically induced anemia. *J. Math. Biology,* **13** (1981), 149158.

[69] LASOTA A., SZAREK T., Dimension of measures invariant with respect to Wazewska partial differential equation, J. Differential Equations, **196** (2004), 448465.

[70] LASOTA A., Stable and chaotic solutions of a first order partial differential equation. *Nonlinear Anal.,* **5**(1981), 11811193.

[71] LEE H. H. , Vector valued Fourier analysis on unimodular groups. *Math. Nachr.,* **279(8)** (2006), 854–874.

[72] LETTENMEYER F., Dissertation, München, 1927, pp. 1-52;

[73] Linear differential equation, http://en.wikipedia.org/wiki/Linear_differential_equation.

[74] LOEB P.A., ROSS D.A., Infinite products of infinite measures. *Illinois J. Math.,* **49 (1)** (2005), 153–158 (electronic).

[75] LIU W., TOLLE J., Existence and uniqueness of invariant measures for stochastic evolution equations with weakly dissipative drifts, *Electron. Commun. Probab.,* 16 (2011), 447–457.

[76] MANKIEWICZ P., On the differentiability of Lipschitz mappings in Fréchet spaces, *Studia Math.*, **45** (1973), 15–29.

[77] MAŘÍK J., The Baire and Borel measure, *Czechoslovak Math. J.* **7** (1957), 248–253.

[78] MARTIN D.A., SOLOVAY R.M., Internal Cohen extensions, *Ann.Math.Logic*, 2 (1970).

[79] MAULDIN R. D., Some problems in set theory, analysis and geometry, Paul Erdös and his mathematics, I (Budapest, 1999), 493–506, Bolyai Soc. Math. Stud., 11, Janos Bolyai Math. Soc., Budapest, 2002.

[80] MAULDIN R. D., Some problems and ideas of Erdös in analysis and geometry, Proceedings of the Erdos Centennial, Budapest, July 1-5, 2013, 10 pages, to appear.

[81] MCKENDRICK A. G., Application of mathematics to medical problems, *Proc. Edin. Math. Soc.,* **44** (1926), 98130.

[82] MYCIELSKI J., SWIERCZKOWSKI S., On the Lebesgue measurability and the axiom of determinateness, *Fund.*, **54** (1964), 67–71.

[83] MYCIELSKI J., Some unsolved problems on the prevalence of ergodicity, instability, and algebraic independence; *Ulam Quart.,* **1 (3)** (1992), 30 ff., approx. 8 pp.

[84] MOORE,C. C. Invariant measures on product spaces, *Proc. Fifth Berkeley Sympos. Math. Statist, and Probability* (Berkeley, Calif., 1965), vol. II, part 2, Univ. of California Press, Berkeley, Calif., (1967), 447-459.

[85] NIKOLSKI S.M., *Cource of mathematical analysis,* no 1, Moscow, 1983. (Russian).

[86] NIKOLSKI S.M., *Cource of mathematical analysis,* no 2, Moscow, 1983. (Russian).

[87] OXTOBY J. C. AND ULAM S. M., ON THE EXISTENCE OF A MEASURE INVARIANT UNDER A TRANSFORMATION, *Ann. of Math.,* **(2)40** (1939),560-566.

[88] OXTOBY J.C., Invariant measures in groups which are not locally compact, *Trans. Amer. Math. Soc.*, **60** (1946), 215–237.

[89] OXTOBY J.C., *Measure and Category*, Springer-Verlag, 1971, second ed. 1980.

[90] PANTSULAIA G.R., Duality of measure and category in infinite-dimensional separable Hilbert space, *IJMMS.* **30(6)** (2002), 353–363.

[91] PANTSULAIA G.R., Relations between shy sets and sets of ν_p-measure zero in Solovay's Model, *Bull. Polish Acad. Sci. Math.* , **52**, No.1, (2004), 63-69.

[92] PANTSULAIA G. R., *Invariant and quasiinvariant measures in infinite-dimensional topological vector spaces.* Nova Science Publishers, Inc., New York, 2007.

[93] PANTSULAIA G.R., GIORGADZE G.P., On analogy of Liouville theorem in infinite-dimensional separable Hilbert space, *Georgian Int. J. Sci. Technol.*, **1 (2)** (2008), 167–180.

[94] PANTSULAIA G.R.,On a generalized Fourier μ-series in some infinite-dimensional Polish topological vector spaces, *Georgian Int. J. Sci. Technol.*, **1 (4)** (2008), 313–326.

[95] PANTSULAIA G.R., GIORGADZE G.P., On Lioville type theorems for Mankiewicz and Preiss-Tiser generators in $\mathbb{R}^{\mathbb{N}}$, *Georgian Int. J. Sci. Technol.*, **1 (2)** (2009).

[96] PANTSULAIA G.R., On generators of shy sets on Polish topological vector spaces. *New York Journal of Mathematics*, **14** (2008), 235–261.

[97] PANTSULAIA G.R., GIORGADZE G.P., On Lioville type theorems for Mankiewicz and Preiss-Tiser generators in $\mathbb{R}^{\mathbb{N}}$, *Georgian Int. J. Sci. Technol.*, **1 (2)** (2009).

[98] PANTSULAIA G.R., On ordinary and standard Lebesgue measures on $\mathbb{R}^{\infty}$, *Bull. Pol. Acad. Sci. Math.* **57 (3-4)** (2009), 209–222.

[99] PANTSULAIA G.R., Change of variable formula for "Lebesgue measures" on $\mathbb{R}^{\mathbb{N}}$. *J. Math. Sci. Adv. Appl.*, **2(1)**, (2009), 1–12.

[100] PANTSULAIA G.R., On Erdős one problem for extensions of Lebesgue measures, Hadronic J., Volume 32, Number 4, August (2009), 425–433.

[101] PANTSULAIA G.R., On a standard product of an arbitrary family of σ-finite Borel measures with domain in Polish spaces, *Theory Stoch. Process,* **16(32)** (2010), no 1, p.84-93.

[102] PANTSULAIA G.R., *Riemann Integrability and Uniform Distribution in Infinite-Dimensional Rectangles* , Georgian International Journal of Science and Technology , Nova Science Publishers, Volume 2, Issue 3 (2010).

[103] PANTSULAIA G.R., On uniform measures in Banach spaces, *Georgian Int. J. Sci. Technol.,* **2(3)** (2010), 243–250.

[104] PANTSULAIA G.R., On T-shy sets in Radon metric groups, *J. Math. Sci. Adv. Appl.* **5 (1)** (2010), 149-186 .

[105] PANTSULAIA G.R., On singularity of non-σ -finite measures and transition kernels, *Georgian Int. J. Sci. Technol.*, **3 (1)** (2011), 21-37.

[106] PANTSULAIA G.R., On ordinary and standard products of infinite family of σ -finite measures and some of their applications. *Acta Math. Sin. (Engl. Ser.)* 27 (2011), no. 3, 477–496.

[107] PANTSULAIA G.,GIORGADZE G., On some applications of infinite-dimensional cellular matrices, *Georg. Inter. J. Sci. Tech., Nova Science Publishers,* Volume 3, Issue 1 (2011), 107-129.

[108] PANTSULAIA G., GIORGADZE G., "Lebesgue "-null sets in $\mathbf{R}^{\infty}$ are not preserved under Lipschitz isomorphisms, *Georg. Inter. J. Sci. Tech.*,**3(1)** (2011), 39–49.

[109] PANTSULAIA G., GIORGADZE G., On an explicit formula of a particular solution of non-homogeneous differential equation of the higher order with real constant coefficients, *Georg. Inter. J. Sci. Tech., Nova Science Publishers,* Volume 4, Issue 1 (2012), in press.

[110] PANTSULAIA G., GIORGADZE G., Description of the behaviour of phase motions defined by generalized von Foersten-Lasota equations in R^∞ in terms of ordinary and standard "Lebesgue measures" (Dedicated to the memory of Professor Andrzej Lasota), *Georg. Inter. J. Sci. Tech., Nova Science Publishers*, Volume 4, Issue 2 (2012), in press.

[111] PANTSULAIA G.R., *Selected topics of an infinite-dimensional classical analysis.* Nova Science Publishers, Inc., New York, 2012, xii -183 p

[112] PANTSULAIA G.R., On existence and uniqueness of generators of shy sets in Polish groups, *J. Math. Sci. Adv. Appl.,* Volume no: 16, Issue no: 1-2, July and August(2012), 61 –78.

[113] PANTSULAIA G., RUSIASHVILI N., On questions of U. Darji and D. Fremlin, *Georg. Inter. J. Sci. Tech.,* Volume 5, Number 3-4 (2013), 89-102.

[114] PANTSULAIA G., RUSIASHVILI N., On a certain version of the Erdös problem, *Georg. Inter. J. Sci. Tech.,* Volume 6, Number 1-2 (2013), pp 1-8(in appear).

[115] PHELPS R.R., Gaussian null sets and differentiability of Lippschitz maps on Banach spaces, *Pac. J. Math.*, **77** (1978), 523–531.

[116] PINCHERLE S., *Memorie Reale Accademia delle Scienze,* Bologna, (4), vol. 9 (1888), pp. 45-71;

[117] PINCHERLE S., Me'moire sur le calcul fonctionnel distributif. (French) *Math. Ann.* **49** (1897), no. 3-4, 325–382.

[118] PINCHERLE S., Sur la re'solution de l'e'quation fonctionnelle $\sum h^\nu \phi(\alpha^\nu) = f(x)$ a' cofficients constantsa' cofficients constants. (French) *Acta Math.* **48** (1926), no. 3-4, 279–304.

[119] PÓLYA G. , *Göttinger Nachrichten,* 1927, pp. 187-195;

[120] PONTRJAGIN L., TOPOLOGICAL GROUPS, PRINCETON, 1939.

[121] POPKEN J., An arithmetical theorem concerning linear differential equations of infinite order. *Nederl. Akad. Wetensch., Proc.* 53, (1950) 1645–1656 = *Indagationes Math.* 12, 522–533 (1950).

[122] PREISS D., TIŠER J., Two unexpected examples concerning differentiability of Lipschitz functions on Banach spaces, *Geometric aspects of functional analysis,* (Israel, 1992–1994), 219–238, *Oper. Theory Adv. Appl.*, Birkhäuser, Basel, **77** (1995).

[123] RITT, J. F., On a general class of linear homogeneous differential equations of infinite order with constant coefficients. *Trans. Amer. Math. Soc.* **18** (1917), no. 1, 27–49.

[124] ROBINSON J.C., *An Introduction to Ordinary Differential Equations,* Cambridge, UK.: Cambridge University Press, 2004.

[125] ROGERS C.A. , *Hausdorff Measures*, Cambridge Univ. Press (1970).

[126] RUDNICKI R., Invariant measures for the flow of a first order partial differential equation, *Ergodic Theory Dynam. Systems,* **5** (1985), 437443.

[127] SCHÜRER F., *Leipziger Berichte* , vol. 70 (1919), pp. 185-246;

[128] SHEFFER I. M., Systems of infinitely many linear differential equations of infinite order, with constant coefficients. *Trans. Amer. Math. Soc.* **31** (1929), no. 2, 281–289.

[129] SHI H., *Measure-Theoretic Notions of Prevalence,* Ph.D. Dissertation (under Brian S.Thomson), Simon Fraser University, October (1997).

[130] Singer I., *Bases in Banach spaces.* II, Springer, Berlin et al. 1981.

[131] SOLECKI S., Haar null and non-dominating sets, Fund. Math. 170 (2001), 197–217.

[132] SOLOVAY R. M. , A model of set theory in which every set of reals is Lebesgue measurable, *Ann. Math.*, **92** (1970), 1–56.

[133] SOWERS R., Large deviations for the invariant measure of a reaction-diffusion equation with non-Gaussian perturbations. *Probab. Theory Related Fields* 92 (1992), no. 3, 393–421.

[134] STONE, A.H. Paracompactness and product spaces. *Bull. Amer. Math. Soc.*, **54** (1948), 977–982.

[135] TOPSØE F., HOFFMANN-JØRGENS J., *Analytic spaces and their application,* in: C. A. Rogers (et al.), Analytic Sets, Academic Press, London (1980), 317–401.

[136] ULAM S., *Zur Masstheorier in der allgemeinen Mengenlehre,* Fund. Math., **16**,(1930) 140-150.

[137] URYSOHN P. , *Zum Metrisationsproblem.* (German) Math. Ann. **94 (1)** (1925), 309–315.

[138] VAKHANYA N.N., TARIELADZE V.I., CHOBANYAN S.A., *Probability distributions in Banach spaces*, Nauka, Moscow (1985) (in Russian).

[139] VINOGRADOW I., *On fractional parts of certain functions.* Ann. of Math. **37 (2)** , (1936), 448–455.

[140] WIENER N., The operational calculus. *Math. Ann.* **95** (1926), no. 1, 557–584.

[141] WILLARD S., *General Topology,* Addison-Wesley Publishing Company, Reading Massachusetts (1970).

[142] WAZEWSKA-CZYZEWSKA M., LASOTA A., Matematyczne problemy dynamiki ukadu krwinek czerwonych, *Roczniki PTM, Mat. Stos.,* VI (1976), 23-40.

[143] WIMMER H. K., Exponential solutions of systems of linear differential equations of infinite order. J. Differential Equations., 33 (1979), no. 1, 39–44.

[144] WEYL H., *Über ein Problem aus dem Gebiete der diophantischen Approximation.* Marchr. Ges. Wiss. GÖtingen. Math-phys. K1., (1914), 234-244.

[145] YAMASAKI Y., Translationally invariant measure on the infinite-dimensional vector space, *Publ. Res. Inst. Math. Sci.* **16(3)** (1980), 693–720.

[146] ZHIDKOV P. E., An invariant measure for the nonlinear Schrodinger equation. (Russian) *Dokl. Akad. Nauk SSSR,* 317(3), 1991, pp. 543546].

Index

G

H

I

L

M

N

O

P

R

S

T

U

V

Y